Invasive and Introduced Plants and Animals

Human perceptions, attitudes and approaches to management

Edited by
Ian D. Rotherham and Robert A. Lambert

First published in paperback 2013

First published 2011
by Earthscan from Routledge
2 Park Square, Milton Park, Abingdon, Oxon, OX14 4RN

Simultaneously published in the USA and Canada
by Routledge
711 Third Avenue, New York, NY 10017

Routledge is an imprint of the Taylor & Francis Group, an informa business

British Library Cataloguing-in-Publication Data
A catalogue record for this book is available from the British Library

Library of Congress Cataloging-in-Publication Data
Invasive and introduced plants and animals : human perceptions, attitudes, and approaches to management / [edited by] Ian D. Rotherham and Robert A. Lambert.
p. cm.
'Earthscan publishes in association with the International Institute for Environment and Development.'
Includes bibliographical references and index.
1. Introduced organisms. 2. Introduced organisms—Case studies. I. Rotherham, Ian D. II. Lambert, Robert A. III. International Institute for Environment and Development.
QH353.I5826 2011
333.95′23—dc22

2010045564

ISBN: 978-1-84971-071-8 (hbk)
ISBN: 978-0-415-83069-0 (pbk)
ISBN: 978-0-203-52575-3 (ebk)

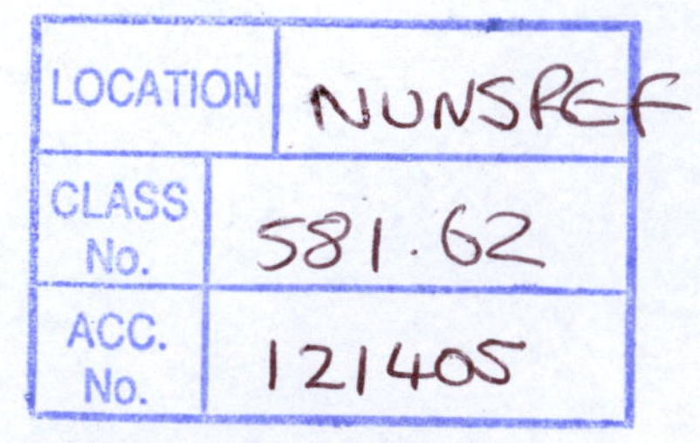

Typeset by Composition and Design Services
Cover design by Andrew Corbett

Printed and bound in Great Britain by
TJ International Ltd, Padstow, Cornwall

Invasive and Introduced Plants and Animals

This book addresses the broader context of invasive and exotic species, in terms of the perceived threats and environmental concerns which surround alien species and ecological invasions. As a result of unprecedented scales of environmental change, combined with rapid globalisation, the mixing of cultures and diversity, and fears over biosecurity and bioterrorism, the known impacts of particular invasions have been catastrophic. However, as several chapters show, reactions to some exotic species, and the justifications for interventions in certain situations, including biological control by introduced natural enemies, rest uncomfortably with social reactions to ethnic cleansing and persecution perpetrated across the globe. The role of democracy in deciding and determining environmental policy is another emerging issue. In an increasingly multicultural society the fundamental questions of ethics and choice. At the same time, in order to redress major ecological losses, the science of reintroduction of native species has also come to the fore, and is widely accepted by many in nature conservation. However, with questions of where and when, and with what species or even species analogues, reintroductions are acceptable, the topic is hotly debated. Again, it is shown that many decisions are based on values and perceptions rather than objective science. Including a wide range of case studies from around the world, his book raises critical issues to stimulate a much wider debate.

Ian D. Rotherham is a leading researcher and writer on ecological history with a long-standing interest in exotic and invasive animals and plants. He is a Reader, Director of the Countryside, Tourism and Environment Research Unit, and International Research Coordinator at Sheffield Hallam University, UK.

Robert A. Lambert has a dual appointment at the University of Nottingham, UK, as Lecturer in Environmental History and Lecturer in Tourism and the Environment. He is also a Senior Honorary Research Fellow at the University of Western Australia and co-editor of the journal *Environment and History*.

'. . . a truly momentous, provocative and, at times, hilarious analysis of the incredibly complex, controversial and conflict-ridden world of invasive and introduced plants and animals. In bridging the gap between objective science and subjective sociocultural fashions and values, this compelling tome dares to go 'where no-one has gone before', with its rich and diverse consideration of the multi-faceted issues surrounding species invasions and introductions. Be warned! After devouring this, you'll never look at plants and animals in quite the same way ever again . . . This is a clarion call – no more excuses.' – *Sarah Simons, Executive Director, Global Invasive Species Programme.*

'An interesting and much-needed book that tackles an important aspect of invasive alien species: how they are perceived, valued and judged by humans. The volume draws together a broad range of fascinating case studies and is very thought-provoking. A must for any serious invasion ecologist.' – *Robert A. Francis, King's College London, Secretary of the British Ecological Society Invasive Species Special Interest Group.*

'For an ecologist it is a salutary experience to examine this excellent record of the widespread confusion and conflict that has surrounded the human response to invasive species. Is this a humiliating testament to the immature and fragmentary state of our subject or should we admit that local, cultural, economic, political and sentimental forces are more important and will often dictate an unpredictable course of action?' – *Professor Philip Grime, University of Sheffield, UK.*

'Abounding with parallels with the human world, *Invasive and Introduced Plants and Animals: Human perceptions, attitudes and approaches to management* is a book to recommend without hesitation.' – *Robert Alexander Hearn, University of Genoa, in* Environment and History *(Vol 19, 2013).*

'It takes a fresh approach to invasive species by asking to what extent popular perceptions and attitudes conflict with objective science and examining how this conflict affects outcomes.' – *Peter Marren, British Wildlife*

'It is gratifying to find the matter so thoroughly dealt with, especially meeting the issues of xenophobia, arbitrary cut off cut-off points in ecological time, and the often ignored fact that in most parts of Britain and Europe at least, one is dealing with a whole landscape shaped by non-native species.' – *Peter Taylor, ECOS*

Contents

List of Figures and Tables

Figures

Tables

Contributors

Dr Stuart K. Allison, Professor of Biology; Director of the Green Oaks Field Research Center, Department of Biology, Knox College, 2 East South Street, Galesburg, IL 61401, USA; email: sallison@knox.edu

Stuart Allison is Professor of Biology at Knox College in the USA, and is a biologist who works on issues of invasive species and how they affect attempts to restore the North American prairie.

Dr John Bailey, Principal Experimental Officer, Biology Department, University of Leicester, Leicester LE1 7RH, UK; email: jpb@leicester.ac.uk

John Bailey has been working for many years on detailed analysis of the genetics of invasive Japanese Knotweed and is the acknowledged authority in this field.

Dr James Beattie, Senior Lecturer, World and Environmental History, Department of History, University of Waikato, Private Bag 3800, Hamilton 2140, New Zealand; email: jbeattie@waikato.ac.nz

James Beattie is a senior lecturer at the University of Waikato in New Zealand and writes on issues of natives and aliens in the ecological history of New Zealand.

Dr Pierre Binggeli, JTC-Design, Plan 3, 2000 Neuchâtel, Switzerland; email: Pierre_Binggeli@hotmail.com

Pierre Binggeli is an international consultant in temperate and tropical ecology and forestry, and a researcher with interests in the issues of existing knowledge and its effects on the human dimensions of introduced invasive plants in tropical regions of continental Africa and surrounding oceanic islands.

Dr Iris Borowy, University of Rostock, August-Bebel-Str 28 R, 6015 18055 Rostock, Germany; email: iris.borowy@uni-rostock.de

Iris Borowy is a lecturer at the University of Rostock and is researching the history of acclimatization societies in relation to invasions and human perceptions of alien species.

Matthew K. Chew, School of Life Sciences, Arizona State University, Tempe, AZ 85287-3301, USA; email: mchew@asu.edu

Matthew Chew lectures at Arizona State University and has a particular interest in the relationships between botany, place and belonging.

Professor Peter Coates, Professor of American and Environmental History, Department of Historical Studies, School of Humanities, University of Bristol, 13 Woodland Road, Bristol BS8 1TB, UK, UK; email: P.A.Coates@bristol.ac.uk

Peter Coates has established a reputation on both sides of the Atlantic as the person who has done most to highlight the issue of perceptions and history in understanding attitudes to exotic species.

Professor Guillaume Decocq, Professor of Botany, University of Picardy Jules Verne, Dynamiques des Systèmes Anthropisés Research Unit (JE 2532 DSA), 1 Rue des Louvels, F-80037, Amiens Cedex, France; email: guillaume.decocq@u-picardie.fr

Guillaume Decocq is a Professor of Botany with long-standing research interests in invasive forest plants.

Dr Francesca Gherardi, Dipartimento di Biologia Evoluzionistica, Università di Firenze, Via Romana 17, 50125 Firenze, Italy. email: francesca.gherardi@unifi.it

Franscesca Gherardi lectures at the University of Florence in Italy, and has been researching issues of invasive aliens in Italy.

Dr Paul H. Gobster, Research Social Scientist, USDA Forest Service, Northern Research Station, 1033 University Place, Suite 360, Evanston, IL 60201-3172, USA; email: pgobster@fs.fed.us

Paul Gobster is research social scientist with the USDA Forest Service, Northern Research Station and has co-led integrated research programmes on mid-western landscape change. He is an authority on North American ecology and land restoration with a particular interest in issues of invasive alien species.

Dr Martin Goulding, 40 Owley Wood Road, Weaverham, Northwich, Cheshire, CW8 3LF, UK; email: mjgoulding@mjgoulding.freeserve.co.uk

Martin Goulding is acknowledged as the leading expert on wild boar in Britain with an interest in attitudes to and perceptions of a native making a comeback or an alien invasive that has escaped from captivity.

Dr Aurélie Javelle, Jules Verne University of Picardy, SupAgro antenne de Florac, Groupe GERME (Gestion des Espaces Ruraux Médiation et Ecologie), 9 Rue Célestin Freinet, F-48400, Florac, France

Aurélie Javelle is a research engineer in environmental anthropology with an interest in perceptions of exotic species.

Professor Bernard Kalaora, University of Picardy Jules Verne, Dynamiques des Systèmes Anthropisés Research Unit (JE 2532 DSA), 1 Rue des Louvels, F-80037, Amiens Cedex, France

Bernard Kalaora is Professor of Sociology with interests in the perceptions of invasive exotic plants.

Dr Robert A. Lambert, Lecturer in Tourism and the Environment, Nottingham University Business School; Lecturer in Environmental History, School of History, University of Nottingham; email: robert.lambert@nottingham.ac.uk

Rob Lambert has been researching issues around both alien species and the reintroduction of supposed natives. Here he addresses this topic with particular reference to the reintroduction of bird species to areas of their ranges from which they have been lost.

Michael Livingston, Agricultural Economist, Economic Research Service, US Department of Agriculture, 1800 M St, NW, Room 4024, Washington, DC 20036-5831; email: mlivingston@ers.usda.gov

Michael Livingston and his colleagues work on the economics of invasion and of management and control.

Dr Jeff McNeeley, Chief Scientist, IUCN, Rue Mauverney 28, 1196 Gland, Switzerland; email: JAM@iucn.org

Jeff McNeeley, formerly Chief Scientist of IUCN, is an acknowledged authority on the global nature of the problem of invasions and the seriousness of the core issues and has researched and written extensively on these issues.

Craig Osteen, Agricultural Economist, Economic Research Service, US Department of Agriculture, 1800 M St, NW, Room 4024, Washington, DC 20036-5831; email: costeen@ers.usda.gov

Craig Osteen and his colleagues work on the economics of invasion and of management and control, relating these to concepts and perceptions of aliens and invasives.

Dr Hazel Petrie, Department of History, Lecturer in New Zealand and Maori history, Department of History, University of Auckland, Private Bag 92019, Auckland 1142, New Zealand; email: h.petrie@auckland.ac.nz

Hazel Petrie lectures at the University of Auckland in New Zealand and has interests in the relationships between alien and native in Maori culture.

Simon Pooley, Modern History, St Antony's College, University of Oxford, 62 Woodstock Road, Oxford, OX2 6JF, UK; email: simon.pooley@sant.ox.ac.uk

Simon Pooley is a researcher with interests in the attitudes to native and exotic species in terms of the management of South African vegetation.

Professor Ian D. Rotherham, Sheffield Hallam University, City Campus, Pond Street, Sheffield S1 1WB, UK; email: i.d.rotherham@shu.ac.uk

Ian Rotherham has worked on the history of invasive species and on problems of their control. His recent work has been on case studies of invasive alien plants and animals and more recently on the human dimensions that are critical to the processes of invasion.

Dr Mark Sagoff, 2101 Van Munching Hall, School of Public Policy, University of Maryland, College Park, MD 20742, USA; email: msagoff@umd.edu

Mark Sagoff is Director and Senior Research Scholar at the Institute for Philosophy and Public Policy, and is a researcher with interests in issues of property rights and perceptions of damage in relation to exotic species.

Dr Daniel Simberloff, Department of Ecology and Evolutionary Biology, University of Tennessee, Knoxville, TN 37996, USA; email: dsimberloff@utk.edu

Dan Simberloff is acknowledged as a global authority on the ecology of animal and plant invasions.

Professor Chris Smout, University of St Andrews and Historiographer Royal, Chesterhill House, Shore Road, Anstruther, Fife, KY10 3DZ, Scotland; email: christopher@smout.org

Chris Smout has been a leading environmental historian in Europe for many years, and as Historiographer Royal to Scotland brings a historian's view to bear on the issue of perceptions of aliens and invasives.

Professor David Trigger, School of Social Science, University of Queensland, St Lucia, Brisbane, Australia 4072; email: d.trigger@uq.edu.au

David Trigger is a leading researcher on Australian wildlife and perceptions of invasions in Australia.

Dr Charles Warren, School of Geography and Geosciences, Irvine Building, University of St Andrews, North Street, St Andrews, KY16 9AL, Fife, Scotland, UK; email: crw2@st-andrews.ac.uk

Charles Warren is a senior lecturer at the University of St Andrews and has a research interest in the specific issues of the Scottish perspective of what might be alien in Scotland.

List of Acronyms and Abbreviations

APHIS	Animal and Plant Health Inspection Service
AS	alien species
BBWF	British Birdwatching Fair
BOU	British Ornithologists' Union
CBD	Convention on Biological Diversity
CDC	Centers for Disease Control
CFK	Cape Floral Kingdom
CCW	Countryside Council for Wales
Defra	Department for Environment, Food And Rural Affairs
DOC	Department of Conservation
DOI	Department of the Interior
ELI	Environmental Law Institute
FWS	Fish and Wildlife Service
GERME	Gestion des Espaces Ruraux Médiation et Ecologie
GISP	Global Invasive Species Programme
GRN	Grassland Restoration Network
IAS	invasive alien species
ICAO	International Civil Aviation Organization
IPPC	International Plant Protection Convention
ISSRs	inter simple sequence repeats
IUCN	International Union for the Conservation of Nature and Natural Resources
mph	multiprimer haplotypes
NAPPRA	not authorized [for importation] pending pest risk analysis
NISC	National Invasive Species Council
NNS	non-native species
PETA	People for the Ethical Treatment of Animals
POET	Preserve Our Eucalyptus Trees
RAPDs	Randomly Amplified Polymorphic DNAs
RISC	Recording Invasive Species Counts
RSPB	Royal Society for the Protection of Birds
RTA	road traffic accident
SCOPE	Scientific Committee on Problems of the Environment

SNH	Scottish Natural Heritage
SNP	Scottish National Party
SOC	Scottish Ornithologists' Club
SSSI	Site of Special Scientific Interest
SWT	Scottish Wildlife Trust
USDA	US Department of Agriculture
WTO	World Trade Organization

Part I
Setting the Scene

1
Balancing Species History, Human Culture and Scientific Insight: Introduction and Overview

Ian D. Rotherham and Robert A. Lambert

Here we set the wider scene and context to the subject matter of this book. This is in terms of issues and perceptions of both alien and introduced (or reintroduced) species. We raise issues of what is native and what is natural, and the ways in which these and our perceptions of nature have changed over time. Our intention is to stimulate the reader to question ideas and received wisdom, and to try to establish the interface between objective science and subjective sociocultural fashions and values.

> *When the late Sir Henry Tizard learnt that I had been asked to write a book about weeds and aliens he remarked in his characteristic manner, 'I do hope you will tell us more than just how to kill them.'* Sir Edward Salisbury (1961)

Writing in 2009, Christopher Lever stated that, 'humans are inveterate and incorrigible meddlers, never content to leave anything as they find it but always seeking to alter and – as they see it – to improve'. He noted that, 'invasive alien species are, after habitat destruction, the most important cause of loss of biodiversity through the extinction or reduction of native species'. Very usefully Lever also considered the dual issues of both non-native introductions and escapes, and native reintroductions. In discussing alien, exotic and invasive fauna and flora it is important to begin with definitions (see Rotherham, 2005a). It is worth considering what the term 'alien' means. Dictionary definitions (Anon., 1983) suggest:

- 'Belonging to another person, place or family, especially to a foreign nation or allegiance. Foreign in nature, character or origin' (1673).

- 'A stranger or a foreigner. A resident foreign in origin and not naturalized' (1330).
- 'One excluded from citizenship, privileges etc' (1549).
- 'A plant originally introduced from other countries' (1847).

Clement and Foster (1994) used 'alien' in a broad sense to denote all plants whether or not they were believed to have arrived as a result of human activities. They include plants referred to by other authors as adventives, casuals, ephemerals, exotics, introductions and volunteers.

Ellis (1993) wrote a very useful pocket-sized introduction to invasive plants in Britain. He suggests that for many an alien plant is essentially one that is not native. In this case a native plant is one that arrived in Britain prior to the closure of the English Channel around 7–8000 years ago, and so an alien is a species arriving after such a date.

An exotic is taken as 'originating in a foreign county' or 'having a strange or bizarre allure, beauty or quality', and invasive is simply 'relating to an invasion' the latter being 'any encroachment or intrusion' (Anon., 2000). Issues such as the changes to, or reinstatement of, national boundaries clearly impact on such interpretations of 'native' or 'non-native' (see Warren, Chapter 5). For most invasive exotic or alien species a key factor is that they occur 'in the wild'. However, there is an often neglected issue (problematic for species subject to legislative controls) of defining exactly what 'in the wild' might really be. While this might seem absolutely obvious to most practising ecologists, when subject to the rigorous inspection of a court of law the precise definition becomes open to interpretation. As Clement and Foster (1994) noted, the word 'naturalized' has been used to describe a wide range of conditions, some records referring to only a single plant among native vegetation. Moreover, the extent to which trees and shrubs reproduce is seldom recorded; nor is it always clear from the records whether an annual species is persisting from self-sown seed or by repeated reintroduction. They also accepted that an 'alien' plant might be 'a single short-lived plant occurring, unintended, in an artificial habitat or many large, long-established colonies overwhelming the native vegetation'. There are further complications with terms such as 'feral' and 'weed', and in his classic book *The New Naturalist Weeds and Aliens* Salisbury (1961) never really tells the reader what these actually are. One suspects that he himself was unclear as to the precise definitions.

However, using accepted concepts and academic traditions it is possible to describe the history of invasions and the impacts of exotic species, species history being a well-established subdiscipline of environmental history, and invasion biology also a well-recognized scientific field. While the general principles and processes are well-known, the more subtle issues of values, perceptions and attitudes are less widely recognized. We argue that human interactions with invasive species, both alien and native, are often of fundamental importance (Rotherham, 2005c, 2009). This may be through a plant or animal being introduced to a potentially new area, and/or in modifications to existing landscapes and ecosystems to precipitate and facilitate bio-invasion. Human sociocultural values and attitudes have shaped many invasions (Coates, 2006). Moreover, perceptions and attitudes towards invasions and invasives are important and often, we suggest, subjective rather than objective; not fixed, but varying with both time and place (Warren, 2002; Coates, 2006). These are key issues in human responses to the undoubted problems generated by some invasive species, and misunderstanding probably compounds, rather than solves, the adverse impacts. Considering what is alien, what is a problem, what should be done in response to damaging bio-invasions, are all reasonable questions, but the answers are not simple and our responses are influenced by entrenched or fluctuating perceptions as well as science. The importance of perception in addressing alien invasions is discussed in this volume for the Mediterranean area by Gherardi (Chapter 12), for the American prairie by Allison (Chapter 17), and for the USA more generally by Gobster (Chapter 16). Javelle, Kalaora and Decocq (Chapter 18) also examine how perceptions of alien species in French forests vary over time and even between stakeholders, with major consequences for control programmes. Peoples' interactions with their local environment are such that they may not even 'see' the problem in the first instance; some invaders are simply invisible to some sectors of the population.

National situations interact across political frontiers with wider global issues of alien species and diverse human cultural perceptions of them, and this vision has often dominated in both ecological and social historical scholarship, influenced through increasing globalization and modern environmental attitudes. Similarly, the problems faced in intensively studied countries such as Great Britain often mirror those which occur in other Western countries around the world. Though conservation organizations are often reluctant or unwilling to enter the societal debate (the non-scientific debate, if you like), perceptions and attitudes towards invasions and invasives are important, and often subjective rather than objective. Discussions about aliens also relate to perceptions of just what is 'natural' and 'wild': loaded concepts with significant subjectivity, that are increasingly open to debate and challenge. This is particularly the case in developed countries where people live mostly in what are, at best, seminatural habitats and largely cultural landscapes shaped both by nature and human history. To address some of these problems botanists have recently adopted the term 'archaeophytes' to cover long-established non-native plants. There may also be a time period after which a species may be accepted almost as an 'honorary native', again a totally subjective non-scientific label.

For plants in particular the mode of arrival or of introduction is central to both definition and to an understanding of the issues. Ellis (1993) took alien plants to be those introduced by people, both deliberately or accidentally. But he also notes that in reality, and with a longer time perspective, most native flora could be considered 'alien invaders'. This is due to the dynamic and fluctuating nature of vegetation in a landscape with long-term changes of key factors such as climate. This latter point may become increasingly important in the years to come. Alien or exotic species can be 'casual', 'persistent', or 'established', and are often also described as either 'introduced' or 'naturalized', the latter implying a self-sustaining and expanding population. This serves to emphasize that much important conservation management is based substantially on subjective human needs, opinions and priorities, not necessarily on hard-nosed ecological science.

Many current conservation problems and issues in terms of exotic and invasive alien plants and animals originate with western European imperial expansion, and thereafter with mostly northern hemisphere-led globalization. Media and scientific expressions of concern have led to some actions, more often a proliferation of policies and strategies but, with some notable exceptions, little effective implementation. This is despite wide recognition and well-supported assertions about the negative impacts of many invaders on both ecology and economy. Professional conservationist Graham Madge (below) was writing specifically about the British experience, but this viewpoint applies globally too.

> *Government documents are long on rhetoric but very thin when it comes down to well-defined actions and accountable responsibility. Invasives are a significant threat to a large proportion of the world's biodiversity.* Graham Madge, Royal Society for the Protection of Birds (RSPB), quoted online 13 October 2008, www.newsforums.bbc.co.uk

Yet these issues are not so simple. The often ignored debate on exactly what is an alien species and what is native has a big impact on what might be considered a problem. Dates and mechanisms of arrival, of human influence and sometimes of extinction too all have a bearing on what we consider is 'in' (acceptable) and what is 'out' (unacceptable). Borowy (Chapter 10) presents an interesting overview of reactions to alien and exotic species and the influences of 19th-century acclimatization societies on attitudes in Europe and around the world. She asserts that invasion biology has yet to make a convincing case for foreign species being inherently more damaging than native species, some of which also behave in ways that contradict peoples' economic and ecological expectations. 'At the end of the day, plants and animals may not be so different from people: there are all kinds of them everywhere, good and bad and mostly in the grey area in between with good sides for some and bad sides for others' (Chapter 10).

Increasingly too in the modern era, conservationists have sought to reverse losses and extinctions through the 'reintroduction' of species to habitats within their former range (see Lambert, Chapter 11). In Great Britain this has led to hugely successful returns (mostly by human hand in reintroduction schemes or active translocation schemes, but also through natural recolonization) by birds such as red kite (Lovegrove, 1990; Carter, 2007) and osprey (Brown and Waterston, 1962), and localized human-instigated recovery by sea eagles in Scotland (Love, 1983). In almost all such cases the species involved are high-profile iconic birds with huge public appeal, and thus sustainable tourism implications that can yield substantial regional economic benefits in remoter rural areas, alongside their conservation value. Yet these anthropogenic interventions are not without their controversies, and proposals to establish the sea eagle in East Anglia triggered media frenzy and a vociferous local campaign opposed to the suggestion. The coastline of Norfolk and Suffolk is now real contested ground.

Other species 'quietly' reintroduced over recent decades include otter (King et al, 1976) and barn owl, both with considerable success and generally without huge public debate. However, even the widespread recovery of the otter has not been totally without issue, with clear evidence of persecution, often with a suggestion of economic interests such as fish farms, for example on the River Don in South Yorkshire. For species such as barn owl, the major concerns were whether or not the issues associated with declines had really been addressed sufficiently to facilitate re-establishment and recovery, after initial reintroduction had taken place.

But perhaps the most controversial proposals for reintroduction have been with regard to large mammals, and sometimes keystone ecological species. Native wild boar (see Goulding, Chapter 19), European beaver (Conroy et al, 1998; Coles, 2000), wolf and lynx (Dennis, 1998) have all been subject to regional proposals, huge media speculation, a degree of public hysteria, and even one romantic novel that we are aware of that markets itself as 'a battle for the hearts and minds of the public' (Plant, 2000). Yet at the same time, as the sociocultural, economic, political and scientific debates rage, the wild boar has established itself unofficially (Goulding, Chapter 19), and there is evidence of breeding wild lynx (David Siddon, personal communication). We even have strong evidence of black leopards deliberately released into the English countryside and suggestions that they have bred successfully. Around Sheffield in South Yorkshire, they were kept as 'guard dogs' for scrap-metal works in the 1970s and let loose into the Peak District when the Dangerous Wild Animals Act of 1976 came into force (Julian Gillott, personal communication). These now present the authorities and scientists with dilemmas as to whether a former native species such as wild boar, absent for some centuries, but ecologically very significant, is native or alien. For the big cats, the issue is often more fundamentally of disbelief.

At the same time as these debates are taking place, exotic species and management techniques devoted to them dominate huge swathes of the landscape. Exotic conifers are imposed on large areas of both upland and lowland as productive forest; and in the lowlands especially, game management for exotic birds such as common pheasant and red-legged partridge has a massive topographical and cultural influence. In the uplands, while game management is mostly for the native (endemic subspecies) red grouse (itself a cultural icon from the Glorious Twelfth to whisky bottle labels), the drainage and burning regimes to produce monoculture heather are a culturally imposed feature of landscape modification to serve an economic purpose, and to benefit one species that is valued over others. Yet many of these landscapes are understood by the wider public to be 'native' and 'natural', and even cherished and embraced as 'wilderness'. Similar issues apply across the planet from Australia and New Zealand, to North America, to Africa and the Mediterranean. In North American prairie restoration for example, there remain serious problems in establishing effective long-term management in the absence of long-extinct large mammals. Since the large herbivores which drove the ecology of the natural prairie are lost to extinction, there are emerging scientific and environmentalist arguments for introducing the nearest extant animals such as the African elephant and other grassland ungulates. Exciting, yes? A touch scary, yes? But how native is that? (see also Allison, Chapter 17, on the prairie and alien species). The chapters in this book address these fundamental matters embedded in the ongoing dialogue about the issues, perceptions and problems surrounding native reintroductions and exotic invaders.

Problem species and their impacts

The scale of impacts of alien species is massive (see Simberloff, Chapter 8), and the relationships between people and invasions are undoubtedly complex (see McNeeley, Chapter 2). According to a recent newspaper report (Bruxelles, 2010) the National Trust estimates the cost of current controls of invasive exotic plants in Great Britain to be around £2.7 billion per year. In response to this threat, the Department for Environment, Food And Rural Affairs (Defra), chiefly responsible for these matters, has announced a new campaign called 'Be Plantwise'. This is the first part of a two-pronged attack on alien invaders and it aims to raise awareness among millions of domestic gardeners of problem species and the consequences of deliberate or accidental release. This is obviously a good idea, and in a nation obsessed by tending the cultural space that is one's garden, it will reach a wide audience, a powerful gardening constituency, but as a campaign it could be criticized as mere gesture-politics: doing something without spending real money. To properly address the long-term problems of invasive exotic species will require finance not forthcoming from central government. It also seems that we are not considering wider landscape management issues and problem species, but focusing on the label 'invasive aliens'. There is little sign of the debate moving towards sustainable conservation land management, supported by a wider community of stakeholders, public and private, scientific and popular. In particular, while an information campaign based on education is especially important in terms of garden escapes, the conservation problems run deeper than just exotics and include aggressively invasive native plants too. Problems such as invasive bracken, birch and even holly are simply ignored because they are 'native', and because in some instances we admire them or associate them with our sociocultural and religious lives (for example, holly at Christmas time). Perceptions of both native and alien species vary over time as does what is or is not a problem (Coates, 2006; Rotherham, 2003). Moreover, linked partly to this ambivalent attitude towards nature, people have often been crucial in both triggering escape and facilitating invasion (Rotherham, Chapter 15; Rotherham, 2001a, 2001b, 2005b).

At a worldwide level the scale of the impacts, and therefore of the challenges facing the delivery of coherent responses, are truly massive. Dr Sarah Simons, Executive Director of the Global Invasive Species Programme (GISP), recently stated that:

Despite the enormous costs, not only to biodiversity but also food security, human health, trade, transport and more broadly, economic development, invasive species continue to receive inadequate attention from policy makers and in 2010, there is simply no excuse for not tackling one of the greatest threats to the environmental and economic well-being of our planet.

Indeed, there is no denying the global impacts of aggressively invasive and often exotic species, especially on once isolated fragile island ecologies and areas of high endemic biodiversity. But problem species include regionally native ones too, and it seems that in many cases the triggers of invasion and of damaging impacts are human-induced environmental changes. These include moves away from traditional land management, and often economically driven controls, and also climate and other environmental changes such as gross eutrophication. To tackle effectively the consequences of aggressive invasions we needed to consider the phenomena 'in the round' and to address wider contextual issues too.

Obvious examples of other natives (along with bracken) causing problems include gorse on many heathland and grassland sites, hawthorn and blackthorn, and even birch, invading grasslands, heaths, moors or bogs. Willow and poplar both cause huge damage to buildings. Both Ted Green of the Ancient Tree Forum and Professor John Rodwell ('godfather' of the British National Vegetation Classification) now argue that sycamore, often the most despised of exotic trees in England, is actually native. Beech, non-native in northern England and Scotland, along with mature larch and other species also not native, are glorious additions to many sylvan landscapes. In England, native clematis (old man's beard) can be a pernicious weed of southern woods, as can native ivy. Wild rhododendron (introduced from Gibraltar in 1764) can be surprisingly good for many wildlife species including winter roosts of birds, breeding nightingales and cover for deer, badgers and otters (Rotherham, 2001b). Moreover, the impacts of exotic invasive rhododendron on ancient woods are not unique to this alien species. The adverse effects are because it is 'invasive', not necessarily because it is 'alien'. Native holly, abandoned and no longer cut for leaf fodder in traditional 'holly hags', now spreads invasively across many ancient coppices. This transforms the woodland ecosystem and eliminates woodland ground flora, yet there is no call to arms to remove it. One wonders why? In urban Sheffield, otters, back on the River Don since the early 21st century, are hiding out under dense stands of Japanese knotweed (Rotherham, 2009).

In terms of mammals, native roe deer can cause similar damage to alien Muntjac deer. Both red squirrel and grey squirrel can damage trees; even badger, one of our most iconic conservation species, can undermine buildings, gardens and occasionally railway lines! Red deer are native but may cause serious overgrazing, damage to trees, woods and forests, and road-traffic accidents. Many species, in the wrong place at the wrong time, can and will cause problems to nature and to people. Alien species are often particularly invasive but then so are many native species. An interesting issue is raised by the culturally significant brown hare, listed as a Red Data Book and Biodiversity Action Plan species but which is alien, a Norman introduction. The rabbit also is alien, but especially following the cultural severance of abandonment of traditional grassland management (Sheail, 1972), is vital in maintaining many species-rich wildflower pastures, and hugely important as food for predators such as common buzzards. And of course, in 1902 Beatrix Potter gave her fictional rabbit a blue waistcoat, a trug with carrots in it, and the name Peter, and single-handedly generated a huge sentimental cultural association between humans and rabbits, particularly from our formative experiences as readers of the genre of children's nature literature (Potter, 1902). This one book has sold 45 million copies worldwide and been translated into 36 languages. The rabbit has many friends. Enemies of the rabbit, most especially the Anti-Rabbit Research Foundation in Australia (since 1998 known as the Foundation for a Rabbit-free Australia), return to the power of children's literature to counter the rampant sentimentality for the British invader down under. Aussie kids are urged (through popular primary school books) to cherish and embrace the native desert marsupial Bilby at Easter time, and to shun the more traditional (but culturally invasive) Easter Bunny (Kessing and Garnett, 1994; 1999). They eat chocolate Easter Bilbies as part of scientifically sponsored cultural ecological restoration. There may be a tendency to chuckle at this evidence, but this is serious stuff in an invaded land such as Australia (Coman, 1999). Severing cultural ties with rabbits needs to be done at a young age, before the powerful and mentally invasive Beatrix Potter-effect can take hold! (See also Coates, Chapter 3).

Controls and controversy

Attempts to remedy damage and to stop or limit invasions can be problematic and control of alien species is frequently controversial, even when based on good science. In the UK, control or removal of planted (alien) conifers (sometimes 100–200 years old) on sand dunes is causing serious concern and even uproar in west Lancashire, and also at Newborough Warren in North Wales. Here after a lot of money has been spent to conserve the 'native' red squirrel, they may be sacrificed to remove planted conifers from the dunes. The Countryside Council for Wales (CCW) has decreed that these aliens should be cleared to free up the ancient sand dune systems; but many locals are dismayed that their squirrels will be lost. There are concerns about the dilution of the genetics of wild daffodils by hybridization with nasty garden escapees in that most cherished of English landscapes, the Wordsworthian Lake District. Similarly in most regions of Britain worries about Spanish bluebells seem totally overstated. Yet the creeping invasion of woodland by variegated yellow archangel has a massive impact on ecology and generates no interest from conservation bodies. Other invasives such as sweet cicely, which is also alien, again often gets no response from conservation bodies. Similarly Norway maple is still widely planted in landscaping schemes and is now colonizing everywhere but triggers no action to control its spread, and Russian vine is another accident waiting to happen, but generates no interest at all.

Shutting the stable door

Wild boar, the European eagle owl, the Monk parakeet and 60 other species have been added to the list of non-native species that pose a threat to Britain's indigenous animals. The Chinese water deer, the snow goose and 13 other birds, the slipper limpet and 7 other invertebrates, 35 plants including 2 kinds of rhododendron, and 2 types of algae have also been included on the list, created jointly by Defra and the Welsh Assembly government. In a statement about the additions, invasive, non-native species were described as 'one of the greatest threats to wildlife worldwide'. Wildlife Minister Huw Irranca-Davies said:

> *It is essential that our native species are given the protection they need to flourish; 2010 is the International Year of Biodiversity and it's more important than ever to do all that we can to halt the loss of biodiversity ... Stopping the spread of invasive non-native species makes a real difference to the survival of our own native plants, birds and animals.*

The Wildlife and Countryside Act (1981) prohibits the introduction into the wild of any animal which does not normally live or visit Britain or any plant or animal on the list, which is detailed in Schedule 9. Doing so carries a maximum punishment of two years in jail and a £5000 fine. Seven animals were removed from the list, including the Mongolian gerbil and the Himalayan porcupine, as these are no longer thought to be a threat. The former native wild boar is back in the wild in Britain in places such as the Weald in Kent and the Forest of Dean about 700 years after being hunted to extinction. It is an ecological agent for good in the management of robust woodland ecosystems, but in more fragile wetland ecosystems can be very destructive and a key predator of waterbird nests. In time, wild boar could be harvested sustainably for food and as an iconic sporting trophy, generating income which could be pumped back into woodland conservation. Would this be acceptable to the public? At the moment, the wild boar is an animal only encountered by landowners, foresters, surprised dog walkers, and naturalists keen to track one down. They have not yet fully invaded the public psyche, despite an elusive TV appearance on the hugely popular BBC Natural History Unit's *Autumnwatch* in November 2007. Calls from enthusiasts for a formal reintroduction into the UK have met with official resistance.

Yet Steve Carver at the University of Leeds comments:

> *It is very worrying to have Defra include native species, albeit largely extinct in the UK save recent reintroductions, in a list of non-native species that threaten UK biodiversity. Where did they get their scientific advice/evidence from on this? I've been trying to get the full list online but can't seem to find it (any hints?). If wild boar is included then it sounds ill thought out ... or is it just a political move? Certainly, wild boar inclusion goes*

> *against Article 22 of the EC Habitats Directive and Article 11 of the EC Birds Directive, doesn't it? Does it include beaver and lynx I wonder? Or is it just a list of species already here and perceived as a problem?*

So it seems that the scientific logic underpinning some of these decisions is very questionable. Furthermore, quite a number of these animals and plants are already here, will be hard or impossible to remove anyway, and in at least some high-profile cases are actually natives. To this must be added the fact that many species now accepted and cherished, like brown hare for example, are clearly not native, and other exotics such as rabbit and common pheasant are hugely influential in domestic ecology and the wider management and imagery of the British countryside. This muddled thinking detracts from the need to take serious action to address those species (like exotic signal crayfish and native bracken) that clearly are very damaging. In most cases, beyond public statements of policy, little real action is taken. Prohibitive legislation has played a role. The Destructive Imported Animals Act of 1932 was domestic declaration of official war on the North American muskrat (musquash) brought to the UK in the 1920s for fur-farming; the preamble mapping out the reason for its promulgation in Parliament is a fascinating insight into cultural values and attitudes towards foreigners in the hard times of the early 1930s: 'an act to make provision for prohibiting or controlling the importation into and the keeping within Great Britain of destructive non-indigenous animals, for exterminating any such animals which may be at large and for purpose connected with the matters aforesaid'. Importing, keeping or releasing a muskrat attracted a £20 fine (around £980 at today's values), and for more than four animals an additional £5 fine per beast (Public Acts, 1932). The Act received royal assent on 17 March 1932 (Hansard PD, 1932). The muskrat was gone from Britain by 1939, by which time 4388 had been killed (Gosling and Baker, 1989), success coming from a combination of good technological planning and rapid response. Sagoff (Chapter 6) considers in detail problems surrounding programmes for controlling damaging aliens in the USA and the degree to which expenditure and enforcement can be justified when human health is not a factor. These are complex sociocultural, economic and political issues and choices.

Specific actions, however apparently laudable in principle, remain controversial in both specialist and public arenas, and problematic in practice. A prime example is the British government-sponsored cull of North American ruddy ducks, one of the less welcome legacies from the Slimbridge Wildfowl Collection in Gloucestershire belonging to Sir Peter Scott (a pioneer of conservation). The British population of wild ruddy ducks descends from Slimbridge escapees after seven adults were brought here in 1948 by Scott. According to *The Observer* (Sunday 7 February 2010):

> *A controversial UK cull of ruddy ducks, a US native that has been compared to a 'feathered lager lout' for its displays of*

> *thuggish and amorous behaviour, has cost the British taxpayer more than £740 for each dead bird. Figures from the Department for Environment, Food and Rural Affairs (Defra) show that shoots of the chestnut-coloured bird have cost taxpayers £4.6m, yet only 6200 have been killed. The disclosure has sparked an outcry from ornithologists and animal activists, who have protested since the cull began five years ago. They say that the bird, targeted because it had interbred with the threatened white-headed duck in Spain, should have been left alone. The cull is due to end in August.*

The newspaper also quoted Lee Evans, founder of the British Birding Association and the radical twitchers' UK400 Club, who is a passionate believer that the cull should be abandoned in the face of poor science and public outrage within the powerful domestic birding community: 'It's appalling and pointless, a complete waste of taxpayers' money. What's the point of it all? Our ruddy ducks don't go to Spain, but the French ducks do, and the French are not culling their birds. These marksmen are getting away with murder.' Data suggests that by the winter of 2008/2009, the UK population of the ruddy duck had been reduced by almost 90 per cent (Henderson, 2009). A section of the British birding community has been very vocal in its fierce opposition to the ruddy duck cull, using the web, blogs and popular birding hobby magazines such as *Birdwatch* and *Birdwatching* or the letters sections of the more scientific *British Birds* and *British Wildlife*.

Without effective international cooperation then surely this project is doomed to failure. One consequence witnessed across much of England is that county bird recorders simply withhold the locational details of ruddy duck breeding and wintering sites in their patch. Regardless of any merits of the case for control, it seems that key arguments with some grassroots ornithologists have yet to be won. Without their support it is highly unlikely that any control programme could be effective. There is now a very subtle ruddy duck information counterinsurgency going on across the UK, with keen British birders and private landowners in a collusion of silence, to keep the cullers at bay. This is a sort of David-versus-Goliath confrontation with government and powerful NGOs being challenged by individual action. Not so much a cry of 'power to the people' as 'power to the birder!' So we question whether the cull has produced any noticeable, long-term, sustainable effects or had any real impact on the white-headed ducks in Spain; or has it been an expensive waste of time? Lessons of history show that to be effective control programmes need to bring together key stakeholders for closely coordinated action. The government-sponsored eradication of coypu from the Fens and Broads of eastern England after 1981 remains one of the few success stories in modern Britain; with landowners, government agencies and conservation bodies working together towards the common objective of removal, founded on a long-term study of population ecology (Gosling et al, 1981) and an incentive scheme for trappers to overcome the basic economic problem that 'trappers would be reluctant to work themselves out of a job' (Gosling and Baker, 1989). Another observation is that too often, even if we want control in order to avoid demonstrable ecological damage, the efforts are too little, too late. No doubt the 'ruddies' could have been dealt with within the first five years of establishment, but that didn't happen. Now, even if doubtful ornithologists (the celebrity television naturalist Bill Oddie is a vocal supporter of the ruddy duck as a charming species of waterfowl) and public could be won round, it is inconceivable that government would fund this level of activity in any sort of sustainable way. This is compounded by evidence that the British cull has been an expensive failure and that the European population is not being controlled. It may be a naive question, but surely the money would have been better spent in Spain helping to control hybrids within the white-headed duck's range, and on associated education and awareness-raising programmes? Many British birders argue that Spain should focus its energies on protecting white-headed duck habitat by halting blind and wholesale economic development of coastal wetlands into mass tourist resorts. They also point out that wild ruddy ducks have reached the Azores archipelago (in the North Atlantic Ocean), so might we be actually standing in the way of the very first wave of emigration as a species seeks to widen its geographical ecological frontiers?

Changing perceptions over time

What is acceptable and what is alien vary with time (Smout, Chapter 4). In the 1930s, the little owl, now a valued and admired diminutive member of British avifauna, yet introduced from France, was considered a serious threat to native (or valuable economic) species, especially chicks of game species, with calls for its eradication. The little owl lived to fight another day, having been publicly found 'not guilty as charged' in an extensive inquiry into its habits and diet (Hibbert-Ware, 1938). Even more challenging today is that eagle owls naturalized in small numbers in Northern England (and harboured on some remote Ministry of Defence estates) were probably once native to the country. We may even receive migrant eagle owls from the Continent from time to time, as natural invaders. Are they welcome or not, and how influential should the tough stance taken against eagle owls in Britain by some conservationists be? What impact are they having on rare upland breeding birds such as the hen harrier? Are they an iconic enough bird of prey species to pull in waves of nature tourists to structured eagle owl nest-viewing opportunities, offering education, interpretation, good PR possibilities for often-criticized big utility companies and income for rural communities? Well, yes, so eagle owl tourism may be a path that we should not be afraid to take. More problematic are internationally rare animals such as Chinese water deer; they are exotic species in Britain, and add to the overabundance of deer at present, but they constitute a significant proportion of the world population. Neither brown hare (UK national Biodiversity Action Plan species) nor rabbit (keystone species of many British ecosystems) are native; they are not accepted despite their minimal ecological impact. Other animals in the spotlight are beavers and wild boar.

Without these two mammals, many of our wildlife habitats lack major functional elements of their natural composition, an absence critically damaging to other key species. Lowland woods without wild boar lack microdisturbance, and the dispersal of important fungi. How do we respond to attempts either to reintroduce these species or to tolerate escapees? There is public outcry and a clamouring for eradication among some quarters; but if wild boar is recognized as native, then logically it should be protected. This might be with a role as game and an associated close season. In Germany and France people and wild boars seem to get along with minimal fuss, so why not here?

As a conservative species and culture, we dislike change and fluidity. They trouble us, and yet environmental and landscape history show constant ebbs and flows in wildlife populations and associated ecosystems. With enhanced global warming, this dynamism increases. The 'Little Ice Age' for example had huge impacts. Yet we do not have reliable data for most species and have little real information about how they have responded. There are few botanical records before about AD 1600 and many 'natives' would have fluctuated dramatically and continue to do so. In South Yorkshire, the prickly lettuce, hemlock and grass vetchling have all extended their ranges since the late 1980s. Yet even today, there is no information on former or current distribution, and no one has really noticed these dramatic changes in status.

Recombinant ecology

Increased urbanization and global climate change mean aliens do and will have increasingly important roles and functions in future landscapes. This new ecology, promoted by George Barker (formerly of the Nature Conservancy Council), has slowly been recognized, with pioneering work in Eastern Europe and more recently in Britain. 'Recombinant ecology' (Barker, 2000) will need to be understood for conservation management to work with this new suite of possibilities and to address actual problems of exotic species. Each generation of ecologists and decision makers has a different set of species that are acceptable to them in their 'natural' environment, but we often do not see this. As individuals passing through spatial and temporal ecologies, we carry personal perceptions of the environment that influence our reactions. Some deeply held precepts might also be wrong. 'English oaks' from which small children gather acorns to nurture, plant and sustain native botanical inheritance often have Dutch or other European parentage from the thousands of oaks and other species imported from continental nurseries during the 18th and 19th centuries. For the genuine article, you may need to gather from a genuine veteran which predates the imports.

We lack scientific rigour on hybridization between related exotic and native species for bluebells and daffodils, for example. We have teams of volunteers roaming woodlands to eradicate alien white bluebells, who are inadvertently removing rare pink and white forms of native bluebell. Where is the evidence of the problem? Variegated yellow archangel continues its spread with barely a murmur; still sold by garden centres with no warning. This seems irresponsible, but where would we stop if it were to be banned? There is a whole list of such plants waiting to jump the garden fence and with accelerated environmental change it is getting longer. Indeed this fence, seen in the past as an enclosing barrier, is now a real conduit for ecological change. Over recent decades landscape architects have created new problems, such as planting Norway maple that are now seeding into ancient woodlands – an accident waiting to happen. Science informs us, but our subsequent decisions are subjective. The new emerging subdiscipline of species history, written by environmental historians, historical geographers, biologists, historical ecologists (Harting, 1880; Beirne, 1952; Lambert, 1998; Yalden, 1999, 2003; Lovegrove, 2007), call them what you will, can only help to provide a fuller and richer and more rigorous understanding of both the distant and more recent ecological past, to be used in shaping and planning the future. This does not mean there is no problem; but it may infer that current approaches will not elicit a solution. At a most fundamental level it is necessary to revisit ideas and values and to consider carefully the lessons of history (a blend of the sociocultural history of humanity with the species history of animals), even before we frame the critical conservation questions for the present day. As plant ecologist Jim Dickson mused at a conference hosted by Scottish Natural Heritage in Perthshire in 1996, 'Good science, good history and pragmatism' may be the way forward (Dickson, in Lambert, 1998, p1) as we begin to understand how nature has changed naturally or has been modified by human action.

How we distinguish alien from native, what belongs from what does not, can be incredibly complicated and yet it lies close to the heart of much of this debate. Chew (Chapter 9), Trigger (Chapter 7), Pooley (Chapter 22) and Petrie (Chapter 21) all raise issues and present case studies that question many basic precepts. In this volume Beattie (Chapter 23) also discusses how these perceptions change over time, and stresses the importance of understanding how historical works may misjudge the evidence to present all-too-rigid dichotomies of alien and native. These are important considerations in understanding why we respond as we do to certain species and not to others. Improved awareness of these influences is vital to informed and effective responses to the problems which undoubtedly exist. Indeed, as argued by Osteen and Livingston (Chapter 20), such perceptions influence government budgets and programmes for control and also lead to potential controversies. In a significant contribution to the debate Binggeli (Chapter 13) discusses how there may be reactions to exotic species that vary according to their utilitarian value to local communities rather than their indigenous nature. Indeed there may be conflicts between stakeholders where useful species had adverse effects on native ecology and nature conservation. Such mundane issues may influence perceptions of what is good and what is bad quite dramatically. A key point raised at a recent conference in Sheffield (Rotherham, 2009) concerned the absence of good science underpinning many assessments of invasion problems, in considering historic changes in species distributions, and in terms of understanding vital aspects of species autecology. Bailey (Chapter 14) in his account of Japanese knotweed as an invader demonstrates the value of meticulous scientific research to help guide and inform our responses to problem species. But for many plants and animals there is a long way to go before we have such insights into their ecology, genetics and physiology. All these factors combine to challenge historians, ecologists, geographers and others to provide a more coherent understanding of the impacts of alien, exotic and introduced species on both people and on Nature.

It is this rich and diverse consideration of the broad sweep of issues relating to species invasions, introductions and reintroduction that this book addresses. In the following chapters we present a range of multidisciplinary contributions with in-depth case studies from around the world and topic-related chapters on particular themes and problems. We believe that only though a more holistic and inclusive approach to this subject, which is both academically interesting, challenging and publicly engaging, can a coherent understanding emerge of this fascinating and complex interaction between people and nature.

References

Anon. (2000) *Collins English Dictionary and Thesaurus,* HarperCollins, Glasgow

Anon. (1983) *The Shorter Oxford English Dictionary,* Guild Publishing, London

Barker, G. (ed) (2000) *Ecological Recombination in Urban Areas: Implications For Nature Conservation,* English Nature, Peterborough, UK

Beirne, B. P. (1952) *The Origin and History of the British Fauna,* Methuen, London

Brown, P. and Waterston, G. (1962) *The Return of the Osprey,* Collins, London

Bruxelles, S. de (2010) 'War is declared on the plant invaders that cost £2.7 billion a year to control', *The Times*, 24 February, p4

Carter, I. (2007) *The Red Kite,* Arlequin Press, Chelmsford, UK

Clement, E. J. and Foster, M. C. (1994) *Alien Plants of the British Isles*, Botanical Society of the British Isles, London

Coates, P. (2006) *American Perceptions of Immigrant and Invasive Species: Strangers on the Land,* University of California Press, Berkeley, CA, USA

Coles, B. (2000) *Beaver in the Western European Landscape,* CEHP, Stirling/St Andrews, Scotland

Coman, B. (1999) *Tooth and Nail: The Story of the Rabbit in Australia,* Text Publishing, Melbourne

Conroy, J. W. H., Kitchener, A. C. and Gibson, J. A. (1998) 'The history of the beaver in Scotland and its future reintroduction', in R. A. Lambert, *Species History in Scotland: Introductions and Extinctions Since the Ice Age*, Chapter 8, Scottish Cultural Press, Edinburgh, pp107–128

Dennis, R. (1998) 'The re-introduction of birds and mammals to Scotland', in R. A. Lambert, *Species History in Scotland: Introductions and Extinctions Since the Ice Age*, Chapter 1, Scottish Cultural Press, Edinburgh, pp5–7

Ellis, G. R. (1993) *Aliens in the British Flora,* National Museum of Wales, Cardiff

Gosling, L. M., Watt, A. D. and Baker, S. J. (1981) 'Continuous retrospective census of the East Anglian coypu population between 1970 and 1979', *Journal of Animal Ecology*, 50(3), October, 885–901

Gosling, L. M. and Baker, S. J. (1989) 'The eradication of muskrats and coypus from Britain', *Biological Journal of the Linnean Society*, 39(1), 39–51

Hansard Parliamentary Debates (1932) *Destructive Imported Animals Act.* 17 March, vol 263, col 502, London

Harting, J. E. (1880) *British Animals Extinct Within Historic Times,* Trubner, London

Henderson, I. (2009) 'Progress of the UK Ruddy Duck eradication programme', *British Birds*, December, 102, 680–690

Hibbert-Ware, A. (1938) *Report of the Little Owl Food Inquiry 1936–37,* H. F. and G. Witherby, London

Kessing, K. and Garnett, A. (1994) *Easter Bilby,* ARRFA/Feral Pests Program, Canberra

Kessing, K. and Garnett, A. (1999) *Easter Bilby's Secret,* Kessing Productions/RFA, Alice Springs

King, A., Ottaway, J. and Potter, A. (1976) *The Declining Otter: A Guide to its Conservation,* Friends of the Earth Otter Campaign, Chard, Somerset, UK

Lambert, R. A. (ed) (1998) *Species History in Scotland: Extinctions and Introductions Since the Ice Age,* Scottish Cultural Press, Edinburgh

Lever, C. (2009) *The Naturalized Animals of Britain and Ireland*, New Holland Publishers, London

Love, J. A. (1983) *The Return of the Sea Eagle*, CUP, Cambridge

Lovegrove, R. (1990) *The Kite's Tale: The Story of the Red Kite in Wales*, RSPB, Sandy, Beds, UK

Lovegrove, R. (2007) *Silent Fields: The Long Decline of a Nation's Wildlife,* OUP, Oxford

Plant, M. (2000) *Project Wolf,* Cairngorm Mountain Press, Edinburgh

Potter, B. (1902) *The Tale of Peter Rabbit,* Frederick Warne, London

Public Acts (1932) *Destructive Imported Animals Act,* ch 12; 22 and 23 George V, UK Parliament

Rotherham, I. D. (2001a) 'Himalayan Balsam: the human touch', in Bradley, P., (ed) *Exotic Invasive Species: Should we be Concerned?* Proceedings of the 11th Conference of the Institute of Ecology and Environmental Management, Birmingham, April 2000. IEEM, Winchester, UK, pp41–50

Rotherham, I. D. (2001b) '*Rhododendron* gone wild', *Biologist*, 48(1), 7–11

Rotherham, I. D. (2003) 'Alien, invasive plants in woods and forests ecology, history and perception', *Quarterly Journal of Forestry*, 97(3), 205–212

Rotherham, I. D. (2005a) 'Invasive plants: ecology, history and perception', *Journal of Practical Ecology and Conservation Special Series*, 4, 52–62

Rotherham, I. D. (2005b) 'Alien plants and the human touch', *Journal of Practical Ecology and Conservation Special Series*, 4, 63–76

Rotherham, I. D. (ed) (2005c) 'Loving the aliens? Ecology, history, culture and management of exotic plants and animals: issues for nature conservation', *Journal of Practical Ecology and Conservation Special Series*, 4

Rotherham, I. D. (2009) 'Exotic and alien species in a changing world', *ECOS*, 30(2), 42–49

Salisbury, E. (1961) *The New Naturalist Weeds and Aliens*, Collins, London

Sheail, J. (1972) *Rabbits and Their History,* Country Book Club, Newton Abbot

Warren, C. (2002) *Managing Scotland's Environment,* Edinburgh University Press, Edinburgh

Yalden, D. (1999) *The History of British Mammals,* T & A D Poyser, London

Yalden, D. (2003) 'Mammals in Britain: a historical perspective', *British Wildlife*, April, 243–251

2
Xenophobia or Conservation: Some Human Dimensions of Invasive Alien Species

Jeffrey A. McNeely

Introduction

The issue of invasive alien species is usually considered primarily a biological concern. But the problem is better seen as an expression of human culture, based on the sense of place that many people have, enlightened self-interest, and basic issues of economics and health. Support for this human perspective comes from four main directions. First, virtually all our planet's ecosystems have a strong and increasing anthropogenic component that is being fed by growing globalization of the economy and society; both people and goods are now moving freely and rapidly across the planet. Second, people are designing the kinds of ecosystems they find productive or congenial, with immigrants often bringing species with them from their native lands. Third, growing travel and trade, coupled with weakening customs' and quarantine controls, enable people to introduce – both inadvertently and intentionally – alien species that may become invasive. Fourth, the issue has important philosophical dimensions, requiring people to examine fundamental ideas, such as 'native' and 'natural'. The great increase in the introduction of non-native species that people are importing for economic, aesthetic, accidental or even psychological reasons is leading to more species invading native ecosystems, often with disastrous results: many become invasive alien species (IAS) that have significant deleterious effects on both ecosystems and economies. This chapter examines some of the important human dimensions of the IAS problem, including historical, economic, cultural, philosophical and political issues. These are addressed in terms of the causes, consequences and responses to the problem of IAS. This introduction shows that successfully addressing the problem of species invasions will call for greater collaboration between economic sectors and among a wide range of disciplines. The Convention on Biological Diversity and many other international agreements offer important opportunities for addressing the complex global problems of IAS through improved international cooperation in what boils down to an ethical issue: the conservation of native biodiversity.

Human impacts on the ecosystems of our planet continue to grow (MA, 2005). The increasing human population and growing wealth mean that more people consume more of nature's goods and services, pushing against the limits of sustainability. Greatly expanding global trade is feeding this consumption, with large containers of goods moving quickly from one part of the world to another by aeroplane, ship, train and truck.

One critical element in this economic and social globalization is the movement of organisms from one part of the world to another through trade, transport, travel and tourism. Many of these movements into new ecosystems where they are alien (also called non-native, non-indigenous or exotic) are generally beneficial to people. But many others have very mixed effects, benefiting some individuals or interest groups while disadvantaging others. In a few cases, especially disease organisms and forest or agricultural pests, the alien species is clearly detrimental to all, or nearly so. This book addresses the latter groups: invasive alien species (IAS), the subset of alien species whose establishment and spread threatens ecosystems, habitats or species with economic or environmental harm (GISP, 2001).

Farmers have been fighting foreign weeds for generations, and disease organisms have been a major focus of physicians for well over 100 years. But the general global problem of invasive alien species has been brought to the world's attention only relatively recently by ecologists concerned that native species and ecosystems are being disrupted (e.g. Elton, 1958; Drake et al, 1989). Much of the work to date on IAS has focused on their biological and ecological characteristics, the vulnerability of ecosystems to invasions, and the use of various means of control against invasives. However, the problem of IAS is above all a human one, for at least the following reasons:

- People are largely responsible for moving eggs, seeds, spores, vegetative parts and whole organisms from one place to another, especially through modern global transport and travel;
- While some species are capable of invading well-protected, 'intact' ecosystems, IAS more often seem to invade habitats altered by humans, such as agricultural fields, human settlements and roadways;
- Many alien species are intentionally introduced for economic reasons (a major human endeavour); and
- The dimensions of the problem of invasive alien species are defined by people, and the response is also designed and implemented by people, with differential impacts on different groups of people.

People introduce organisms into new habitats unintentionally (often invertebrates and pathogens), intentionally (usually plants and vertebrates), or inadvertently when organisms imported for a limited purpose subsequently spread into new habitats (Levin, 1989). Many deliberate introductions relate to human interest in nurturing species that are helpful to people for agricultural, forestry, ornamental or even psychological purposes (Staples, 2001). The great bulk of human dietary needs in most parts of the world are met by species introduced from elsewhere (Hoyt, 1992); it is difficult to imagine an Africa without cattle, goats, maize and cassava, or a North America without wheat, soy beans, cattle and pigs, or a Europe without tomatoes, potatoes and maize – all introduced species. Species introductions, therefore, are an essential part of human welfare and local cultures in virtually all parts of the world. Further, maintaining the health of these introduced alien species of undoubted net benefit to humans may sometimes require the introduction of additional alien species for use in biological control programmes; for example the importation of natural enemies of agricultural pests (Waage, 1991; Thomas and Willis, 1998); and these biological controls may themselves become invasive.

Considerable evidence indicates a rapid recent growth in the number and impact of IAS (Mooney et al, 2005). Trade and more general economic development lead to more IAS; Vilà and Pujadas (2001), for example, found that countries more effectively tied into the global trading system tend to have more IAS, being positively linked to development of terrestrial transport networks, migration rates, tourists visiting the country and trade in commodities (Dalmazzone, 2000). The global picture shows tremendous mixing of species, with unpredictable long-term results but a clear trend toward homogenization (Bright, 1998; Mooney et al, 2005). The future is certain to bring considerable additional ecological homogenization as people continue to introduce species. This ecological shuffling will enable some species to become more abundant and others to decline in numbers (or even become extinct), but the overall effect will probably be a global loss of biodiversity at species and genetic levels (McNeely, 2001). How is the great reshuffling of species being driven by human interests and how will it affect them? How should people think about the issue? What stakes are involved? Whose interests are affected? How can the human dimensions best be addressed by scientists, resource managers and policy makers?

These are not trivial questions, because the issue of IAS has ramifications throughout modern economies. It involves global trade, settlement patterns, agriculture, economics, health, water management, climate change, genetic engineering and many other fields and concerns. It therefore goes to the very heart of problems that policy makers spend much time debating, ironically usually without reference to IAS. This chapter examines some ramifications of IAS through many dimensions of human endeavour. It shows that IAS are deeply woven into the fabric of modern life, so more effective responses to the problems they pose must incorporate the kinds of human dimensions that are discussed in this chapter and elsewhere in this book.

Historical dimensions

Because of a long geological and evolutionary history, our planet has very different species of plants, animals and microorganisms on the various continents, and in the various ecosystems (Wallace, 1876). As a broad illustration, Africa has gorillas, Indonesia has orang-utans, South America has monkeys but no apes, and Australia has no non-human primates at all. Even within continents, most species are confined to particular types of habitats: gorillas live in forests, zebras mostly in grasslands, and addax in deserts. Oceanic islands and other geographically isolated ecosystems often have their own suites of species, many found nowhere else (termed 'endemic species'); about 20 per cent of the world's flora is made up of insular endemics found on only 3.6 per cent of the land surface area. Geographical barriers have ensured that most species remain within their region, thus resulting in much greater species richness across the planet than would have been the case if all land masses were part of a single continent. This historical biogeographical framework provides the basis for defining concepts of native and alien species. Of course biogeography has always been dynamic, as species expand and contract their ranges and the contents of ecosystems change as a result of factors such as climate change (Udvardy, 1969), so some movement of species is natural, just as climate change is. But just as climate change is accelerating due to human factors (IPCC, 2007), so are people speeding the movement of species around the world.

Homo sapiens apparently evolved in Africa, spreading to Europe and Asia over 100,000 years ago, Australia 40–60,000 years ago, the Americas about 15–20,000 years ago, and the far reaches of the Pacific less than 1,000 years ago. Our species is a good example of a naturally invasive species, spreading quickly, modifying ecosystems through the use of fire, and driving other species to extinction (Martin and Klein, 1984). Wherever people have moved they have also carried other species with them. The Asians who first peopled the Americas, for example, were accompanied by dogs, and Polynesians sailed with pigs, taro, yams and at least 30 other species of plants (with rats and lizards as stowaways).

Trade is known far back in human prehistory, judging from the discovery of stone tools at a considerable distance from where they were quarried. But as long-distance travel became more regular, trade became more important. The Chinese have traded with Southeast Asia for at least several thousand years, and trading routes between India and the Middle East stretch back at least as long. As sailing craft became larger and more reliable, trade increased further and was given a great boost with the voyages of Christopher Columbus that opened up entirely new sources of species (Crosby, 1972), and led to the replacement of the rigid moral strictures of medieval Europe by a new set of merchant values that stressed consumption (Low, 2001).

For at least several thousand years, armies have been an important pathway for moving species from one region to another, with at least some of these becoming invasive (like the armies). The spread of new diseases by armies is well known. For example, measles was carried into the Americas from Europe by the early conquistadors and perhaps syphilis went in the opposite direction (McNeill, 1976). Rinderpest, a virus closely related to measles and canine distemper, is native to the steppes of Central Asia, but it frequently swept through Europe, carried by cattle moved to feed armies during military campaigns. Africa remained free of this disease until 1887, when it appeared in Eritrea at the site of the Italian invasion, spreading through Ethiopia in 1888, and conquering the entire continent in less than a decade. In some parts of Africa, rinderpest was followed by wars and cattle raids as the tribal pastoralists sought to maintain their herd. Another result was that rinderpest led an ecological revolution against people and cattle and in favour of wildlife species resistant to the disease.

The period of European colonialism ushered in a new era of species introductions, as European settlers sought to recreate the familiar conditions of home (Crosby, 1986). They took with them species such as wheat, barley, rye, cattle, pigs, horses, sheep and goats, but in the early years their impacts were limited by available means of transport. Once steam-powered ships came into common use, the floodgates opened and between 1820 and 1930 more than 50 million Europeans emigrated to distant shores. They were carrying numerous plants and animals that were added to the native flora and fauna (Reichard and White, 2001). More recently, emigrants from Asia and Africa have carried familiar species with them to grow in their new homelands in Europe, Australia and the Americas.

The era of European colonialism also encouraged the spread of plant exploration, in the quest for new species of ornamental plants for botanical gardens, nurseries and private individuals back home, some of which escaped and became invasive (Reichard and White, 2001). The spread of global consumerism was given a significant boost in the early decades of the 20th century through advertising and marketing that was strategically designed to motivate the public to buy more goods (Staples, 2001). This ultimately led to an accelerating search to find new species to grow and market, creating consumer demand for products that previously were unknown. The invasive characteristics of the newly introduced species often came as a surprise, because those responsible for the introduction were unaware of possible negative ecological ramifications.

Many invasive species were carried by the colonial military, especially to Pacific and Indian Ocean islands that had numerous endemic species vulnerable to such invasives. In the 17th and 18th centuries, navies introduced many plants and animals to remote islands as future food sources, and these frequently became invasive (Binggeli, 2001). The military sometimes brought in exotic species of plants to form barriers. For example, the French introduced a cactus (*Opuntia monacantha*) to Fort Dauphin in southeast Madagascar in 1768 to provide what they hoped was an impregnable barrier around the fort. Later, the military also introduced a spineless variety (*O. ficus-indica*) to feed oxen (Decary, 1947). Both these cacti have now colonized much of Madagascar, though most of the French have returned home. The role of the military in the spread of IAS has continued. World War II was a particularly active time for the introduction of weeds in the Pacific. Some species, such as

Bermuda grass (*Cynodon dactylon*), were deliberately introduced to re-vegetate islands that were devastated by military activity. Other species spread by accident, clinging to military equipment and supplies or sticking to wheels of planes. Some grass species were carried from one island to another as seeds adhering to soldiers' clothing. Because many weeds do best on bare or disturbed ground, war helped to prepare a fertile ground for them. The brown tree snake (*Boiga irregularis*) came to Guam from New Guinea or the Solomon Islands during the war, apparently hitching a ride on either Allied or Japanese ships or planes. Arriving in Guam, the brown tree snake found that local birds and lizards were not well adapted to such an agile predator with poisonous fangs. As a result, the populations of native birds and lizards have plummeted on Guam. Now the brown tree snake threatens Hawaii, possibly carried by military transports from Guam.

Thus the faunal and floral assemblages found in any particular location have been profoundly influenced by past human activities, and people are likely to have an even greater impact in the future. This leads us to consider whether the current episode of globalization might lead to increased diversity in at least some places, after the dust settles on the current extinction spasm (Parker et al, 1999). One example is that New Zealand has twice as many plants today as it did when humans first arrived, as well as a whole suite of new mammals; one tragic cost was the loss of an extensive unique fauna of birds, including at least 12 species of flightless birds known as moas, some of which were larger than ostriches. On the other hand, relatively few species of native plants have become extinct, so some argue that the net effect on New Zealand ecosystems has been enrichment. Further development of biotic communities as climates change will depend on organisms invading novel habitats, sometimes hybridizing with native species, sometimes replacing them, and sometimes adding to ecosystem diversity by new species interactions. Through introducing species, humans are creating entirely new ecosystems, often more or less by accident, and disrupting ecosystems that had evolved over millions of years.

How people help species invade

Global trade has enabled modern societies to benefit from unprecedented movement and establishment of species around the world. Agriculture, forestry, fisheries, the pet trade, the horticultural industry and many industrial consumers of raw materials today depend on species that are native to distant parts of the world. The lives of people everywhere have been greatly enriched by their access to a greater share of the world's biological diversity, and expanding global trade is providing additional opportunities for further enrichment. Most people warmly welcome this globalization of trade, and growing incomes in many parts of the world are leading to increased demand for imported products. North American nursery catalogues, for example, offer nearly 60,000 plant species and varieties to a global market, often through the Internet (Ewel et al, 1999). A generally unrecognized side effect of this globalization is the introduction of alien species, at least some of which may be invasive.

Global mobility has undermined the sense of place that previously provided a psychological anchor to most societies. Modern information technology leaves us

> *with the disorienting experience that there is no single or universal context with clearly defined borderlines within which we can appeal to reason to settle our differences, but rather a multiplicity of mini-contexts that are not universally shared, within which numerous incommensurable interpretations co-exist alongside one another.* (Hattingh, 2001)

The speed at which many modern people live combined with great mobility seriously compromises their ability to develop a sense of place, a clear vision of the future or a sense of origin. This makes it much more difficult for people to distinguish between native and alien species, or to be concerned about the difference.

Linked to the global marketplace, the world is becoming increasingly urban, with half the world's population now living in cities and losing their links to nature. Cities tend to be focal points of the global economy and entry points for many invasives, which are most prolific in urban and urban-fringe environments where long histories of human disturbance have created abundant bare ground and many opportunities for invasion. Many urban dwellers seek ornamentals from a wide range of sources, and these may become invasive. For example, Berlin has 839 native species of plants and 593 aliens (Kowarik, 1990). Urbanization involves large and mobile populations that can easily escape the environmental penalties from misuse of resources. Further, they are seldom aware of the problems of invasive species because they have essentially lost their connections to the natural environment (Staples, 2001).

Many people who seek to introduce a non-native species into a new habitat do so for an economic reason (McNeely, 1999). They may wish to increase their profits from agriculture; they may believe that the public will be attracted to purchase a newly discovered flower from a distant part of the globe; or they may think that non-native species will be able to carry out functions that native species cannot carry out as effectively. But few of those introducing alien species have carried out a thorough cost–benefit analysis before initiating the introduction, ignoring ('externalizing') the negative impacts that may follow from species introductions because they have not been required to recognize them. They might also be worried that they would be expected to compensate those who are negatively affected. Similarly, those responsible for inadvertently introducing species into new habitats may not be willing to make the investment necessary to prevent such accidents from occurring. They may not realize the dangers, and in any case the dangers would be unlikely to affect their own welfare. Costs of such accidents are borne disproportionately by people other than those allowing them to happen. So the costs of introducing potentially invasive alien species into new habitats are externalized in consideration of the costs of global trade. The line of responsibility is insufficiently clear to bring about the necessary changes in behaviour, and the general public and future generations pay most of the costs of damage to native ecosystems.

In the early 1990s, Serbian scientists discovered the western corn rootworm (a beetle, *Diabrotica vigifera*, whose worm-like larvae feed on the roots of maize plants) near Belgrade airport, apparently inadvertently flown in on military aircraft from the USA. Vigorous international action might have curbed this pest's first known venture outside North America, but the turmoil of war prevented such collaboration and now it is too late. By 1995 the pest had spread into Croatia and Hungary, subsequently invading Romania, Bosnia-Herzegovina, Bulgaria and Italy (Enserink, 1999). It may well spread to every maize-growing country in Europe, and perhaps eventually to Asia, forcing farmers to use more chemical pesticides and kill harmless insects in the process. A problem that would have been relatively easy and cheap to solve if addressed promptly was not controlled due to the war that blocked the necessary collaboration, and the problem now has serious economic effects.

One limitation of human perception of the costs of IAS is that invasions often happen almost invisibly, without any clear responsibility, and with very limited initial impact. Further, monitoring, early detection and containment of invaders before they cause widespread damage are unlikely to be considered to have a positive cost–benefit ratio because the expense is required now while the main benefits (at least in terms of future costs avoided) remain speculative. On the other hand, where sound cost–benefit studies have been done, they demonstrate the value of control and that prevention is the best strategy (Jenkins, 2001, 2002).

Practically all human cultures actively modify their surroundings to achieve an environment that they find pleasing. At least part of the world's cultural diversity is due to the local patterns of distribution of plants and animals, because locally available resources and how their use help to define the character of a cultural group. Some IAS become part of the local culture. For example, Australian Aborigines today hunt mammals that have invaded Australia (such as water buffalo, rabbits and camels) and they argue that the government programmes to control these invasive aliens in the name of protecting native fauna and flora affect an important food source for them. They contend that in this way they are being forced to turn their hunting attention to native species already under threat.

The Maori people of New Zealand have similar concerns. For example, some Maori leaders initially opposed the eradication of invasive Pacific rats (*Rattus exulans*) from some islands, claiming that they were 'a treasure' brought to New Zealand by their ancestors in the course of their migrations (Veitch and Clout, 2001). While neither Maoris nor Aborigines have a single perspective on IAS, they share a concern about the use of poisons to control IAS, the potential for pollution of water supplies, and the introduction of yet more alien species for biological control.

Some suggest that people have an innate tendency to focus on life and life-like processes, a condition Wilson (1984) called 'biophilia'. This leads many people to value diversity for its own sake, perhaps seeking to enhance the options available for improving their physical or social well-being. One manifestation of this tendency may be a need or desire to have other, non-human, species living close by (Mack, 2001; Staples, 2001). In the USA, China, Europe and elsewhere, a thriving pet trade that answers this human need also poses continuous risks due to intentional or accidental releases by pet owners (see, for example, Genovesi and Bertolino, 2001). Even people who are professional resource managers, such as the staff at South Africa's Kruger National Park, can be remarkably resistant to the idea of limiting their cultivation of potentially invasive garden plants (Foxcroft, 2001). Thus human preference rather than biological traits may determine whether a plant or animal species is intentionally introduced.

How invasive species affect people

IAS have many negative impacts on human economic interests. Weeds reduce crop yields, increase control costs and decrease water supply by degrading catchment areas and freshwater ecosystems. Tourists unwittingly introduce alien plants into national parks, where they degrade protected ecosystems and drive up management costs. Pests and pathogens of crops, livestock and trees destroy them outright, or reduce yields and increase pest control costs. The discharge of ballast water introduces harmful aquatic organisms, including diseases, bacteria and viruses, to both marine and freshwater ecosystems, thereby degrading commercially important fisheries and recreational opportunities. Recently spread pathogens continue to kill or disable millions of people each year, with profound social and economic implications. While considerable uncertainty remains about the total economic costs of invasions, estimates of the economic costs of particular invasives to particular sectors indicate the seriousness of the problem (Perrings et al, 2000). Pimentel et al (2000), for example, estimate the economic losses in the USA due to invasive species at US$137 billion per year, a figure disputed by some economists.

Globalization is bringing with it a series of new medical threats, many of which can be considered a subset of the IAS problem. The global changes that affect many parts of the world are expected to lead to expansion of the ranges of many viruses potentially dangerous to humans. When people invade formerly unoccupied wilderness areas, this brings them into contact with a wider range of viruses and bacteria, while air travel carries them around the globe before their symptoms become apparent, as in the case of SARS.

Infectious disease agents often, and perhaps typically, are invasive alien species. Unfamiliar types of infectious agents, either acquired by humans from domesticated or other animals or imported inadvertently by travellers, can have devastating impacts on human populations. Pathogens can also undermine local food and livestock production, thereby causing hunger and famine. Examples include bubonic plague (caused by *Pasturella pestis*) spread from central Asia through North Africa, Europe and China using a flea vector on an invasive species of rat (*Rattus rattus*) that came originally from India. Viruses carrying smallpox and measles spread from Europe into the western hemisphere during European colonization. The low resistance of indigenous peoples to these diseases helped bring down the mighty Aztec and Inca empires, as well as devastating many tribal peoples. The Irish potato famine in the 1840s was caused by a fungus (*Phytophtora infestans*) introduced from North America, which attacked potatoes, with devastating impacts on the health of local people and driving many of them to emigrate; a similar fungus is now exterminating the native oak trees of California, fundamentally changing a significant native ecosystem based on these plants.

The influenza A virus has its origins in birds but multiplies through domestic pigs, which can be infected by multiple strains of avian influenza virus and then act as genetic 'mixing vessels' that yield new recombinant-DNA viral strains. These strains can then infect pig-tending people, who then infect other people, especially via rapid air transport. Avian influenza and swine flu have become a global scourge, dramatic examples of the impact of IAS, yet seldom recognized as such.

Dynamism among invasive pathogens, human behaviour and economic development is complex and depends on interactions between the virulence of the disease, infected and susceptible populations, the pattern of human settlements and their level of development. Large development projects, such as dams, irrigation schemes, land reclamation, road construction and population resettlement programmes, have contributed to disease invasions such as malaria, dengue fever, schistosomiasis and trypanosomiasis. Forest clearing in tropical regions to extend agricultural land has opened up new possibilities for wider transmission of viruses carrying haemorrhagic fevers that previously circulated benignly in wild animal hosts. Invasive species combined with variations in interannual rainfall, temperature, human population density, population mobility and pesticide use all contribute to one of the most profound human dimensions of invasive species: the threat to human health.

Components of biological diversity that are threatened or lost as a result of IAS can lead to the loss of traditional knowledge, innovations and practices. Likewise, customary uses of biological resources in accordance with traditional cultural practices may be inhibited or, in the worst case, discontinued completely. As intimate users of local biological resources, indigenous and local communities are potentially best-qualified to monitor the impacts of alien species on local ecosystems and their components (Article 7 of the Convention on Biological Diversity), to identify when those species become invasive, and to be involved in eradication and mitigation programmes (Article 8h of the CBD).

However, this depends on awareness of the problem. In China, Vietnam, Malaysia, Thailand, Korea and Cambodia (at least), people 'make merit' by releasing captive animals, especially birds, fish and turtles; but one study in Taiwan found that 6 per cent of birds released were exotic, and most of the fish and turtles were captive-bred exotic species that could become invasive (Severinghaus and Chi, 1999). Clearly, the cultural process of 'making merit' does not intentionally include deleterious impacts on native ecosystems, but occurs largely because the people involved have no concept of IAS.

How people respond to IAS

Addressing the problems of IAS uses four main management approaches: first, subject all alien species proposed for introduction to expert consideration, following the precautionary principle; second, improve the scientific basis for predicting which species proposed for deliberate introduction are likely to become invasive and which are likely to be beneficial; third, improve control of pathways for unplanned introductions (through ballast water, international trade, wooden packing material and so on); and fourth, improve management techniques to eradicate or control invasive alien species once prevention has failed or become impractical.

Governments have a responsibility to provide regulations in the public interest, but current economic orthodoxy argues that global trade is fostered through removing regulations that may constrain such trade, such as restrictions that may prevent the introduction of a potentially invasive alien species. These contradictions help to underline the conflict of interests between global trade and the control of IAS, and the challenges to current management measures and legal frameworks. The human dimension is the most unpredictable variable in any management programme to control IAS. Reaser (2001) and Mack (2001) go into considerable detail about the psychological factors motivating people to import or use alien species that sometimes become invasive, and show how a more thorough understanding of these psychological factors can slow further invasions and promote the control of the existing ones. They demonstrate that IAS are a by-product of human values, decisions and behaviours, suggesting that a focus on human beliefs and resultant behaviour might be more effective than focusing primarily on IAS themselves as the problem. Resource managers must therefore generate public support and understanding for any control programme before it begins. Thus, social embedding of management actions, as through the 'Working for Water Programme' in South Africa (Noemdoe, 2001), can foster effective management intervention.

Economic arguments have much to contribute to programmes that address the problems of IAS (Perrings et al, 2000). Decision makers often find arguments couched in economic terms to be more convincing than those cast in emotive or ethical terms, and economics-based arguments of costs and benefits can be used to support stronger programmes to deal with invasive species.

But while it is important to identify costs and benefits of IAS, such determination does not automatically lead to a decision because politically charged value judgements and issues about distribution of benefits are nearly always involved. Further, the magnitude of the costs may sometimes be so great as to render an action politically unacceptable, even when the benefits are likely to be even greater; part of the problem is that the benefits may be widely spread throughout the public over a period of many years, while the costs of control may need to be paid rather quickly by taxpayers. It appears that conflicts of interest between various sectors of society regarding the costs and benefits of IAS are an inevitable fact of modern life. Such conflicts might be mediated through a more thorough identification of the full costs of the IAS. However, the value of an alien species to any particular interest group may change over time, complicating the determination of costs and benefits.

Cultural factors also affect the perceptions different people have of the benefits and costs of IAS. Luken and Thieret (1996), for example, report that within less than a century after the deliberate introduction of Amur honeysuckle (*Lonicera maackii*) into North America to improve habitat for birds, serve ornamental functions in landscape plantings, and stabilize and reclaim soil, the shrub had become established in at least 24 states in the eastern USA. While many resource managers perceive the plant as an undesirable element, gardeners and horticulturalists consider it useful. Similarly, fishermen value the cultural importance of alien species of trout (*Oncorhynchus* spp, *Salmo* spp) in the Sierra Nevada mountains of the American West, despite the powerful negative impact they have on native frog species. St John's wort (*Hypericum perforatum*), a noxious weed with harmful effects on livestock in North America, is also gaining popularity in the natural pharmaceutical trade as an antidepressant and is being grown legally as an agricultural crop in northwest USA (Reichard, 2001). Thus the noxious invasive of one cultural group is the desirable addition of others.

The perception that local people have of introduced species may be different from that of conservationists, affecting how they respond. For example, in recent years, people living on Pitcairn Island – descendants of the *Bounty* mutineers – have not considered *Lantana camara* as a major weed as conservationists have done, but believe the shrub to be a soil improver. On the other hand, they view the tree *Syzygium jambos* as a major pest, not because of its impact on the native flora and fauna, but rather because of its heavy shading and its spreading, shallow and dense rooting system which renders cultivation of gardens an arduous task. Thus the weed status of a species relates to the way it interferes with day-to-day activities and will change through time as society develops (Binggeli, 2001).

The words that people use to articulate their concepts and values are often taken for granted, but their linguistic framework contains many assumptions, unarticulated values, implications and consequences that need to be critically scrutinized if people are to be effective in airing their concerns about IAS (Hattingh, 2001). The vocabulary used to describe IAS often implies conceptual oppositions, such as native–alien, pure–contaminated, harmless–harmful, original–degraded and diversity–homogeneity. Ideals like ecological integrity and authenticity associated with the values implied by such oppositions are undermined by the modern forces of globalization, which use an even more powerful set of oppositions, such as wealth–poverty, freedom–constraint, private–public, and connected–disconnected.

Because globalization tends to promote homogenization, Hattingh argues that conservationists are unable to provide a sufficiently strong conceptual framework that can incorporate the basic concepts that are required to respond to the problem of IAS, like uniqueness, ecological integrity and biological diversity. Some methods of controlling IAS may carry health hazards as well. For example, pesticides can have serious effects on both people and ecosystems.

Between 1975 and 1985, forests in Atlantic Canada were sprayed with the insecticide Matacil to control spruce budworm (*Choristoneura fumiferana*). In the late 1990s, fisheries and environmental scientists inferred that the declines in the Atlantic salmon (*Salmo salar*) stocks in the Restigouche River that occurred at that time were related to exposure of the smolt to nonylphenol used as an inert solvent in the pesticide (Fairchild et al, 1999).

Once public enthusiasm to control IAS has been generated, it must be channelled in the right direction. For example, gorse (*Ulex europeus*) has become invasive in montane grasslands of Sri Lanka following its introduction about 150 years ago. Recently, several local NGOs have launched volunteer programmes to remove gorse. However, some species of endemic reptiles and amphibians have found gorse a congenial habitat providing food and cover. When eradication programmes removed this habitat virtually overnight, endemic species were exposed to native opportunistic predators such as crows (Marambe et al, 2001). Therefore, programmes to eradicate invasive plant species should consider restoring its ecological functions.

More than 40 international conventions, agreements and guidelines have been enacted for addressing the problem of IAS, at least in part, and many more are being prepared (Shine et al, 2000). Governments have expressed their concerns about the problem of IAS especially through the Convention on Biological Diversity (CBD), which calls on the Parties to 'prevent the introduction of, control or eradicate those alien species which threaten ecosystems, habitats, or species' (Article 8h). But the expanding impact of IAS on both global economies and the environment implies that these international instruments have been insufficient to prevent and combat IAS effectively, suggesting that additional measures, such as a protocol under the CBD, are advisable.

At the national level, those opposed to eradicating IAS on ethical grounds are often prepared to argue their case in court, where litigation can be effective. This challenge calls for a legal framework that clearly recognizes the need to eradicate IAS when they threaten the greater public good, and education for judges to ensure that they understand the issues before them. National and even local legislation also needs to recognize the human dimensions of species invasions identified here, including such elements as ethical concerns, human health, trade, cultural considerations and even international obligations. Human dimensions are an essential element in determining what existing regulatory, financial and penal disincentives could be adjusted to deter high-risk trade and transport activities, and to determine specific levels of disincentives that will deter invasives (Jenkins, 2001, 2002).

Political support is clearly essential to implement coherent policies, laws and regulations to address the problem of IAS. This depends in turn on the support of the public, which ultimately depends on the quality of information provided on the issue and the effectiveness with which it is transmitted. Advocates need to convince the general public that controlling an invasive species is worthwhile. For example, the programme in New Zealand to control brush-tailed possum (*Trichosurus vulpecula*) was called 'Operation ForestSave' and the promotional information showed lovely flowering native trees, without a cute furry possum in sight. In Europe, one issue is presented as 'Save the Red Squirrels', not 'Kill the Grey Squirrels'. In Queensland, Australia, the endangered cassowary (*Casuarius casuarius*) is used as the front for controlling feral pigs, fully involving community groups in the programme (Low, 2001).

The current South African effort to eradicate, or at least substantially reduce, the population of invasive European rabbits on Robbin Island is made more acceptable because the island is a World Heritage Site that has iconic status as the prison home of Nelson Mandela for 18 years. Even there though, some animal rights advocates argue that turning what is almost a holy place into a killing field is immoral (ignoring the ecological damage caused by the rabbits).

The US National Park Service has encountered considerable public resistance to efforts to eradicate wild burros (*Equus asinus*) from the Mojave National Reserve, mountain goats (*Oreamnos americanus*) from Olympic National Park, and sheep (*Ovis aries*) from Santa Cruz Island. At least part of this resistance arises because these IAS are large mammals that some interest groups find attractive, but the negative ecological and economic impact of these species could be used to influence the political process. For example, heavy grazing of native vegetation by feral populations of horses and donkeys allows non-native annual plants to displace native perennials, and costs the nation an estimated US$5 million per year in forage losses, implying that these species eat forage worth US$100 per animal per year. They also diminish the primary food sources of native bighorn sheep (*Ovis canadensis*) and seed-eating birds, reducing the abundance of these natives (Pimentel et al, 2000). By making people aware of the ecological and economic impact of the invasive large grazers, public attitudes to control operations may be changed. But this issue remains controversial even among scientists, with some arguing that since North America had several species of equids before humans arrived, these now-invasive species should instead be seen as part of 'rewilding' the American west (Foreman, 2004).

Different interest groups may have different ethical positions. At least some animal rights groups, for example, argue that the intrinsic right to exist rests at the level of individual animals, not only of the species as a whole, and therefore strongly resist any measures to control them, much less eradicate entire populations even if they are IAS (though few extend this to pathogenic microorganisms or even insects). Animal rights advocates contend that nature will find its own solution to the new situation, and that any human intervention is immoral because people have no right to select one species or individual over another. (Though of course people do interfere in this way when introducing an alien species into a new habitat.) Public acceptance can be significantly influenced by animal rights groups, whose views must be considered when assessing the feasibility of eradicating an invasive species. Unfortunately, their ethical concerns cannot be answered by scientific evidence of conservation threats, or even by economic arguments. Some widely held ethical values have unrecognized ramifications for IAS. For example, the world has become increasingly interconnected over both time and space, where individuals have come to expect great freedom of individual behaviour (Low, 2001). But their behaviour when introducing alien species has significant, though undefined, influences on many other people. Most of these people are unknown to those who are affecting that behaviour.

Ethics of obligations and responsibilities are not always easily understood against the backdrop of the ethic of 'consumer freedom' to grow exotic plants or keep exotic pets. This is pertinent to the keeping of species that might escape captivity and become invasive, such as Burmese pythons spreading through Florida's Everglades National Park.

Thus the invasive alien species issue can be seen as ultimately an ethical concern. If people seek to maximize their material welfare, or even the diversity of species with which they surround themselves, alien species might well be a part of their rational response. But when alien species become invasive, destabilizing ecosystems and reducing diversity, then control is a far more acceptable, even necessary, response. Since invasions invariably involve trade-offs, the determination of costs and benefits of IAS becomes paramount (though this too has its ethical components).

Conclusions

IAS are able to invade new habitats and constantly extend their distribution, thereby representing a threat to native species, human health or other economic or social interests. One remarkable human dimension is the fact that a strong consensus can be built that many specific invasions are harmful, including killer bees, water hyacinth, kudzu, spruce budworms, various pathogens and agricultural weeds. The issue of IAS, therefore, can bring together interest groups that might otherwise be in opposition, such as farmers and conservationists. Bringing in the human dimensions can shift the focus from the IAS itself onto the human actions that facilitate its spread or manage its control. It implies that focusing directly on the invasive species is likely to provide only symptomatic relief. A more fundamental solution requires addressing the ultimate human causes of the problem, often the economic motivations that drive or enable species introductions.

As Pfeiffer and Voeks (2008) conclude, 'An understanding of the processes by which invasive biota become culturally enriching, facilitating, or impoverishing can contribute to articulating interdisciplinary programmes aimed at simultaneously conserving biological and cultural diversity.'

Gould (1998) argues that preference for native species provides 'the only sure protection against our profound ignorance of the consequences when we import exotics', because of the lack of certainty about the behaviour of an alien species imported into a new environment. Thus everything possible should be done to prevent unwanted invasions, carry out careful assessments before intentionally introducing an alien species into a new environment, build a stronger awareness among the general public about the problems of IAS, mobilize conservation organizations to address the problems, and build an ethic of responsibility among those most directly involved in the problem. The global trading system brings many benefits but it needs to be managed in a way that minimizes any deleterious impacts of invasive alien species on ecosystems, human health and economic interests. Human dimensions are central in doing so.

References

Binggeli, P. (2001) 'The human dimension of invasive woody plants', in McNeely, J. A. (ed) *The Great Reshuffling: Human Dimensions of Invasive Alien Species*, IUCN, Gland, Switzerland, pp145–160

Bright, C. (1998) *Life Out of Bounds: Bioinvasion in a Borderless World*, W. W. Norton, New York

Crosby, A. W. (1972) *The Colombian Exchange: Biological and Cultural Consequences of 1492*, Greenword Press, New York

Crosby, A. W. (1986) *Ecological Imperialism: The Biological Expansion of Europe, 900–1900*, Cambridge University Press, Cambridge, UK

Dalmazzone, S. (2000) 'Economics factors affecting vulnerability in biological invasions', in Perrings, C., Williamson, M. and Dalmazzone, S. (eds) (2000) *The Economics of Biological Invasions*, Edward Elgar, Cheltenham, UK, pp17–30

Decary, R. (1947) 'Epoque d'introduction des Opuntias monocantha dans le Sud de Madagascar', *Revue Internationale de Botanique Appliquée et d'Agriculture Tropicale*, 27, 455–457

Drake, J. A., Mooney, H. A., di Castri, F., Groves, R. H., Kruger, F. J., Rejmanek, M. and Williamson, M. (eds) (1989) *Biological Invasions: A Global Perspective*, Wiley, Chichester, UK

Elton, C. S. (1958) *The Ecology of Invasions by Plants and Animals*, Wiley, New York

Enserink, M. (1999) 'Biological invaders sweep in', *Science*, 285, 1834–1836

Ewel, J. J., O'Dowd, D. J., Bergelson, J., Daehler, C. C., D'Antonio, C. M., Gomez, D., Gordon, D. R., Hobbs, R. J., Holt, A., Hopper, K. R., Hughes, C. E., Lahart, M., Leakey, R. R. B., Lee, W. G., Loope, L. L., Lorence, D. H., Louda, S. M., Lugo, A. E., Mcevoy, P. B., Richardson, D. M. and Vitousek, P. M. (1999) 'Deliberate introductions of species: research needs', *BioScience*, 49(8), 619–630

Fairchild, W. L., Swansburg, E. O., Arsenault, J. T. and Brown, S. B. (1999) 'Does an association between pesticide use and subsequent declines in catch of Atlantic salmon represent a case of endocrine disruption?' *Environmental Health Perspectives*, 107, 49–358

Foreman, D. (2004) *Rewilding North America: A Vision for Conservation in the 21st Century*, Island Press, Washington, DC

Foxcroft, L. C. (2001) 'A case study of human dimensions in invasion and control of alien plants in the personnel villages of Kruger National Park', in McNeely, J. A. (ed) (2001) *The Great Reshuffling: Human Dimensions of Invasive Alien Species*, IUCN, Gland, Switzerland, pp127–134

Genoviesi, P. and Bertolino, S. (2001) *Piano di Azione Nazionale per il controllo dello Scoiattiolo grigio (Sciurus carolinesis)*, Ministry of Environment, National Wildlife Institute, Bologna, Italy

Genovesi, P. and Sandro B. (2001) 'Human dimension aspects in invasive alien species issues', in McNeely, J. A. (ed) (2001) *The Great Reshuffling: Human Dimensions of Invasive Alien Species*, IUCN, Gland, Switzerland, pp113–120

GISP (Global Invasive Species Programme) (2001) *Global Invasive Alien Species Strategy*, GISP, Cambridge, UK

Gould, S. (1998) 'An evolutionary perspective on strengths, fallacies, and confusions in the concept of native plants', *Arnoldia*, vol 58, no 1, pp3–10

Hattingh, J. (2001) 'Human dimensions of alien invasive species in philosophical perspective: toward an ethic of conceptual responsibility', in McNeely, J. A. (ed) (2001) *The Great Reshuffling: Human Dimensions of Invasive Alien Species*, IUCN, Gland, Switzerland, pp183–194

Hoyt, E. (1992) *Conserving the Wild Relatives of Crops*, IBPGR, IUCN, and WWF, Rome, Italy

IPCC (Intergovernmental Panel on Climate Change) (2007) *Fourth Assessment Report*, UNEP, Nairobi

Jenkins, P. T. (2001) 'Who should pay? Economic dimensions of preventing harmful invasions through international trade and travel', in McNeely, J. A. (ed) *The Great Reshuffling: Human Dimensions of Invasive Alien Species*, IUCN, Gland, Switzerland, pp79–88

Jenkins, P. T. (2002) 'Paying for protection from invasive species', *Issues in Science and Technology*, Fall, pp67–72

Kowarik, I. (1990) 'Some responses of flora and vegetation to urbanization in Central Europe', in Sukopp, H. J., Mejny, S. and Kowarik, I. (eds) *Urban Ecology: Plants and Plant Communities in Urban Environments*, SPB Academic Publishing, The Hague, pp45–74

Levin, S. A. (1989) 'Analysis of risk for invasions and control programmes', in Drake, J. A., Mooney, H. A., di Castri, F., Groves, R. H., Kruger, F. J., Rejmanek, M. and Williamson, M. (eds) (1989) *Biological Invasions: A Global Perspective*, Wiley, Chichester, UK, pp425–432

Low, T. (2001) 'From ecology to politics: the human side of alien invasions', in McNeely, J. A. (ed) (2001) *The Great Reshuffling: Human Dimensions of Invasive Alien Species*, IUCN, Gland, Switzerland, pp35–42

Luken, J. O. and Thieret, J. W. (1996) 'Amur honeysuckle, its fall from grace', *BioScience*, 46(1), 18–24

MA (Millennium Ecosystem Assessment) (2005) *Ecosystems and Human Well-Being: Synthesis*, Island Press, Washington, DC

Mack, R. N. (2001) 'Motivations and consequences of the human dispersal of plants', in McNeely, J. A. (ed) (2001) *The Great Reshuffling: Human Dimensions of Invasive Alien Species*, IUCN, Gland, Switzerland, pp23–34

Marambe, B., Bambaradeniya, C., Pushpa Kumara, D. K. and Pallewatta, N. (2001) 'Human dimensions of invasive alien species in Sri Lanka', in McNeely, J. A. (ed) (2001) *The Great Reshuffling: Human Dimensions of Invasive Alien Species*, IUCN, Gland, Switzerland, pp135–144

Martin, P. S. and Klein, R. G. (eds) (1984) *Quaternary Extinctions: A Prehistoric Revolution*, University of Arizona Press, Tucson, AZ, USA

McNeely, J. A. (1999) 'The great reshuffling: How alien species help feed the global economy', in Sandlund, O. T., Schei, P. J. and Viken, A. (eds) *Invasive Species and Biodiversity Management*, Kluwer Academic Publishers, Dordrecht, The Netherlands, pp11–31

McNeely, J. A. (ed) (2001) *The Great Reshuffling: Human Dimensions of Invasive Alien Species*, IUCN, Gland, Switzerland

McNeill, W. H. (1976) *Plagues and Peoples*, Anchor Press, Garden City, New York

Mooney, H., Mack, R., McNeely, J. A., Neville, L., Schei, P. and Wagge, J. (eds) (2005) *Invasive Alien Species: A New Synthesis*, Island Press, Washington, DC

Noemdoe, S. (2001) 'Putting people first in an invasive alien clearing programme: Working for Water programme', in McNeely, J. A. (ed) (2001) *The Great Reshuffling: Human Dimensions of Invasive Alien Species*, IUCN, Gland, Switzerland, pp121–126

Parker, I. M., Simberloff, D., Lonsadale, W. M., Goodell, K., Wonham, M., Kareiva, P. M., Williamson, M. H., Von Holle, B., Moyle, P. B., Byers, J. E. and Goldwasser, L. (1999) 'Impact: Toward a framework for understanding the ecological effects of invaders', *Biological Invasions*, 1, 3–19

Perrings, C., Williamson, M. and Dalmazzone, S. (eds) (2000) *The Economics of Biological Invasions*, Edward Elgar, Cheltenham, UK

Pfeiffer, J. and Voeks, R. (2008) 'Biological invasions and biocultural diversity: linking ecological and cultural systems', *Environmental Conservation*, 35(4), 281–293

Pimentel, D., Lach, L., Zuniga, R. and Morrison, D. (2000) 'Environmental and economic costs of non-indigenous species in the USA', *BioScience*, 50, 53–65

Reaser, J. K. (2001) 'Invasive alien species prevention and control: The art and science of managing people', in McNeely, J. A. (ed) (2001) *The Great Reshuffling: Human Dimensions of Invasive Alien Species*, IUCN, Gland, Switzerland, pp89–104

Reichard, S. H. and White, P. (2001) 'Horticulture as a pathway of invasive plant introductions in the United States', *BioScience*, vol 51, pp103–113

Severinghaus, L. L. and Chi, L. (1999) 'Prayer animal release in Taiwan', *Biological Conservation*, 89, 301–304

Shine, C., Williams, N. and Burhenne-Guilmin, F. (2000) *Legal and Institutional Frameworks on Alien Invasive Species*, IUCN Environmental Law Programme, Bonn, Germany

Staples, G. W. (2001) 'The understorey of human dimensions in biological invasions', in McNeely, J. A. (ed) (2001) *The Great Reshuffling: Human Dimensions of Invasive Alien Species*, IUCN, Gland, Switzerland, pp171–182

Thomas, M. B. and Willis, A. J. (1998) 'Biocontrol: Risky but necessary?' *TREE*, 13(8), 325–329

Udvardy, M. D. F. (1969) *Dynamic Zoogeography*, Van Nostrand Reinhold, New York

Veitch, C. R. and Clout, M. N. (2001) 'Human dimensions in the management of invasive species in New Zealand', in McNeely, J. A. (ed) (2001) *The Great Reshuffling: Human Dimensions of Invasive Alien Species*, IUCN, Gland, Switzerland, pp63–74

Vilà, M. and Pujadas, J. (2001) 'Socio-economic parameters influencing plant invasions in Europe and North Africa', in McNeely, J. A. (ed) (2001) *The Great Reshuffling: Human Dimensions of Invasive Alien Species*, IUCN, Gland, Switzerland, pp75–78

Waage, J. K. (1991) 'Biodiversity as a resource for biological control', in D. L. Hawksworth (ed) *The Biodiversity of Micro-organisms and Invertebrates: Its Role in Sustainable Agriculture*, CAB International, Oxford

Wallace, A. R. (1876) *The Geographical Distribution of Animals*, Macmillan, London

Wilson, E. O. (1984) *Biophilia*, Harvard University Press, Cambridge, MA, USA

Part II
Attitudes and Perceptions

3
Over Here: American Animals in Britain

Peter Coates

Introduction

In his seminal work on the impact of European flora and fauna intentionally and inadvertently brought to the Americas, Alfred Crosby emphasized the asymmetry of biotic exchange between the old and new worlds (Crosby, 1972, 1986). Europe had received few problem species in return for the plethora of notorious invasive non-natives inflicted on North America, not least the 'English' sparrow and the starling, which had 'dispossessed millions of American birds'. However, as John MacKenzie has pointed out, the flow of troublesome transatlantic traffic was not quite so one-directional. For MacKenzie, Crosby's perspective was a 'strikingly American one, for anyone who lives in the British Isles is almost daily brought face to face with the fact that ecological colonialism has been a two-way process'. Plenty of extra-European species caused headaches for wildlife and habitat conservationists (MacKenzie, 2001a, 2001b). Crosby remained unmoved. 'How often', he rejoined, 'have American species swamped and driven to the verge of extinction native species in Great Britain? Do you have over there, for instance, millions of American rats as we do millions of Old World brown and black rats? If there are equivalently successful aliens in the UK, I would like to know about them' (MacKenzie, 2001b).

There may be no American faunal species whose British exploits are comparable to Crosby's rats (which, in Britain, were themselves non-native species from the Eurasian mainland). Nonetheless, various American animals have flourished in the UK. This chapter shifts attention from European species over there to American species over here. In particular, it challenges the frequent assumption in US literature on invasive non-native species that British consciousness of the distinction between native and non-native species has been far lower than that of Americans and that Britons have been far more relaxed about their activities (Sagoff, 2002). In Britain, controversial immigrant animals abound, and many of the most heavily contested are North American introductions: the grey squirrel (1870s), mink (1920s), ruddy duck (1940s), signal crayfish (1970s) and the bullfrog (1990s).

I emphasize the cultural dimensions of British debates over the desirability and impact of American immigrant animals, especially the role of anti-American sentiments. Furthermore, I suggest that the issue of British attitudes to American 'problem' animals is the least studied aspect of anti-Americanism and Americanization. Not one of the many recent books on anti-Americanism published within the climate of exacerbated anti-American feeling that has marked the first decade of the new millennium, mentions the phenomenon of ecological nationalism or a single disruptive animal of American provenance (Revel, 2004; Hollander, 2004; Sweig, 2006; Markovits, 2007; Berman, 2008). Scrutinizing animals resident in Britain without faunal citizenship – particularly the grey squirrel – this chapter is a preliminary attempt at a natural history of British anti-Americanism.

Oversexed and over here

The strong American representation among animals having 'major negative environmental effects' was highlighted in an audit of non-native species by English Nature, the then statutory government agency responsible for wildlife and nature conservation. Of 12 non-native animals wreaking the most serious havoc, five were North American: signal crayfish, Canada goose, ruddy duck, mink and grey squirrel (Hill et al, 2005). The Environment Agency's recent compilations of the 'Top Ten Most Wanted Foreign Species' confirm this high American profile; the American signal crayfish and American mink ranked second and third on the list (Environment Agency, 2006). They also loom large in journalistic critiques of British 'hatred' of non-native animals (Liddle, 2002; Johnston, 2009).

The signal crayfish exemplifies characteristics that have equipped various American species for competitive success. Brought to southern England in the early 1970s to establish a crayfish farming industry, the freshwater crustaceans readily escaped their tank, and many were deliberately released. Relative to its native counterpart, the white-clawed crayfish, the American species is bigger and more aggressive, breeds earlier, lays more eggs and hatches sooner, eats a wider range of foods, is more tolerant of poor water quality, and spreads a fungal disease to which it is immune. Able to crawl for considerable distances and climb substantial heights, it colonizes watercourses at the rate of 1 km a year. Moreover, unlike native crayfish, it burrows, and its excavations honeycomb river banks, rendering them liable to collapse. As a result, indigenous crayfish have been virtually supplanted in rivers south of a line drawn between the Severn and Humber. The signal crayfish has also begun to colonize tributaries of the Tweed in southern Scotland, where its taste for salmon eggs threatens recovery of salmon stocks (Amos, 2003; Kirkup and Johnston, 2007).

The pejorative humanization of these American species by government scientists, natural resource managers, journalists and members of the public is incorrigible. The Environment Agency explains that signal crayfish and mink have 'taken advantage of Britain's welcoming living conditions' and 'overstayed their environmental visa' (Environment Agency, 2006). The agency's press office personifies them as 'wanted' persons and describes their activities as 'crimes'. The British National Party (whose website features stories about the misdeeds of alien species and the plight of various native creatures) has also entered the fray, one of its officials characterizing the signal crayfish as 'the Mike Tyson of crayfish ... a diseased, psychotic, evil, illegal immigrant colonist who displaces the indigenous crayfish, colonizes their territory and then reproduces until it totally devastates the indigenous environment and indigenous crayfish' (Monbiot, 2009).

Metaphors of the vicious, fast-breeding, all-conquering alien saturated the earlier mink debate of the 1970s and 1980s (Linn and Chanin, 1978a, 1978b). A semi-aquatic member of the weasel family, the mink was imported to stock fur farms in the late 1920s. However, it proved to be an accomplished escape artist, and was also released by bankrupt mink farmers and 'liberated' by animal rights activists. By the early 1950s, it was putting down roots in the English countryside and naturalized in Scottish waters within a decade.

Few animals are more vilified in British sporting and conservation circles, where the foreign mink 'menace' is blamed for the depletion of native game fish and waterfowl populations. Its infestation is also heavily implicated in the decline of otter and water vole (Dunstone, 1993; Dunstone and Ireland, 1989; Lever, 1978).[1] The water vole, enshrined in the affections of generations of Britons in the endearing shape of Kenneth Grahame's fictional character Ratty from *The Wind in the Willows*, has apparently been pushed to the brink by what is routinely lambasted as a 'deadly' 'American invader' (Marks, 2007). Tales of the mink's infamy have even entered children's fiction. In Alan Lloyd's *Kine* (1982), the unruffled world of marshland and woods inhabited by noble native species is disrupted by the brutal raid of a ferocious overseas aggressor. Seeking revenge for the incarceration of their ancestors on fur farms, the mink are 'monstrous aliens' with insatiable appetites that slay 'promiscuously' and 'terrorize' the coastal flats and valleys where Kent and Sussex meet. The hero, Kine, is a male weasel who rallies the native creatures and spurs them on to the ultimately successful reconquest of their homeland (Lloyd, 1982).

Crosby was right to point out that Britain lacks rats of American origin. An American species is, however, routinely compared to the rat: 'tree rat' has been a well-established (if zoologically incorrect) term for the grey squirrel since the early 20th century. This is the American species that, if not quite 'equivalently successful' to the 'old world' rat in the new world, still generates the most heated public debate: a typical denunciation targets a 'vicious tree pest' with a 'vile temper' (Drummond, 1984). The grey squirrel's misdeeds in Britain are often compared to the exploits of the uncouth and bullying 'English' (house) sparrow and 'European' starling in the USA. In a letter to the editor of *The Week* (23 February 2008), a Pennsylvanian suggested a swap of 'their'

Figure 3.1 *Grey squirrel*

starlings for 'our' grey squirrels. The oldest of American faunal immigrants arrived during an era of innocent faith in the virtues of acclimatization. The first recorded release to enrich Britain's impoverished fauna (as distinct from the escape of a pet) was at Henbury Park, Cheshire, in 1876. The most important release, however, took place at Woburn Park, Bedfordshire, in 1890; eight secondary releases were drawn from Woburn's flourishing stock. The main phase of releases began in 1902, and though importation ended before 1914, internal transplantations persisted until 1929 (Middleton, 1930, 1935).

The grey squirrel quickly became naturalized and rapidly expanded its range. In 1930, it occupied grids covering 9920 square miles; by 1935, the area had virtually doubled. By the 1950s, its occupation of central England was complete and the species was prospering across much of the southeast, colonizing fresh territory at an annual rate of 1000 square miles (Middleton, 1930; Shorten, 1953, 1954). For the pioneering figure in invasion biology, Charles Elton, the 'American' grey squirrel was a prime example of an ecological population 'explosion' (Elton, 1958).

Criticism emerged in tandem with proliferation. The grey was described as pugnacious, greedy and opportunistic. The belief that its success was inimical to the native red squirrel's welfare took hold while the grey squirrel was still being released. Sir Frederick Treves complained (1917) that pestiferous greys evicted reds and 'eat everything than can be eaten, and destroy twenty times more than they eat' (Ritchie, 1920). The leading squirrel researcher of the 1920s, A. D. Middleton, reported the deep-seated belief among residents of areas colonized by greys that the red 'has been reduced in numbers or driven out by the introduced aliens'. On a more sinister note, he also mentioned the widespread rumour – completely unsubstantiated – that the larger, more powerful and more aggressive grey male castrated its red counterpart by biting off its testicles (Middleton, 1930). Equally fanciful rural folklore held that male greys were killing red squirrel litters and interbreeding to produce a hybrid offspring that was more grey than red (Tittensor, 1975). In 1936, 26 county councils distributed a total of more than 5000 posters beseeching the public to 'kill the tree rat' (*The Times*, 7 April 1936). In 1937, it was designated a pest because of the damage it caused to hardwoods, cereal crops and fruit trees (Kenward, 1987; Roots, 1976). A bounty system was subsequently introduced, lasting until 1958 (Thompson and Peace, 1962; Sheail, 1999). It hardly made a dent, a forester commenting a few years later on the 'alarming progress' of the 'grey invasion' (Seymour, 1961).

A nationalistic element has long been more or less inseparable from this consternation. It was not just a grey squirrel but an American (sometimes North American) grey squirrel, and not just a tree rat but an American tree rat (Ritchie, 1920; Watt, 1923; Lancum, 1947; Leyland, 1955). On the opening page of his book about the grey squirrel, Middleton remarked that, 'I know of more than one patriotic Englishman who has been embittered against the whole American nation on account of the presence of their squirrels in his garden' (Middleton, 1931). In a representative example of demonization from the mid-1950s, an eminent British geographer called it a 'villain of the worst order … steadily replacing the lovely little native red squirrel' (Stamp, 1955). And in a letter to the editor of *The Times* (21 October 1971), the owner of a woodland garden terrorized by the grey squirrel admitted that 'in respect of the alien grey immigrant I plead guilty to racial discrimination'.

This hostility did not escape American attention. The *New York Times*, reporting on the Forestry Commission's investigation in the late 1920s of the grey squirrel's impact on timber resources – and drawing a direct comparison with earlier American debates over the desirability of wiping out the 'English' sparrow – reckoned that prejudice against the animal was so strong that it was unlikely to receive a fair appraisal. A year later, the paper drew attention to the many 'outbursts' against the 'gray invader' triggered by 'patriotic' naturalists, foresters and agriculturalists (*New York Times*, 11 November 1929, 3 August 1930).

Integral to the grey's vilification has been the rehabilitation of its alleged victim, the smaller, gentler, 'cuter,' more timid and less prolific red squirrel (Kean, 2000). The red squirrel is distributed across Eurasia, from Spain to Japan, and there are seventeen subspecies. But one of these is endemic to Britain (Tittensor, 1975). As various commentators pointed out in the 1920s, the 'attractive American stranger' was mischievous and destructive, but no more so than the pesky red squirrel (Watt, 1923). However, with a little help from Beatrix Potter, and one of her most beloved animal creations, Squirrel Nutkin, the red has been converted from forester's scourge to national animal emblem (Potter, 1903).[2] (Potter tried to follow up the popularity of Squirrel Nutkin with her story about a 'little fat comfortable' grey squirrel, Timmy Tiptoes (Potter, 1911; Lear, 2007). Timmy's national origin is not mentioned, nor is the story's North American setting, but since the cast of animal characters includes a bear and a chipmunk, this is clearly not the Lake District, where at the time there were no greys. Though Timmy and his wife, Goody, drink tea like good British squirrels, the story was not a success.) The red squirrel is a charming, feisty little chap, 'truly British', who, as Hilda Kean notes, has assumed a hallowed position in the pantheon of Englishness with red telephone boxes, warm beer and cricket bats (Leyland, 1955; Kean, 2001). Once known as the 'common' squirrel, by 1945 the red was highly uncommon. In 1979, 'our island race' was already confined to 'islands' of conifer plantations amid a 'sea of grey squirrel country' (Tittensor, 1979).

The Cumbrian Lake District, home of Potter and Squirrel Nutkin, was one of those islands, a besieged redoubt of a species whose numbers have plunged to 160,000, while the grey population now stands at an estimated 3.3 million (Skelcher, 1997; Carrell, 2009). During an impassioned squirrel debate in the House of Lords, Lord Inglewood of Cumbria identified the creature as the 'most lovable and loved of our British native animals' (*Hansard*, 25 March 1998). The same patriotic sentiments were displayed during the Lords' most recent squirrel debate (2006), when peers who spoke lost no opportunity to emphasize the grey squirrel's national origins. Embracing the red squirrel as an iconic national animal that was firmly British (rather than merely English), Lady Saltoun of Abernethy pronounced its innate superiority over the grey. Cloaking reds in the admirable qualities of a better class of person, she observed that they are 'rather like quiet, well behaved people, who do not make a nuisance or an exhibition of themselves, or commit crimes, and so do not get themselves into the papers in the vulgar way grey squirrels do'. For Saltoun, the bottom line was that Squirrel Nutkin was not a grey but a red (*Hansard*, 23 March 2006). The distinction between squirrels in Britain and British squirrels is subtle but profound. Speaking in the Lords debate of 1998, Lord Rowallan observed that 'we should not encourage them merely because they are beautiful animals ... They ... have another advantage – they are British' (*Hansard*, 25 March 1998).

By 2006, the beleaguered red was even more embattled than it had been in 1998. Appeasement of the grey, according to Lord Inglewood, had failed to ease the red's predicament. Desperate times demanded drastic measures. Therefore, he urged the celebrity chef, Jamie Oliver, who had recently launched a high-profile campaign to revolutionize British school dinners, to add the grey squirrel to the school menu. This was because 'unless something radical and imaginative is done ... Squirrel Nutkin and his friends and relations are "going to be toast"' (*Hansard*, 23 March 2006; Massie, 2007). Lord Redesdale vigorously concurred with this elegantly simple policy of 'eat a grey to save a red' (which can also be summed up as 'if you can't beat them, eat them').[3] Redesdale, whose seat in Northumberland enjoys a remnant population of reds, has founded the Red Squirrel Protection Partnership, whose mission is also elegantly simple: to kill grey squirrels (Max, 2007; Adams, 2008). Red squirrel defence groups are spreading at the rate of the grey squirrel itself. Save Our Squirrels has received National Lottery funding to operate 16 squirrel reserves established in the north of England in 2006.[4] Drawings of red squirrels by more than 100 well-known personalities from the worlds of music, comedy, sport, television and radio, politics, stage and screen were auctioned on the Internet in September 2000 to raise money for Red Alert (Paul McCartney's sketch, apparently, bore the closest resemblance to Squirrel Nutkin) (Ingham, 2000). For many British nature and countryside enthusiasts, the grey squirrel is inauthentic and out of place. The nimble creature raiding the bird feeder in our back gardens 'should be the red original, not this grey impostor' (Baker, 2005).

The popular case against the grey squirrel is so entrenched that it is largely impervious to research findings suggesting that its initial spread was not responsible for the red's decline. Red squirrel populations have a long history of fluctuation. Against a background of partial woodland clearance, reds had virtually vanished from Scotland by the late 18th century, surviving only in a few relics of native woodland in the Cairngorm area (Harvie-Brown, 1881; Ritchie, 1920). The period immediately preceding the grey's importation was marked by red abundance; extensive late-18th-century conifer plantings matured between 1860 and 1890, offering congenial conditions. But red populations were under stress in many parts of England before the grey's advent due to disease (coccidiosis) and the early 20th-century retreat of coniferous woodland, not least due to extensive felling due to wartime demand. Grey squirrels, rather than displacing reds, often advanced into empty territory (MacKinnon, 1978; Usher et al, 1992; Kenward and Holm, 1993). After World War I conditions for the red became even more unfavourable with the decline of hazel coppicing; reds are partly dependent on pine cones and hazelnuts whereas greys draw on a more extensive range of foods.[5] The remaining sanctuaries of the red, which makes no distinction between native trees and non-natives such as Sitka and Norway spruce, are coniferous forests in northern England and Scotland (Garson and Lurz, 1992). Ironically, the firm conservationist preference for native deciduous woodland works to the grey's advantage.

A surprising outlet for this revisionist data was a children's book. *The Further Exploits of Mr. Saucy Squirrel* (1977) by Woodrow Wyatt, a former Labour MP and well-known public figure who became a Conservative life peer in 1987, features a dapper and jaunty squirrel that lives in a beech tree near Devizes, Wiltshire. Despite his impeccably English manner, behaviour and pursuits (he sports a fancy waistcoat and patronizes a London club), Mr Saucy Squirrel was in fact a grey squirrel. Asked if he was 'a good English squirrel' when he goes to buy a bowler hat, Mr Saucy Squirrel replies, 'Dear me. I'm always being asked that question. I'm sorry to have to tell you that my family came from America at the end of the nineteenth century. We are not the old kind of English squirrel people call the red squirrel. But we've been here a long time now.' Nor does he suffer from an identity crisis or conflicting allegiance: 'We came from America so long ago that we are now properly naturalized English. I don't sing "The Star-Spangled Banner" ... I sing "God Save The Queen".'

Mr Saucy Squirrel is also at pains to exonerate his kind from the alleged crime of having marginalized the red squirrel. The 'English' squirrel, he explains, was already in trouble when he and his fellow Americans arrived: 'We just took over in various places that they'd already left.' He also maintains that the two types co-exist happily: 'We don't quarrel with English squirrels at all. We eat the same food but we don't fight over it' (Wyatt, 1977).[6] Though Mr Saucy Squirrel concedes that grey squirrels 'probably put the red squirrels off by living in the same areas', this charming picture of Anglo-American harmony is rather overdone. Greys may not entirely have caused the reds' decline, but they probably prevented their recovery (Tittensor, 1975). Moreover, most greys carry a virus, squirrel pox, to which they are immune, but which kills reds within two weeks.

Regardless of the evidence accumulated by even-handed researchers, who frown on the term 'tree rat' and suspect that the grey's alien nature is itself part of the objection (Shorten, 1954), the grey squirrel continues to serve as a scapegoat. Here are a few recent examples of the cultural construction of an unsavoury Yankee critter that, in the view of Andrew Tyler, director of Animal Aid, reflects the mentality that 'the only good squirrel is a red squirrel' and fuels the 'pogrom' against the grey (Tyler, 2001). An entry on the Internet blog of a self-described Tennessee expatriate and long-term London resident reflects on the pervasive feeling that the grey squirrel is the UK's Public Enemy Number One. The immediate prompt was the aforementioned squirrel debate in the Lords (March 2006). At that time, Americans who had been living peacefully in London for years were reporting their first experiences of abusive treatment, regardless of whether they voted for Bush or supported the invasion of Iraq ('some people just fly off the handle without even talking to me – it's as if they had been waiting to run into an American all day to let their feelings out', a 29-year-old woman reported) (BBC, 2006). Within this climate of opinion, the Tennessean wondered whether the assault on the grey squirrel is a 'further sign of rampant anti-Americanism among the British elite' (Blogspot, 2006).

A feature in the *New York Times* about international couples negotiating differences with their respective families supplied another instance of perceived hostility to Americans associated with the greys' exploits. One of the profiled couples was an American woman and her British husband who live in London. Whereas her staunchly Anglophile family readily accepted her British boyfriend, her future mother-in-law made a point, at their first meeting, of bringing to her attention the 'awful American gray squirrels' that were 'chasing all the lovely English red squirrels out of Britain' – something she attributed to generic anti-Americanism, until she read up about the exploits of grey squirrels (Emling, 2006).

A strategy perhaps less vulnerable to accusations of knee-jerk anti-Americanism is to package the case against the grey and for the red as part of a desire – socially and culturally progressive as well as ethically sound – to preserve national and regional distinctiveness and embattled minorities in a world of rampant globalization that threatens us with suffocating sameness. In other words, the campaign involves the same commitment to cultural survival, community identity and diversity that fuels the championing of local cheeses and apples against the tasteless universalism of the products of international agribusiness. In a recent article in a conservative British magazine welcoming the tea shop's plucky reappearance in London amid the sprawling American empire of a coffee-shop chain, the author confessed that Starbucks is 'liable to bring out unworthy feelings of anti-Americanism in me'. Casting around for an evocative analogy, he portrayed Starbucks as 'the Yankee grey squirrel of the high street, which has reduced our native tea and coffee shops to a few redoubts in the North' (Trefgarne, 2007).

Anthropomorphism of this stripe riddles debates about American bio-villains. The most familiar critical device is to humanize them through analogy with the 'friendly invasion' of American troops during World War II (3 million in total between January 1942 and December 1945). GIs stationed in Britain were famously described as 'oversexed, overpaid, overfed, and over here' (Reynolds, 1995). This celebrated phrase encapsulated the resentment of British males toward their strapping American counterparts who, bearing gifts of nylon stockings, lipstick, chocolate, cigarettes and chewing gum, allegedly enjoyed a huge advantage in the competition for British women.[7]

Succumbing to this glib rhetorical strategy, an Environment Agency press release remarked that, from the standpoint of white-clawed crayfish and water vole, signal crayfish and mink are 'oversized, oversexed and over here' (Environment Agency, 2006). Publicizing its depredations on ground-nesting birds, Steve Sankey, chief executive of the Scottish Wildlife Trust, characterized the 'American' mink as 'oversexed and, unfortunately, over here' (Edwards, 2001). This tendency has been most pronounced, though, with reference to the ruddy duck, the analogy with randy GIs reinforced by the arrival of the 'over-plumed' (Ryder, 1995) and 'pushy American settler' (BBC, 2003) in the UK in the same decade (Johnston, 2009).

Introduced in the late 1940s to the Wildfowl and Wetlands Trust collections at Slimbridge, Gloucestershire, the first young ruddy ducks escaped around 1954. By the mid-1990s, the UK population had reached 3,500. In the early 1980s, the bird flew east into Europe and particularly south towards Spain. Whereas male crayfish and grey squirrels have not mated with their British counterparts, by the early 1990s male ruddy ducks were interbreeding with Spain's endangered white-headed duck. Responding to international pressure and with the support of the Royal Society for the Protection of Birds (RSPB), the UK government ordered a three-year trial cull from April 1999. The British animal rights lobby protested that the ruddy duck was a scapegoat for Spanish hunters and wetland drainers who had already driven white-heads to the brink of extinction by the late 1970s. The West Midland Bird Club had even adopted the ruddy duck as their mascot, indicating how far it had entered at least some British affections. Nonetheless, the trial went ahead. By its end in March 2002, some 2,500 birds had been killed. The subsequent eradication campaign had reduced numbers to 500 by the spring of 2008, despite a degree of protest from both birdwatchers and the general public (Smout, 2009).

Uncertainty over the duck's national identity (Spain regards it as a double alien) and the fact that the site of its misbehaviour is Spain rather than Britain has generated a more complex set of analogies than those involving the mink and the grey squirrel. For some commentators, the notoriety of a certain type of young British male tourist and his beery misdeeds in Spanish coastal resorts has provided an irresistible point of comparison: the ruddy duck is the lager lout of the bird world (*Sunday Express*, 24 January 1993, in Milton, 2000). Yet the default action is to trot out variations on the tired old cliché: 'over-plumed, oversexed and over here' (Ryder, 1995).

Those who insist on the bird's innocence lament what they see as an irrational and obsessive interest in native species and denounce fears of 'genetic impurity' as racist (Tyler, undated). Far from waging war on the Spanish duck, argued David Cox, the ruddy duck's only crimes were being American and making love: 'though they may be oversexed, it is not their fault they are over here'. As well as suspecting that the bird had fallen foul of anti-American 'fervour' ('you might have thought that in multicultural Britain such origins would prove no bar'), Cox believed that concern over the dilution of racial purity was also wrong-headed from a scientific standpoint (as well as racist). The ruddy duck was a problem animal for conservationists, not for the white-headed duck. To bolster his argument that interbreeding had boosted the latter's ailing genetic viability, he fleshed out the analogy with Americans and World War II. The ruddy duck's arrival in Spain constituted a 'natural version' of the Marshall Plan's post-war economic assistance to impoverished Europe. He adds that conservationists should give the white-heads more credit for knowing for themselves what was best in an evolutionary sense. Maybe this was why they chose ruddy ducks rather than their own males, 'just as our grandmothers favoured brash GIs over the weedy, local males' (Cox, 2003).

Conclusions

Not all invasive and troublesome species are non-native. The red squirrel's champions tend to forget that it too was once treated as a major pest, incurring the forester's wrath by eating the seeds and shoots of young pine and stripping their bark. Scotland's Highland Squirrel Club killed 82,000 reds between 1903 and 1929 (Middleton, 1930). However, the red's native status offered more protection and restricted its critics' rhetorical options. An invader that is also a foreigner seems more sinister and its threat can be more easily sexed up. Even a thoroughly dispassionate biologists' account of the respective status of the two types of squirrel in Britain bore the provocative title, 'An American invasion of Great Britain' (Usher et al, 1992).

How seriously should we take the language of anti-Americanism that flavours British perceptions of the grey squirrel and its fellow American species? How closely do responses to their exploits mirror attitudes to Americans and the activities of the USA? We can chart the peaks and troughs of anti-Americanism in Britain with some degree of accuracy. The Vietnam War represents a fairly unambiguous high point; and during the Clinton presidency the profile of anti-American feeling was comparatively low. On a British tour in 1997 to promote his book, *A History of the American People*, Paul Johnson, a great admirer of the USA and its achievements, encountered minimal anti-Americanism. He was pleased to report that 'overpaid, over-sexed and over here' was no more than a 'historical curiosity', incomprehensible to Britons under 40 (Johnson, 1997). His verdict on the imminent death of anti-Americanism proved premature. The George W. Bush presidency and the invasions of Afghanistan and Iraq precipitated a powerful resurgence. Yet this most recent and particularly virulent strain of anti-Americanism does not explain the high degree of hostility encountered by the grey squirrel over the past decade.

It is tempting to conclude that British attitudes to the mink and grey squirrel tell us as much about British stereotypes of Americans (and therefore of American animals) as they do about the creatures themselves. This would fit with the main thrust of recent studies of anti-Americanism in Europe, which contend that the phenomenon is a general orientation with deep and extensive cultural and ideological roots, possessing its own momentum, rather than a response to specific events and policies. Yet what these animals do matters as much as who they are. Anti-Americanism is not usually the root source of objections to the squirrel and other US invasive species. For the most part, anti-Americanism supplements and embroiders critiques anchored in ecological (and economic) realities. For rhetorical purposes, the grey squirrel has become a furry embodiment of US multinationals and imperious US foreign policy. One suspects, though, that it would have been as unpopular lately if Al Gore had been president, thanks to its own continuing success and the hapless red's continuing decline. The accumulating power of grey squirrel and signal crayfish is not connected to the global advance of American military and economic might, or presidential popularity. These processes have their own independent momentum. It is also worth remembering that fortunes decline as well as rise. As the otter stages a comeback, thanks to cleaner rivers, there are signs that mink are in retreat (O'Neill, 2006).

Still, responses to the grey squirrel's 'policy' of expansion and nefarious activities (barking trees, preying on garden plants, digging up bulbs, raiding birds' nests, and chewing telephone wires and electricity cables), for better or worse, are influenced by and sometimes conflated with attitudes to human Americans and American behaviour abroad. The metaphors and similes through which criticism of creatures like the grey squirrel is expressed would certainly have been less engaging for the cultural historian if Walmart or Starbucks had never crossed the Atlantic. For while anti-squirrel rhetoric is partly reflexive and generic, with enduring and predictable points of reference ('American cousin' being the least judgmental of labels), it also develops and shifts in response to particular conditions. It remains to be seen whether British delight with the Obama presidency eventually rubs off on the grey squirrel and other bêtes noires that appear to be forever American in the British mental and physical landscape.

Notes

1 The adoption of organochlorine pesticides (dieldrin and aldrin) for sheep dipping and seed dressing from 1955 was arguably a more important cause of the precipitous decline in otter numbers during the mid-1950s, but the otter's disappearance undoubtedly assisted the mink's spread. The mink is also blamed for preying on fish farms and game birds.

2 Of somewhat less enduring fame than Squirrel Nutkin was Tufty Fluffytail, who, through the Tufty Club, taught schoolchildren about road safety between the early 1950s and 1980s.
3 Various restaurants of repute have started to include grey squirrel on their menus. At St John Bar and Restaurant at Smithfield Market, London, diners are offered roast squirrel or squirrel liver pate, and at the Famous Wild Boar Hotel at Crook, Cumbria (of which Lord Inglewood was a director), squirrel meat is served in Peking duck-style pancakes (canapés), with chutney made from locally sourced damsons. See Linzi Watson, 'Game Dealer rustles up recipe for red squirrel conservation', *Cumberland News*, 17 October 2008; Nigel Burnham, 'Eating the enemy', *The Guardian*, 18 March 2009, p7.
4 I should declare a strong personal interest in the red's fate. I grew up on the Lancashire coast north of Liverpool, at Formby, site of a National Trust squirrel reserve. Red squirrels were part of my boyhood.
5 The survival of reds on islands like the Isle of Wight suggests, however, that the hazel coppicing may not in fact have been such an important consideration; hazels, after all, still fruit when they are not coppiced.
6 Wyatt's daughter, Petronella (for whom this and his first squirrel book, *The Exploits of Mr Saucy Squirrel* [1976], were written), puts the case for accepting the grey squirrel in 'The red and the grey', *The Spectator*, 29 December, 2001, vol 287, no 9047, p49.
7 American servicemen may have been overfed, but some who hailed from rural parts of the eastern seaboard nevertheless craved an authentic taste of home – so much so that they were willing to offer as much as 5 shillings for a grey squirrel: Monica Shorten (1954), *Squirrels*, Collins, London, p58. Shorten, who inherited Middleton's mantle as the pre-eminent British squirrel researcher, was herself a keen consumer of squirrel pie and fried squirrel. Some Britons boiled up grey squirrel for dog food during World War II (Stephen Tallents, letter to the editor, *The Times*, 2 December 1946, p7).

References

Adams, T. (2008) 'They shoot squirrels, don't they?', *The Observer*, 19 October, p24

Amos, S. (2003) 'If you go down to the woods today…', *The Independent*, 23 February, pp28–29

Baker, N. (2005) 'Wild year', *National Trust Magazine*, 106 (Autumn), p84

BBC News (2003) 'R.I.P. ruddy duck', 3 March, http://news.bbc.co.uk/1/hi/uk/2814293.stm, accessed 16 September 2009

BBC News (2006): 'Anti-Americanism "feels like racism"', 16 April, http://news.bbc.co.uk/1/hi/uk/4881474.stm, accessed 20 September 2009

Berman, R. A. (2008) *Anti-Americanism in Europe: A Cultural Problem*, Hoover Institution Press, Stanford, CA, USA

Blogspot (2006) http://thevolabroad.blogspot.com/2006_03_19_archive.html, accessed 17 September 2009

Burnham, N. (2009) 'Eating the enemy', *The Guardian*, 18 March, p7

Carrell, S. (2009) 'Sure he's cute … but not cute enough to save him from the great squirrel cull', *The Guardian*, 10 February, p3

Cox, D. (2003) 'Sex and the single duck', *New Statesman*, 30 June, www.newstatesman.com/200306300009, accessed 30 September 2009

Crosby, A. W. (1972) *The Columbian Exchange: Biological and Cultural Consequences of 1492*, Greenwood Press, Westport, CT, pp210–211

Crosby, A. W. (1986) *Ecological Imperialism: The Biological Expansion of Europe*, Cambridge University Press, New York, pp164–165

Drummond, H. (1984) 'The unequal status of the red squirrel,' *The Field*, 1 (December), 48

Dunstone, N. (1993) *The Mink*, T & A D Poyser, London, p187

Dunstone, N. and Ireland, M. (1989) 'The mink menace? A reappraisal', in Putman, R. J. (ed) *Mammals as Pests*, Chapman and Hall, London, pp225–241

Edwards, R. (2001) 'Beastly aliens taking over the countryside,' *Glasgow Sunday Herald*, 1 July, p3

Elton, C. S. (1958) *The Ecology of Invasions by Animals and Plants*, Methuen, London, pp15, 73, 123, 148

Emling, S. (2006) 'Meet the family: Complexity and stress at holiday time,' *New York Times*, 10 January, www.nytimes.com/2005/11/25/style/25iht-ameet.html?_r=1, accessed 29 September 2010

Environment Agency (2006) 'Top Ten Most Wanted Foreign Species', 3 August, www.environment-agency.gov.uk/news/1444976, accessed 17 September 2009

Garson, P. J. and Lurz, P. W. W. (1992) 'The distribution of red and grey squirrels in northeast England in relation to available woodland habitats', *Bulletin of the British Ecological Society*, vol 23, no 2, pp133–139

Hansard, Lords (1998) Session 1997–98, vol 587, Column 1318, 25 March www.parliament.the-stationery-office.co.uk/pa/ld199798/ldhansrd/v0980325/text/80325-10.htm#80325-10_head0, accessed 17 September 2009

Hansard, Lords (1998) Session 1997–98, vol 587, Column 362, 23 March, www.parliament.the-stationery-office.co.uk/pa/ld199798/ldhansrd/v0980325/text/80325-10.htm#80325-10_head0, accessed 17 September 2009

Hansard, Lords (2006) Session 2005–2006, vol 680, Columns 362, 370–71, 23 March, www.publications.parliament.uk/pa/ld200506/ldhansrd/v0060323/text/60323-04.htm#60323-04_head1, accessed 29 September 2010

Harvie-Brown, J. A. (1881) *The History of the Squirrel in Great Britain*, McFarlane & Erskine, Edinburgh

Hill, M., Baker, R., Broad, G., Chandler, P. J., Copp, G. H., Ellis, J., Jones, D., Moore, N., Parrott, D., Pearman, D., Preston, C., Smith, R. M. and Waters, R. (2005) *Audit of Non-Native Species in England*, English Nature Research Report 662, English Nature, Peterborough, p30

Hollander, P. (ed) (2004) *Understanding Anti-Americanism: Its Origins and Impact at Home and Abroad*, Ivan R. Dee, Chicago

Ingham, J. (2000) 'Stars go nuts to help save the red squirrel', *Daily Express*, 15 September, http://atgbcentral.com/dasquirrels.html, accessed 29 September 2010

Johnson, P. (1997) 'The decline and fall of anti-Americanism in Britain', *The Spectator*, 1 November, 279(8831), p30

Johnston, I. (2009) 'UK accused of "racism" towards invaders from across the pond', *The Independent*, 26 January, p2

Kean, H. (2000) 'Save "our" red squirrel: Kill the American grey Tree Rat', in Kean, H., Martin, P. and Morgan, S. (eds), *Seeing History: Public History in Britain Now*, Francis Boutle, London, pp51–64

Kean, H. (2001) 'Imagining rabbits and squirrels in the English countryside', *Society & Animals*, vol 9, no 2, 164

Kenward, R. E. (1987) 'Bark stripping by grey squirrels in Britain and North America: Why does the damage differ?', in Putman, R. J. (ed) *Mammals as Pests*, Chapman and Hall, London, pp144–154

Kenward, R. E. and Holm, J. L. (1993) 'On the replacement of the red squirrel in Britain: A phytotoxic explanation', *Proceedings of the Royal Society of London, Series B, Biological Sciences*, vol 251, pp187–194

Kirkup, J. and Johnston, I. (2007) 'Alien invaders face being exterminated', *The Scotsman*, 10 March, http://news.scotsman.com/hedgehogs/Alien-invaders-face-being-exterminated.3353349.jp, accessed 2 October 2009

Lancum, F. H. (1947) *Wild Animals and the Land*, Crosby Lockwood, London, p69

Lear, L. (2007) *Beatrix Potter: A Life in Nature*, Allen Lane, London, p245

Lever, C. (1978) 'The not so innocuous mink', *New Scientist*, 78, 481–484

Leyland, E. (1955) *Wild Animals*, Edmund Ward, London, p63

Liddle, R. (2002) 'Our hatred of certain furry foreigners', *The Guardian*, 4 September, p5

Linn, I. L. and Chanin, P. R. F. (1978a) 'Are mink really pests in Britain?', *New Scientist*, 78, 560–562

Linn, I. L. and Chanin, P. R. F. (1978b) 'More on the mink menace', *New Scientist*, 79, 38–40

Lloyd, A. R. (1982) *Kine*, Hamlyn, Feltham, Middx, UK; republished as *Marshworld* (1990)

MacKenzie, J. M. (2001a) 'Editorial', *Environment and History*, vol 7, no 3, p253

MacKenzie, J. M. (2001b) 'Editorial', *Environment and History*, vol 7, no 4, p380

MacKinnon, K. (1978) 'Competition between red and grey squirrels', *Mammal Review*, vol 8, no 4, pp185–190

Markovits, A. S. (2007) *Uncouth Nation: Why Europe Dislikes America*, Princeton University Press, Princeton

Marks, R. (2007) 'Rafts put mink in the clink to save Ratty', *Western Mail*, 13 February, www.walesonline.co.uk/countryside-farming-news/equestrian-news/2007/02/13/rafts-put-mink-in-the-clink-to-save-ratty-91466–18613417/#, accessed 2 October 2009

Massie, A. (2007) 'The ugly American abroad: Animal version', 9 October, www.spectator.co.uk/alexmassie/3256806/the-ugly-american-abroad-animal-version.thtml, accessed 30 September 2009

Max, D. T. (2007) 'The squirrel wars', *New York Times*, 7 October, www.nytimes.com/2007/10/07/magazine/07squirrels-t.html, accessed 1 October 2009

Middleton, A. D. (1930) 'The ecology of the American grey squirrel in the British Isles', *Proceedings of the General Meetings for Scientific Business of the Zoological Society of London*, Part 2, p812

Middleton, A. D. (1931) *The Grey Squirrel*, Sidgwick & Jackson, London

Middleton, A. D. (1935) 'The distribution of the grey squirrel in Great Britain in 1935', *Journal of Animal Ecology*, vol 4, pp274–276

Milton, K. (2000) 'Ducks out of water: Nature conservation as boundary maintenance', in Knight, J. (ed) *Natural Enemies: People–Wildlife Conflicts in Anthropological Perspective*, Routledge, London, p229

Monbiot, G. (2009) 'How British nationalists got their claws into my crayfish', *The Guardian*, 1 October, www.guardian.co.uk/environment/georgemonbiot/2009/oct/01/crayfish-bnp, accessed 2 October 2009

New York Times (1929) 'American squirrel on trial for his life in England: He is charged with killing his English cousins and with destroying bird life', 11 November, p171

New York Times (1930) 'Gray squirrel in England gains faster than in American home', 3 August, p106

O'Neill, B. (2006) 'Backstory: A fierce and furry fight on the banks of the Thames', *Christian Science Monitor*, 27 September, p20 www.csmonitor.com/2006/0927/p20s01-sten.html, accessed 1 October 2009

Potter, B. (1903) *The Tale of Squirrel Nutkin*, Frederick Warne, London

Potter, B. (1911) *The Tale of Timmy Tiptoes*, Frederick Warne, London

Revel, J.-F. (2004) *Anti-Americanism*, Encounter Books, London

Reynolds, D. (1995) *Rich Relations: The American Occupation of Britain, 1942–1945*, HarperCollins, London, xxiii

Ritchie, J. (1920) *The Influence of Man on Animal Life in Scotland: A Study in Faunal Evolution*, Cambridge University Press, Cambridge

Roots, C. (1976) *Animal Invaders*, David & Charles, Newton Abbot

Ryder, R. (1995) 'Hands off our ruddy ducks', *The Independent*, 30 June, p20

Sagoff, M. (2002) 'What's wrong with exotic species?', in Gehring, V. and Galston, W. (eds) *Philosophical Dimensions of Public Policy*, Transaction Publishers, New Brunswick, NJ, pp327–340

Seymour, W. (1961) 'Grey squirrels', *Quarterly Journal of Forestry*, vol 55, no 4, p293

Sheail, J. (1999) 'The grey squirrel (*Scirius carolinensis*): A UK historical perspective on a vertebrate pest species', *Journal of Environmental Management*, vol 55, pp145–156

Shorten, M. (1953) 'Notes on the distribution of the grey squirrel (*Scirius carolinensis*) and the red squirrel (*S. vulgaris leucourus*) in England and Wales from 1945 to 1952', *Journal of Animal Ecology*, vol 2, pp134–140

Shorten, M. (1954) *Squirrels*, Collins New Naturalist Series, London

Skelcher, G. (1997) 'The ecological replacement of red by grey squirrels', in Gurnell, J. and Lurz, P. (eds) *The Conservation of the Red Squirrel: Scirius Vulgaris L.*, People's Trust for Endangered Species, London, p76

Smout, T. C. (2009) 'The alien species in twentieth-century Britain: Inventing a new vermin', in Smout, T. C., *Exploring Environmental History*, Edinburgh University Press, Edinburgh, pp169–181

Stamp, L. D. (1955) *Man and the Land*, Collins, London, p210

Sweig, J. E. (2006) *Friendly Fire: Losing Friends and Making Enemies in the Anti-American Century*, PublicAffairs, New York

Thompson, H. V. and Peace, T. R. (1962) 'The grey squirrel problem', *Quarterly Journal of Forestry*, vol 56, no 1, pp33–34

Tittensor, A. M. (1975) *Red Squirrel*, Forestry Commission, London, p16

Tittensor, A. (1979) 'What future for the reds?', *Country Life*, 166(4294), 25 October, pp1394–1395

Trefgarne, G. (2007) 'Is it time for tea?', *The Spectator*, 303(9310), 20 January, p65

Tyler, D. (2001) BBC Radio 4, *Today Programme*, 29 January

Tyler, D. (undated) 'Scapegoating the Aliens', www.animalaid.org.uk/images/pdf/factfiles/aliens.pdf, accessed 18 September 2009

Usher, M. B., Crawford, T. J. and Banwell, J. L. (1992) 'An American invasion of Great Britain: The case of the native and alien squirrel (*Sciurus*) species', *Conservation Biology*, vol 6, no 1, p108

Watson, L. (2008) 'Game dealer rustles up recipe for red squirrel conservation', *Cumberland News*, 17 October, www.cumberland-news.co.uk/news/1.256329, accessed 2 October 2009

Watt, H. B. (1923) 'On the American grey squirrel (*Sciurus carolinensis*) in the British Isles', *The Essex Naturalist*, vol 20, pp189–205

Wyatt, P. (2001) 'The red and the grey', *The Spectator*, 287(9047), 29 December, 49
Wyatt, W. (1976) *The Exploits of Mr. Saucy Squirrel*, Allen & Unwin, London
Wyatt, W. (1977) *The Further Adventures of Mr. Saucy Squirrel*, Allen & Unwin, London, pp78–80

4
How the Concept of Alien Species Emerged and Developed in 20th-century Britain

Chris Smout

Introduction

Alien species, we are told with good reason, are the biggest threat to biodiversity on the planet after habitat loss and climate change. Earlier ages would not have known what we are talking about; we think very differently about nature as the centuries pass. The modern preference for biodiversity over bio-uniformity, like the privileging of rare or unusual taxa over common ones, is a cultural construct of recent times. It has little to do with science per se, though from the 18th century and the time of Linnaeus, collecting and examining the differences between species was a critical part of science. The dominant attitude towards species even today, moreover, does not consider that biodiversity should override bio-utility. Animals, plants and insects either have a value to people, or they do not, and those that do are held to be more worthwhile than those that do not. Scientists and conservationists who want to be taken seriously by governments must still disguise biodiversity as bio-utility and play up the value of, for instance, tropical habitats as a gene pool of unknown treasures which might come in handy to people one day, or stress that reintroduction of the European beaver into Scotland will benefit the tourist trade. The most important taxa, in this traditional and enduring world view, have clear economic value (e.g. salmon, cows, cabbages, bees), although some have only decorative or amusement value (e.g. goldfinches or monkeys in a cage).

Some past attitudes towards species have been discarded. It was believed that some species were morally instructive to man ('go to the ant, thou sluggard'), were intrinsically evil (serpents, toads) or were scourges sent by God (locusts, rats). None at all were good only because they were native, or bad because they were alien. So the preference for biodiversity and the notion that alien species are deadly enemies of biodiversity are relatively modern ideas.

Transporting nature

The movement of species from place to place through human agency, however, is not at all new. It was often accidental, witness the successive rodent invasions of Europe and Britain by the house mouse, black rat and brown rat, all alien species arriving from the east between the Neolithic and the 18th century. These movements were often deliberate, for food, such as spelt, apples and goats, for traction, like horses, for skins, like rabbits, or for hunting, like pheasants and fallow deer. This has of course continued – mink and signal crayfish are two of too many 20th-century examples. Then from the late 18th century, things were moved about more for non-economic reasons too. Other continents had their naturalization societies that brought house sparrows to America, rabbits to Australia, gorse to New Zealand because they reminded the settlers of home – with incalculable consequences. In Britain it was landowners wishing to add variety and curiosity to their estates who introduced alien species, innumerable rhododendrons of which only *R. ponticum* became aggressively invasive, and also Japanese knotweed, mandarin ducks, Canada geese, grey squirrel and little owl. Some only partially succeeded as invaders – the red-necked wallaby is still there in a few places but never exactly swept the ground, and many failed. Introduction of the American beaver for example was tried by the Marquess of Bute but failed.

The London area at the end of the 19th century was a popular focus for introductions. No fewer than six separate attempts were made between 1890 and 1916 within 20 miles of St Paul's to bring in grey squirrels, the first two by American citizens who thought they were doing us a favour. One of these introduced 100 animals at Kingston Hill in Surrey. The background to this was that London and the southeast was at that time a squirrel-free zone, the native red squirrels having died out following an attack of squirrel plague. The inhabitants felt the need for a squirrel no less frisky but more robust, and it succeeded only too well. There was also an attempt to liven up the London waterways in the 1890s with edible frogs from Belgium and Germany. They persisted, but the experiment was rather less disastrous (Fitter, 1945).

These examples were the recent exotics, but in the organic countryside of the early 20th century alien species were commonplace and often, though not always, of ancient lineage. Many of these were what botanists have come to call archaeophytes, present for more than 500 years. Whole swathes of economically valuable habitats and many of their wild inhabitants were composed of alien species. Agriculture depended on alien species, and the alien corn was bright with alien poppies and cornflowers. Alien cattle and sheep grazed fields of mixed native and alien grasses. Native barn owls were extraordinarily abundant, feeding on alien rats before the age of warfarin. Alien rabbits, as yet unscathed by alien myxomatosis, kept the Brecklands vegetation short and open for rare native herbs and birds such as the stone curlew. Alien pheasants amused native lords shooting in the autumnal woods. The native woods themselves harboured many nationally alien trees such as sycamores and wellingtonias, or regional aliens like Scots pines in the south of Britain, and beech in the north, though estate forestry had as yet adopted commercially few alien conifers aside from the Norway spruce and Corsican pine. These trees sheltered hosts of native birds, like treecreepers that loved to roost in wellingtonia bark and migrant goldcrests whose favourite forage tree on October coasts was the sycamore. They also held a further host of native insects, as well as still more (and more worrying) alien insects that had come along for the ride.

Then there were the new urban ecosystems, singularly biodiverse with aliens. Some escaped from the gardens of middle-class villas, such as privet, buddleia and Japanese knotweed. Some arrived uninvited at the docks as part of Victorian global free trade, and settled down for a while on wasteland nearby, or caught the train north and west in bales of wool and cotton. Few of these were much trouble. Figs came to Sheffield to relieve Victorian bowels of their endemic constipation, and floated their seeds down the sewers to the river. Here they germinated on the banks warmed by the waste hot water from Bessemer steel converters. The figs outlived the converters and the Victorians, and they are now a valued part of the cultural heritage of this city.

It was not that the 19th century did not care which wild species were alien and which were not. Obviously in the age of Darwin there were a growing number of both amateur and professional naturalists with an interest in collecting and examining biodiversity, and first attempts to define a species as native or alien (and to suggest the alien was not so interesting or authentic as the native) come as early as the 1830s, when John Stevens Henslow (Darwin's friend and mentor) and Hewett Cottrell Watson (father of British botanical geography) were trying to define British flora. As a matter of classification and an attempt at definition, the alien species starts here (Chew, 2006). Amateur botanists found that henceforth they would find an asterisk placed against their records of non-native species (they still are), so they were marked as inferior.

The concept of vermin

The 19th century also had a very well developed concept of species as vermin. The trades of vermin-killers began: less respectable rat-catchers and rabbit-catchers alongside admired gamekeepers, employing thousands of trained men to war against nature. Gamekeepers and their employers defined as vermin many species which we would now consider precious as native biodiversity to be protected. These included eagles, ospreys, dippers, kingfishers, bullfinches and polecats, as well as the jays, magpies, stoats and so on which it is still legal to kill. The game-preserving movement in the 19th century inflicted enormous harm to biodiversity, from which we have still not recovered. They also killed aliens, making no distinction between long-established introductions like rats and rabbits and recent ones like grey squirrels.

It might be as well to set this against the attitude of the British towards human aliens. It was towards the end of the 19th century that British society became for the first time considerably exercised about foreigners settling in Britain. This is not to say that xenophobia and hostility towards minorities did not exist earlier; though there were often minorities present, especially the Irish. Up to this point foreigners had been free to settle in Britain and to apply in due course for British nationality; entry documents were normally unnecessary, and Britain prided herself on giving asylum both to the politically persecuted and the economically unfortunate. British passports as documents of national identity were first issued in 1858 as matters of convenience for those travelling abroad.

Things began to change in the 1890s with the influx of large numbers of Jews from Eastern Europe. They were fleeing the pogroms of the Czars, and they aroused considerable antisemitism and xenophobia, particularly in the East End of London where tens of thousands settled, but not only there. Coloured migrants also began to feel the prejudice of politicians – Keir Hardie, the pioneer labour leader and MP, spoke of the yellow peril. There were parliamentary enquiries and commissions, and in the 1905 Aliens Act, immigration controls were instituted for the first time, though only to prevent paupers and criminals entering Britain. Antisemitic demands were completely defeated and Britain continued a liberal policy towards granting asylum to victims of persecution. Nor at this stage was there any question of a cap on numbers. At the time of World War I restrictions were tightened for security reasons. After the war the League of Nations convened an international meeting to regularize passports in international travel, the origin of our present 32-page passport system.

It was as the language of anti-immigrant protest in the streets became strident, and the press spoke of 'the dirty, destitute, diseased, verminous and criminal foreigner who dumps himself on our soil and rates simultaneously' (*Manchester Evening Chronicle*, 1905), that we first began to hear in Britain of anxiety about non-native animal and plant species. 'Should alien stock be introduced with success the animal life of a country alters appreciably', warned James Ritchie in his path-breaking *Influence of Man on Animal Life in Scotland* of 1920. He devoted 60 pages to the problem, concluding with rebuking those guilty 'of many thoughtless introductions' and pointing a 'warning finger at the naturalist and reformer who, by introducing animals would revise nature's order, and by short cuts and unimaginative experiments tends to make a wilderness where he had looked for a paradise' (Ritchie, 1920). This sounds all rather measured and sensible, and no overt connection is made between the foreign person and the foreign animal. Anxiety is fuelled more by the fashion for releasing alien species into the wild than by racist suspicion of all immigrants, human and animal alike.

Ritchie pointed out two mammals in particular whose introduction had had disastrous consequences, the rabbit and the squirrel. It comes as a shock to realize that the squirrel he had in mind was the red squirrel, which in the 19th century had all but died out in Scotland and had been reintroduced from northern English and European stock. 'The country of their adoption has favoured them; they have multiplied so enormously that they have come to be regarded as one of the prime pests of the forester', he says, mentioning 14,123 killed in 16 years on one estate, Cawdor, and the destruction by squirrels of 1000 trees in 16 years in Glentanar (Ritchie, 1920). These hostile remarks are echoed by Anderson in his *History of Scottish Forestry*, published in 1967, who speaks of the 'devastating menace' of the reintroduced red squirrel in the northern pinewoods, and describes 60,000 killed in 16 years by the Highland Squirrel Club (Anderson, 1967).

Here we have a species defined today as a native that was only 40 years ago considered in Scotland to be an introduced pest. Now it is expensively and carefully protected because it is being displaced by another species, the grey squirrel, which is undoubtedly not native, and which has almost identical habits. The red squirrel has not changed its ways, but we have certainly changed ours. To be explicit, we are defending what used to be considered a major forestry pest because it is of British stock and is threatened by an alien.

Modern natural history

After World War II, the New Naturalist book series started to celebrate the new world of nature conservation and inform a public thirsty for information about biological science: the world of Julian Huxley and Max Nicholson. It is interesting to see how the different volumes dealt with alien species. Richard Fitter in *London's Natural History* wrote essentially an early environmental history of what was at the time, in his words, 'the largest aggregation of human beings ever recorded in the history of the world as a living in a single community' (Fitter, 1945). He treated natives and aliens as equally interesting, devoting space to alien rats, mice, cockroaches and bed bugs as unwelcome pests, but no worse in that category than native common houseflies and bluebottles. Canada geese were as valid as mallard. Max Nicholson in *Birds and Men* was sympathetic to alien species like the red-legged partridge and the little owl, and described the latter with affection. It had been introduced successfully in the late 19th century, but he told how, 'as the wave of little owls rolled across England, a wave of hysteria sprang up in its wake, and eventually overtook it'. Aristotle himself had described its diet of mice, lizards and beetles, but, says Nicholson:

> *A number of gentlemen whose powers of observation and logic would have done them more credit 2300 years before Aristotle than 2300 years after him carried on a virulent campaign of emotional abuse against the little owl, which they pictured as emptying our coverts of game and our coverts of songsters.* (Nicholson, 1951)

The scientific evidence was entirely supportive of Aristotle, and the little owl was included on the schedule of protected birds in postwar legislation.

A different attitude was that of Sir Edward Salisbury, in *Weeds and Aliens* (1961). This was a book dealing with the ecology of plants considered pests of agriculture and horticulture. He assumed that weeds and aliens were virtually synonymous. 'Many, and indeed most weeds', he said, 'are either known to be introductions or are under suspicion of having been such ... the subject of weeds cannot be naturally separated from that of alien species ... the most aggressive weeds are in fact usually those known to have been introductions' (Salisbury, 1961). That skims over certain notorious natives, such as bracken, common ragwort and couch grass, but it certainly encapsulates the truth about the invasive tendencies of a number of non-native species.

The defining book in the emerging study of the ecology of alien species, however, was Charles Elton's *The Ecology of Invasions by Animals and Plants*, which originated as three talks broadcast on the BBC Third Programme in 1958 (Elton, 2000). It is a remarkable book, sharing with *The Origin of Species* brevity, lucidity and brilliance. The author was a distinguished mammal ecologist at Oxford, and his work was extremely farsighted. He identified the main characteristics of the subject, including the special problem invasions posed to island populations, the difficulty in identifying which of innumerable species would eventually pose a problem, and the risks of introducing as biological controls other non-native species ('counterpests') to combat existing invaders.

But his perspective was a measured one. He believed that the best defence against damaging invasions was a robust native ecosystem with many existing native species, but this is unfortunately not true. The example of Australia alone shows this to be wrong – the rabbit, cat and fox ruined half a wonderfully biodiverse continent. He believed that in time some invasions might moderate their initial destructive force as new predators and diseases emerged to control them. Such had indeed been the case with the pondweed *Elodea canadensis* that had choked British canals and rivers in the 19th century, growing so profusely that bathers caught in it and drowned. The Thames was rendered impassable in places, and on the Trent fishermen could not operate their nets. Yet by the 1950s it was growing quietly 'in moderate and permanent occupation of many waters' without doing much harm. He even believed that there might be a place for the scientist to encourage non-native species in a healthy ecosystem:

> *I believe that conservation should mean the keeping or putting in the landscape of the greatest possible ecological variety – in the world, in every continent or island, and so far as is practicable in every district. And provided the native species have their place, I see no reason why the reconstitution of communities to make them rich and interesting and stable should not include a careful selection of exotic forms, especially as many of these are in any case going to arrive in due course and occupy some niche.* (Elton, 2000)

At the same time he emphasized the enormous problems that some invaders caused across the world, citing as three examples (among many) the African mosquito in Brazil, Asiatic chestnut blight in the eastern United States and the sea lamprey in the Great Lakes.

It is interesting that he could not find many examples within Britain of very damaging effects, though he found a few rampant invasive alien species. One example was sea cordgrass *Spartina townsendii* (a hybrid with an alien), which was altering the ecology of estuaries but had bio-utility for land reclamation, and the American slipper limpet on neglected oyster beds. One has the impression that Elton thought the ecology of the British countryside comparatively little affected because he saw it as naturally diverse and therefore robust. But many species that appear problematic now, like the Canada goose, the Japanese knotweed, the giant hogweed and the signal crayfish were not a concern then: the last-named had not arrived, and the first three had not yet exploded their populations to the degree they have today. He was not even too concerned about the grey squirrel, noting that it had replaced the red squirrel in the English Midlands and parts of the south of England, but also that there were large areas with no recent records of red squirrels where the grey had first become established. Beyond this zone, he said, 'there are plenty of red squirrel populations still, though they have fluctuated, often severely' (Elton, 2000).

Recent times

In the years up to the early 1990s there were several developments which in Britain changed the tone of the discussion about alien species. I am not sure what significance one should attribute to the rising tide of racial tension of these years, but it has to be mentioned: 1958 was the year of the Notting Hill riots, 1968 of Enoch Powell's 'rivers of blood' speech, 1981 of the Brixton riots and 1993 of the racially provoked murder of Stephen Lawrence. I would not suggest that those who battled against the human-made tide of what they called 'alien conifers' equated Sitka spruce and lodgepole pine with West Indians and Pakistanis, even subliminally. Nor were those who founded the Woodland Trust in 1972 to defend what they described as 'native woodland' eco-fascists. Yet the choice of language then and now can make minority groups sensitive and uneasy about conservation. They asked, for instance, about the pointed contrasts between native and alien, and whether 'rhodo-bashing' resonated with 'Paki-bashing' (Wong, 2005). 'Alien' after all is not a value-neutral word in the English language: when my Danish wife first came to Edinburgh in 1959 she had to register at intervals with the police beneath a sign that read 'Aliens, firearms and dangerous drugs', which indicates what we thought of her. Even the arrival on our screens of E T in Steven Spielberg's movie of 1982, though it improved the image of an alien and gave it a new, whimsical, even cuddly, twist, hardly made it more like one of us.

But the ecological reasons why the problems of non-native species came to the fore in the years 1960–1990 were real, substantive and pressing. First, the Forestry Commission and its private-sector clients were doing untold damage to valued ecosystems. This was by planting Sitka spruce forests as a new non-native habitat; it was hard to argue this was in any sense a good replacement for open country and broad-leaf native woodlands. Second, the Nature Conservancy Council and the Wildlife Trusts, charged with the defence of an ever-increasing number of nature reserves and Sites of Special Scientific Interest, were finding problems of predation and invasion seriously threatening and worrying. Often the problem was caused by a native species, like crows, or the foxes that destroyed the famous Ravenglass gullery in Cumbria, or by the withdrawal of the alien rabbit after myxomatosis which led to invasions by native scrub of heaths that harboured rare native species. It took time, sometimes, to realize that nature did not reach a balance but had to be constantly managed, and there was at first some reluctance to tackle native predators and invaders. We had become very much more squeamish about identifying any of our native taxa as vermin, after the Victorian slaughter and the arrival in the 20th century of comprehensive protection.

In cases where the menace was alien, however, no one had any hesitation in figuring out what to do. American mink were devastating wildfowl colonies and driving water vole to the edge of extinction. Muntjac deer were destroying bluebells and other ground flora of ancient oak woodland in the south of England, devastating the National Nature Reserve at Monks Wood and many besides. Giant hogweed and Himalayan balsam were shading out stream sides and though the effect was slightly unclear, they certainly altered wetlands that had been declared special for other reasons. Sycamore proved invasive in some woods, and beech (where it was an alien) in others. There seemed no alternative other than to cull and to chop, and indeed if the sites were to be left unaltered there was none. Often the action failed, but at least wardens had the satisfaction of knowing it had been attempted.

Public reaction varied. Sometimes the actions of conservationists called forth local community protests, as at the threat to fell much-loved beech trees at Twentyshilling Wood near Comrie, central Scotland, and then Scottish Natural Heritage (SNH) had to withdraw. Sometimes culling of an alien species was applauded by the public, as when the brown rats on Ailsa Craig were destroyed and the puffins flourished again. On the other hand not everyone in authority was absolutely purist about native and alien species. When it was proposed to do the same to the black rats on the Shiants it was vetoed by the local officer of SNH, on the grounds that there were millions of puffins in Scotland but only a handful of black rats left in the country. The landowner complained that he was left as owner of the only rat reserve in Britain, but the deputy chairman of SNH argued that the rats had cultural interest as vectors of the medieval Black Death.

What was happening in Britain was also happening all over the world, and indeed British problems over sycamore and squirrels were relatively minor compared to, for example, prickly pear in South Africa, Nile perch in the lakes of East Africa, cats and foxes in Australia, possums in New Zealand or water hyacinth in Florida. The science of invasive species and how to control them exercised many countries, for good economic reasons as well as for biodiversity considerations.

The emergence of modern genetics

At the same time, science developed and changed. The discovery of DNA in 1953 placed genetics on a new footing, and raised questions and explanations about how differences and similarities between and within species were transmitted. In the late 1960s and 1970s, the expressions 'natural diversity' in Britain and 'biological diversity' in America arose, and did not necessarily refer only to the numbers of species. In 1982 biological diversity was defined by Bruce Wilcox for the IUCN at the World National Park Conference in Bali as the 'variety of life forms ... at all levels of biological systems', including the molecular (Wilcox, 1984). The term was contracted to biodiversity, and by the time of the 1992 Rio Convention it was defined as 'diversity within species, between species and of ecosystems', that is to say, including genetic diversity within a species. At the same time it was being argued that biodiversity was threatened by global homogenization. A few superspecies would come to dominate the world's ecosystems, just as McDonald's, Burger King and Kentucky Fried Chicken seemed to be taking over the world's high streets. Most nations, though not the USA, signed the Rio Convention to protect global biodiversity.

This began to move the debate about invasive species on to another plane of finesse. The concept of 'genetic pollution' began to emerge, applied both to possible gene flows from GM crops to wild species and from invasive species into native species. Conservationists began to worry not only that one species might displace another, but that one might pollute another by hybridization. Native genes would be irreversibly diluted. It was not a case of 'make love not war', but make neither love nor war if it could be prevented.

In Britain this led to anxiety over the amorous proclivities of two alien species in particular, the sika deer and the ruddy duck. The sika deer in Scotland are descended from several deliberate releases and park escapes between 1893 and 1918. Much smaller than red deer, they favour coniferous woodland habitat rather than open moor: the big mature male reds did not care for the undersized female sika, and the puny male sika did not impress the female reds, so that the risk of hybridization did not at first seem large. However, frustrated young male reds driven off their own females by more mature stags did rather fancy and impress the female sikas, and the two species proved completely interfertile (Yalden, 1999).

The sika population did not expand much until after 1970, but the planting of Sitka spruce forests suited it perfectly. Soon it was all over the north and west of the Highlands, deep into the range of red deer. Now on the mainland even pure-looking red deer living far from known sika colonies have sika genes in them. Derek Yalden comments that the process has probably gone too far to be reversed: 'Many of the hybrids are unrecognizable as such, so culling them is not an option. This seems a very sad way to lose our largest native land mammal.' It is not clear to me exactly what is lost, especially if you cannot tell the difference by looking at them. The red deer has possibly acquired some genes that will help it adapt to hiding in dense woodland, certainly a disadvantage from our perspective in a sporting animal and one already judged too numerous for the good of the ecosystem. But from the deer's point of view, what is the matter with that? Should we not consider the deer's point of view?

The case of the ruddy duck is better known and more widely discussed, but it will bear recapitulation. It is a North American species of stifftail, introduced into Peter Scott's wildfowl collection at Slimbridge in Gloucestershire after World War II. It escaped from captivity in 1952, and by 1960 birds began to breed regularly in the west and the Midlands; by 1975 there was a population of around 50–60 breeding pairs and they were expanding at the rate of 25 per cent a year. In those innocent days, so pleased was the West Midland Bird Club to have acquired such an attractive newcomer (it has a bright blue bill which it clatters in courtship, dashing in circles like a toy wind-up duck in a bath), that they made it their logo. In due course, it occupied most lowland counties in Britain without causing obvious trouble to the native fauna and began to emigrate to Europe, where it is now found in 20 countries. By 1982, some had reached Spain, where they encountered the closely related but native white-headed duck. This was a species that the Spaniards, with great trouble and expense, had brought back from the brink of local extinction, from only 22 in 1977 to about 1,000 today. The ruddy duck and the white-headed duck hybridized and proved interfertile, producing a number of cross-bred ducks that could themselves produce young.

Considering this a threat to the white-headed duck's genetic integrity, the Spanish culled what hybrids they could find, and complained to Britain that it was a reservoir of genetic pollution. The RSPB took up the case with government, and the government, which was anxious for various diplomatic reasons to placate the Spanish, eventually agreed in 1999 (at that time against the advice of English Nature, their own conservation advisory agency) to instigate a trial cull 'to investigate the feasibility of eradication'. By 2003, 2,651 ducks had been killed and the government announced it would spend a further £5 million and go ahead with a programme to eradicate the remaining birds, reckoned to be about as many again. In 2008, the British population had been reduced to 400 individuals. There have been well-founded anxieties about animal welfare in the shooting process and about disturbance to native wildfowl, but they have been brushed aside.

Scientifically, it is likely that left to themselves the two species would interbreed extensively, and possible that in time the white-headed duck hybrids would replace the pure-bred white-headed ducks. I don't see the problem myself. If the new hybrid is better adapted to the environment than the old one, it will, by natural selection, succeed; if not, it will fail. So either way what is the issue? To use the language of genetic pollution seems to me to be dangerously racist. We are condemning creatures for breeding together and producing something less 'pure', which is only to say different from what it was when science first described it. If we apply this rule to birds then why should we not also include man in this restriction? Man is part of nature, and the logic that culls the ruddy duck could equally well apply to humans of another race. Others see it differently, but for me the application of the concept of genetic pollution is a step too far in invasion ecology.

Some concluding thoughts

With the important exception of the application of genetic purism, it seems to me that at the start of the 21st century the matter of alien species is being handled rather well. Purists who would have all aliens driven out are a small minority; science and politics concentrate on what they term 'invasive aliens', and do not in fact bother about all these. Nothing could have invaded more successfully than the New Zealand willowherb, a small creeping pink perennial of damp places that was first recorded in 1904. In the north and west it now reaches from sea level to mountain top, but interferes with no one and no one interferes with it. Science is rightly much more bothered by what an earlier age more readily called weeds and vermin – Japanese knotweed and grey squirrels, because they are deemed to displace native flora and fauna and cause economic damage. We have at length come to a definition of alien species in Britain, as those who have invaded since the Mesolithic with the assistance of man (deliberate or otherwise). But under the guidance of botanists we have also come to a sensible division between archaeophytes and neophytes, plants which arrived before AD 1500 and those which came later. The former get special protection if they are threatened in Britain, notably including ancient weeds of arable land, such as cornflower, blue pimpernel and Venus's looking glass. No neophytes get such protection, though some animals at large in Britain like the Chinese water deer and Lady Amherst's pheasant are threatened in their native land, and deserve guarding in Britain. We should recognize, though, that these definitions are not scientific ones, but arbitrary and devised for the convenience of nature management. To be an alien is not a biological characteristic like being blue or having a square stem; it is a character imputed by man. That being the case, commonsense and not scientific dogma should be the guide when it comes to deciding whether or not to treat an alien species as vermin.

Nature needs managing, but it also demands study, and as a final plea I would ask for a more serious study of the ecology of those fascinating and rich habitats where aliens form such an interesting component, our cities. I need not make such a plea in Sheffield, where study of the urban ecosystem is well advanced, nor in Glasgow, where botanists have also been busy tracing the rise and fall immigrants, but as a general rule I think there is prejudice against study of urban ecology as somehow less natural. Where man is part of nature that is not true at all. Environmental historians and ecologists could well follow where Richard Fitter led in 1945 with his book on *London's Natural History*.

References

Anderson, M. L. (1967) *A History of Scottish Forestry*, Nelson, London, pp290, 404

Chew, M. K. (2006) 'Ending with Elton: Preludes to invasion biology', unpublished PhD thesis, Arizona State University

Elton, C. S. (2000) *The Ecology of invasions by Animals and Plants* (edition with foreword by D. Simberloff), Chicago University Press, Chicago, pvii

Fitter, R. S. R. (1945) *London's Natural History*, Collins New Naturalist, London, pp211, 214

Manchester Evening Chronicle (1905), quoted online without publication details, see http://en.wikipedia.org/wiki/Aliens_Act_1905

Nicholson, E. M. (1951) *Birds and Men: The Bird Life of British Towns, Villages, Gardens and Farmland*, Collins New Naturalist, London, p68

Ritchie, J. (1920) *The Influence of Man on Animal Life in Scotland: A Study in Faunal Evolution*, Cambridge University Press, Cambridge, pp241–300

Salisbury, E. J. (1961) *Weeds and Aliens*, Collins New Naturalist, London, p18

Wilcox, B. A. (1984) '*In situ* conservation of genetic resources: Determinants of minimum area required', in J. R McNealy and K. R. Miller (eds) *National Parks, Conservation and Development, Proceedings of the World Congress on National Parks*, Smithsonian Institution Press, Washington, pp18–30

Wong, J. L. (2005) 'Cultural aspects: The "native" and "alien" issue in relation to ethnic minorities', *Journal of Practical Ecology and Conservation Special Series*, vol 4, pp94–96

Yalden, D. (1999) *The History of British Mammals*, Poyser, London, pp197–198

5
Nativeness and Nationhood: What Species 'Belong' in Post-devolution Scotland?

Charles Warren

Introduction: Natives, aliens and home rule

On 12 May 1999, the new Scottish Parliament met for the first time. After an interval of 292 years, Scots once again possessed a key signifier of nationhood, emphasizing the differences which set it apart from the rest of the British Isles. Those differences encompass history, geography, culture, law and, in places, language, but they also include distinctive biodiversity. In the context of devolution, it is natural to celebrate and protect those things which are most evocative of the country. In Scotland, this most certainly includes not only the cherished landscapes of mountain, loch, glen and moor but the native fauna and flora which inhabit them – the red deer, Scots pine, wild salmon, red grouse, heather and, of course, the Scottish thistle, the national emblem for 700 years. In a recent poll, red deer were voted Scotland's most iconic animal species, followed by roe deer, red squirrel and golden eagle, and one factor influencing people's choices was an explicit recognition that such species are symbols of national identity (Stewart, 2006). In most constructions of 'Scottishness', nature looms sufficiently large that it becomes 'a semiotic player in politics' (Toogood, 1996). Accordingly, a decade after devolution, it is interesting to reflect on the significance and symbolism of Scotland's native plants and animals, and the evolving understandings of the terms 'native' and 'alien'.

In doing so, it is worth noting that the political devolution settlement of 1999 does not encompass all the devolutionary trends which have transformed the context of nature management. The earliest significant act of devolution came in 1991 when the UK-wide Nature Conservancy Council was broken up and replaced with new conservation agencies in England, Wales and Scotland.

Equally, some devolution came after 1999; the Scottish part of the Forestry Commission, formerly a UK-wide organization, gained its 'independence' in 2003, for instance. This steady, one-way devolution of powers and responsibilities from the UK to Scotland has progressively heightened and sharpened the sense of Scottish distinctiveness.

These political and perceptual trends represent the culmination of a longer process. The Scottish 'home rule' movement grew slowly but surely during the post-war decades, and these were also the decades which witnessed the steady rise to prominence and orthodoxy of 'native-only' policies in nature conservation. Simultaneously, there was a growing awareness and promotion of 'national identity'. During the heyday of the British Empire, the distinctive identities of the constituent nations and regions of the UK were partially subsumed within a wider Anglo-British narrative, but the demise of empire removed a central linchpin of UK unity and watered the seeds of devolution (Devine, 1999). 'National' increasingly came to refer to Scotland rather than Britain. Of course, national identity is a many-faceted thing, as richly demonstrated by the diverse essays on 'being Scottish' in Devine and Logue (2002). However, it is interesting to note that the phrase 'national identity' itself is of surprisingly recent provenance (Condor and Abell, 2006) and that its growing currency was a new component of political rhetoric in post-imperial Britain. The fact that concerns about nationhood and nativeness developed in parallel is no coincidence. It is therefore ironic that when devolution finally delivered nationhood, heightening the perceived significance of all things 'authentically Scottish', it came at a time when the conceptual foundations of the native/alien framework were coming under increasingly persistent attack, thereby potentially undermining one of the most rhetorically powerful arguments for the protection of 'our' nature.

To understand why the idea and practice of classifying species as native and alien is now contested, it is necessary to outline a range of issues concerning the definition of 'native' and 'alien' and the way in which these terms have come to be understood and applied. The classification scheme which categorizes species in this way is an almost universally adopted framework within contemporary conservation science and popular perception, utilized to set agendas and priorities. The fact that the term 'alien' has come to be strongly associated with the problem of invasive species is unsurprising given the enormous ecological and economic damage which invasive alien species can cause (Macdonald et al, 2007). Well-known Scottish examples include the dwindling range of the red squirrel following the arrival of its grey cousin, the damage being caused by American mink and hedgehogs to ground-nesting birds on the Western Isles and the spreading infestations of giant hogweed, Japanese knotweed and rhododendron (Warren, 2009). These are merely the headline species; at least 992 alien species are established in Scotland (Welch et al, 2001). Governments the world over have developed policies to counter such 'alien invasions', and the Scottish Parliament is no exception; in 2009 it began considering a Wildlife and Natural Environment Bill which includes proposals for stronger powers for dealing with invasive alien species.

Such species are almost universally vilified as a major ecological evil – as one of the 'mindless horsemen of the environmental apocalypse' (E. O. Wilson in Baskin, 2002). The employment of the alien–native distinction as a key criterion in our management of species has become so familiar that it goes largely unchallenged. Similarly, the underlying value hierarchy which ranks those things which are (perceived to be) natural and native above those which are wholly or partially anthropogenic mostly goes unquestioned in contemporary public discourse. However, a simple dualistic notion of 'native good, alien bad' is demonstrably false (Kendle and Rose, 2000). Not only are some native species damagingly invasive (e.g. bracken and common ragwort) but society accords a high value to many alien species. This is partly because of their contributions to economic and social well-being through agriculture, forestry, horticulture, fisheries and medicine, but also partly because of 'the lure of the exotic' – the aesthetic pleasure and/or cultural significance which many people attach to numerous introduced species. Thus the popular notion that 'alien' and 'native' are polar opposites is woefully misguided. These terms actually represent the ends of an extended continuum along which many complex entanglements and grey areas exist, and their application is anything but straightforward (Macdonald et al, 2007).

Critiques: Defining and implementing the indefinable?

It is clear even from the foregoing that society's approach to so-called native and alien species is not wholly based on the principled application of rigorous science but is shot through with contradictions, anomalies and cultural contingencies. But there are, in fact, some fundamental difficulties with the very classification of some species as native and others as alien. This is because the concepts themselves are intrinsically geographical, being defined according to spatial and temporal boundaries which can be – and are – constructed in many different and changing ways. In a recent review, Warren (2007) divides the difficulties into problems of definition and of implementation. What follows is an outline summary of these problems, setting the stage for the subsequent discussion.

Rigorous definitions of 'native' and 'alien' are elusive because they are essentially relative terms, both in time and space. No species is inherently alien but only in relation to a particular environment at a particular time. The difficulty lies in trying to pin down that time and place because, in this context, both timescale and spatial scale are socially constructed, not naturally given. The choice of scale is crucial but deeply problematic because, in Townsend's (2005) provocative words, 'by playing around with the temporal and spatial criteria almost anything can be native or non-native'. The terms are not discrete and unambiguous but cluster concepts with overlapping boundaries (Woods and Moriarty, 2001). Consequently, critics have argued that they are arbitrary and indefensible.

If the definitions are elusive, implementing conservation policies which depend on them is doubly fraught. Warren (2007) discusses a range of challenging issues. These include:

- the partial and evolving nature of scientific knowledge about species' status;
- the clash between scientific prescriptions and social preferences (as witnessed in the public outcries over the hedgehog eradication campaign on the Hebrides (Webb and Raffaelli, 2008)) and over conservation-inspired plans to fell alien beech trees in central Scotland (Smout, this volume, Chapter 4);
- debates about whether we humans should be classified as native or alien, and the implications of that choice for our dealings with non-human nature;
- the prospect of rapid climate change which destabilizes the concept of a fixed endowment of native species.

These all represent real difficulties for the practical application of the native/alien distinction, but there is a further challenge, one which is arguably the most contentious and the most relevant for this chapter. It might be labelled the taint of bio-xenophobia, referring to the accusation that a policy of eradicating alien species is xenophobic, racist and tantamount to ethnic cleansing (Gröning and Wolschke-Bulmahn, 2003). Some strongly refute this charge (Simberloff, 2003), but even if the motivations of individual conservationists are entirely innocent, there are disturbing links and parallels with 'racial purity' discourses (Wong, 2005; O'Brien, 2006). A desire to preserve native species may be positively represented as an expression of ecological patriotism, but just as patriotism can slide into racism, so the language concerning the 'righteousness' of native species can all too easily bleed into the claims made by racists and xenophobic nationalists (Olwig, 2003). The demonizing of alien species clearly represents a value system that is reprehensible when applied in human society. Although the scientific discourse concerning native and alien species is not inherently xenophobic, there is a significant risk of it being interpreted as such, not least because the terminology has created a hybrid language which melds value judgements with scientific concepts (Davis, 2009). This risk is considerably exacerbated when the region selected for the definition of native status is a nation state. It is hard, therefore, to disagree with Brown's (1997) verdict that applying the native/alien distinction at a country scale 'reflects a bizarrely nationalistic view of biogeography and has neither ecological nor practical value'. The questionable value of using national borders for ecological classification is highlighted by the fact that political change such as Scottish devolution or the break-up of former nation states such as Czechoslovakia and Yugoslavia can rewrite nationally based taxonomies overnight.

Because of these widely ranging conceptual and practical difficulties with the native–alien framework, there has been growing criticism of the terminology itself and particularly of its sometimes overzealous, uncritical application. For example, Davis (2009) is critical of what he dubs the 'simple-minded "nativism" paradigm', and he recommends that species should be distinguished simply on the basis of how long they have been resident in a region, without having recourse to the loaded labels of 'native' and 'alien' with their normative implications. Does this add up to a case for jettisoning the terms 'alien' and 'native' altogether, as some (e.g. Kendle and Rose, 2000; Fischer, 2009) have argued?

In answering this, it is important to differentiate between the use of these terms (or concepts) for description and prescription. Thus Aitken (2004) suggests that, although they may be colloquially valuable in a descriptive sense, it does not seem acceptable to apply them in a prescriptive sense. Preston (2009) makes a strong case for continuing with the descriptive differentiation of native and introduced species for the purposes of biogeographical explanation, showing persuasively that today's species distributions simply cannot be understood if human agency is ignored. That an understanding of the history of species arrivals is valuable is undeniable, but it is the next step – the progression from description to prescription – which is controversial. Understanding when and how a species arrived in an area is a straightforwardly scientific exercise; deciding whether it now belongs there is anything but. Such decisions stretch beyond science to include values and ethics, and the value judgements – such as the critical decision whether *change* constitutes *harm* – cannot be read directly from science.

It can therefore be argued that, in place of the alien–native distinction, a more workable and ethically defensible criterion for guiding conservation choices would be a species' potential for causing damage in a particular place and time (Lodge and Shrader-Frechette, 2003; Rotherham, 2009). The defining issue would then cease to be a species' 'immigrant status' but whether it is a well-behaved citizen of the ecological community (Callicott, 2002; Aitken, 2004). This suggestion leads Warren (2007) to draw a parallel between ecological and human communities, suggesting that 'terrorists' – whether home-grown or foreign – deserve to be rooted out while immigrants which make a positive contribution deserve to be welcomed as citizens. Combating and preventing damage is arguably a better strategy than waging war on alien species. As Davis (2009) comments, 'as long as the harm is real, it should not be necessary to promote a native vs "alien" dichotomy to get society to respond'.

Do these arguments imply that the whole edifice of protecting natives and resisting aliens, which lies at the heart of conservation policy, should be abandoned forthwith? It may be that it is the language and justifications for conservation policies which are most in need of a critical overhaul rather than the practical conservation activities which those policies support. The use of native–alien terminology imposes a sharply dichotomous paradigm on a sociobiological context which is actually a continuum – and a contested, dynamic and messy continuum at that. The reliance on pro-native arguments to justify conservation action appears to be a strategy resting on unsound foundations, and which can, moreover, prove to be counterproductive by offending the public on whose support conservation depends. However, although conservation arguments in recent decades have come to be framed within the compromised native–alien construct, the fundamental rationale of conservation – the need to care for the non-human living world around us – is not in question. Caring for nature can be supported with numerous arguments which do not need to employ the language of natives and aliens or the rhetoric of naturalness and authenticity.

Natives and aliens in post-devolution Scotland

Where does this discussion leave us with regard to the cherishing of Scotland's plants and animals? The zeal for protecting and promoting native species and for weeding out those which don't belong is entirely understandable in the context of a newly distinct Scotland. It is the wish to reassert 'an authentic version of Scottish natural heritage concomitant with Scottish national identity' (Toogood, 1996). The desire to resist species which threaten to dilute or change the distinctive character of Scotland, and to save species such as the red squirrel, which have become conservation icons, has obvious emotional appeal. However, Smout (2009, this volume) is one of a growing number of critics who believe that this zeal has been taken to unjustified lengths. For a start, if genetic purity is taken to be a criterion of native status, then even animals as quintessentially native to Scotland as red squirrels and capercaillie do not qualify because they are actually genetically alien, having been reintroduced in the late 18th and early 19th centuries from England and Scandinavia (Kitchener, 1998). Equally, scientific uncertainty surrounds the status of many species; pedunculate oak may turn out not to be a native Scottish species after all whereas the much-vilified 'alien' sycamore may actually prove to be native (Newton et al, 2001; Leslie, 2005). As such cases show, it is often the case that pro-native conservation causes turn out to be much less clear-cut than

Figure 5.1 *Red squirrel and capercaillie: Cherished native species or alien intruders?*

official conservation narratives would suggest. Perhaps what society values 'is not necessarily a native ecology but one perceived to be' (Rotherham, 2009). A similar but wider point is made by Kendle and Rose (2000) who argue that 'conservation and preservation of nature are really value-laden activities that get confused with objective science'.

A revealing illustration of this is provided by the current debates over hybridization between alien and native species. The prevention of such hybridization is routinely championed as a cause worth fighting for on the grounds of protecting the 'genetic integrity' of native species such as red deer and wildcats. Consequently, much anguish is expressed over the ongoing interbreeding between red deer and Sika deer, introduced from eastern Asia. However, Scottish red deer are actually far from being thoroughbreds. This is a consequence of the Victorians' penchant for 'improving' stock using imported deer such as North American wapiti, and, in fact, red deer populations in some parts of the country are entirely the result of introductions following local extinction (Pérez-Espona et al, 2009). Consequently, it is already centuries too late to preserve the genetic integrity of Scottish red deer. Long before the emergence of 'the Sika threat', the 'monarch of the glen' had already become the 'mongrel of the glen' (Ratcliffe, 1995). The inexorable spread of Sika has also made a mockery of official policies of containment, and this, combined with the fact that Sika stags have become a valuable sporting asset, leads Warren (2009) to propose pragmatically that we should 'live and let live' and treat Sika as a naturalized species.

More fundamentally, Smout (Chapter 4) questions the values which underlie efforts to prevent hybridization, arguing that the explicit objective of resisting genetic pollution trespasses disturbingly close to racism, as discussed above. It is, he contends, not at all clear what we are trying to save or what would be lost if conservation efforts fail. Taylor (2005) is equally forthright in his iconoclasm:

> *We are not losing our largest native land mammal, but simply watching it being transformed to a species better adapted to the ... environment. ... This is a classic example of scientific conservation values drawn from an old paradigm of an almost nationalistic taxonomy, rather than from functional ecology.*

Ironically, Scotland may now have one of the largest and healthiest populations of Sika deer in the world, while in its native lands it is increasingly endangered. Does this international perspective argue for conservation rather than eradication of Scottish Sika populations? This situation echoes that of the rabbit, a notoriously invasive alien in Scotland but an endangered species in its native Iberia (Lees and Bell, 2008). The existence of such conservation paradoxes illustrates the need for policies to be formulated using broad perspectives, incorporating both time depth (history) and broad spatial awareness (geography), and to lean more towards case-specific pragmatism than dogmatic purism.

Although a concern for genetic purity and site-specific 'authenticity' is now an established part of official conservation policies, both Rotherham (2009) and Smout (2003, Chapter 4) show how comparatively recently such native-only policies emerged, and how quickly attitudes to particular species can change. Thus the red squirrel, killed by the thousand as a pest in the early 20th century and still regarded as a 'devastating menace' in the 1960s (Anderson in Smout, Chapter 4), is now championed as a threatened native species and protected at considerable cost. In the context of contemporary Scotland, where it is a commonplace to point out that the 'natural environment' is anything but natural – consisting as it does of a cultural landscape resulting from many millennia of human management – it must be questioned whether striving for genetic purity and 'naturalness' makes much sense. 'If there is no real purity, why be purist?' asks Taylor (1996). Of course, the Scottish population itself is far from genetically pure. How would one set out to define a 'pure-bred Scot' in the context of the successive waves of invaders and immigrants throughout ancient and recent history, all of whom have contributed to the richly diverse genetic make-up of today's Scottish people? As the Scottish novelist William McIlvanney put it, 'Scottishness isn't some pedigree lineage, it's a mongrel tradition' (in Ascherson, 1993). This being so, it is hard to justify the imposition of genetic purity policies on other species.

How then, has Scottish devolution affected perceptions of nativeness? The creation of a new parliament, whether in new sovereign nation states (such as the constituent nations of the former Yugoslavia) or in a nation within a nation such as Scotland, inevitably intensifies the sense of nationhood and distinctiveness. Condor and Abell (2006) confirm that many people in post-devolution Scotland employ different lay geographies and historiographies to those in England, and that their understandings of national identity differ too. One corollary of this heightened sense of separateness is the desire to preserve and celebrate everything which is unique and/or most emblematically distinctive. The redrawing or reweighting of political boundaries is thus typically accompanied by a tightening of perceptual boundaries to emphasize the here and now – the 'here' of this particular nation and the 'now' of its present endowments. This spatial and temporal focus gives added impetus to efforts to conserve those elements of the natural heritage which are considered to be 'national treasures' – in the newly defined sense of national (e.g. Scottish instead of British, Croatian instead of Yugoslavian).

Since 1999 this trend has been very evident in Scotland. It has been seen, for example, in the creation of the first two national parks – Loch Lomond and the Trossachs (2002) and the Cairngorms (2003) – which was accompanied by overtly nationalistic rhetoric, and also in the active reframing of environmental decision-making to assess it using Scottish rather than UK criteria. This was strikingly apparent in the decision to reject a proposed superquarry on the Isle of Harris, a decision which turned partly on the definition of 'national interest' (Warren, 2002); the Reporter at the predevolution Public Inquiry in 1995 interpreted 'national' to refer to the UK and recommended that the quarry should go ahead, whereas the final decision was made after devolution by the Scottish Government which interpreted 'national' as Scottish and rejected the proposal.

Such political, emotional and psychological reactions to devolution are understandable and probably inevitable. They are simply an accentuation of the almost universal human tendency to differentiate 'us' from 'them', 'to identify with a homeland, a home tribe, a home religion, a home team, and to declare someone else the opposition or the enemy' (Davis, 2009). Cherishing native species and rooting out aliens is, in part, an extension to the natural world of this declaration of home identity. However, in the light of the above discussion this promotion of all things native at the expense of all things alien (or, at least, of those things perceived to be native and alien) can be seen to rest on decidedly dubious foundations. Instead of focusing on the here and now, the arguments above suggest that informed management of environments and species requires a broader view to incorporate a longer-term historical perspective and a spatial perspective which stretches far beyond Scottish shores. Otherwise, 'natives only' or 'natives best' policies can justly be accused of being based on a narrow nativism which is first cousin to xenophobic nationalism. Of course, such nationalism is a cousin loudly disowned by native species enthusiasts, but both impulses spring, at least in part, from the same wells of love of country and beliefs about what (or who) belongs where. Such attempts at distancing echo the distinction made in political rhetoric between 'ethnic' and 'civic' forms of nationalism; the Scottish devolution campaign strongly emphasized the inclusive, civic sense of being Scottish while frowning on ethnic nationalism (Kiely et al, 2005). The familiar aphorism 'think global, act local' is very relevant in this context. Devolution brings with it an insidious temptation to 'think local, act local', but a narrow concern for the here and now is hard to justify. Even the Scottish National Party (SNP) has been careful to balance its campaigning for independence with appeals to a broader European identity (Toogood, 1996).

Given the need to adapt to predicted climate change, there is a faultless logic behind arguments that the species composition of ecosystems must be allowed (or encouraged) to change, even to the extent of welcoming species which are currently labelled alien. Thus Davis (2009), for example, suggests that we should 'learn to love 'em'; and Trudgill (2008) wonders whether 'invasive aliens' which arrive as a consequence of climate change should be rebranded as 'refugee species'. This directly parallels the SNP's policy of positively reconceptualizing immigrants as 'new Scots' (Kiely et al, 2005). However, the reality is that learning to love species which we have long persecuted as unwelcome aliens will be as difficult as immigration policy is politically charged, because people's sense of place creates strong beliefs about 'what should be where'. For this reason Trudgill (2008) believes that there are likely to be 'rearguard actions to preserve the status quo for a long time to come'.

Importantly, this 'zooming out' to adopt broader spatio-temporal horizons does not rule out the celebration, promotion and protection of valued species. It simply means that the basis on which they are valued cannot rest on their supposed nativeness and naturalness. Nativist purism does not withstand close scrutiny. Tellingly, nor does it chime with the perceptions and values of the general public. For example, surveys of the Scottish public's attitudes towards changes in biodiversity and towards the relative value of native and non-native species reveal that nativeness is not, in fact, a particularly highly valued attribute, and that non-native species are judged primarily by their impacts on habitats and other species rather than by their foreign nature (Fischer and van der Wal, 2007; Fischer, 2009). Perhaps, then, we should simply acknowledge that, for a complex mixture of reasons – ecological, sociocultural, emotional, economic, historical and aesthetic – society wishes to grant privileged status to particular species. Smout (2009) argues in favour of such an approach and against the proscription of 'aliens', concluding uncompromisingly that, 'there is no logical reason whatever not to celebrate, preserve and respect what came about, or was favoured, by human cultural activity'. Official acknowledgement that conservation policy springs from such a diverse mix of value-laden motivations would, of course, be acutely difficult, given society's avowed post-Enlightenment commitment to scientific rationalism. Being able to claim objective science as your ally enormously strengthens your case because science is widely (if improperly) held to be an impartial arbiter of truth, whereas appealing to subjective preferences and cultural attachments is perceived as infinitely weaker. However, the arguments summarized here show that classifying species as native and alien is not as scientific an exercise as is commonly supposed but is itself a culturally contingent process shot through with contested human values.

The sooner this is admitted, the sooner we will be able to adopt a more honest approach to our management of nature and abandon the pretence that conservation is, purely and simply, a scientifically motivated, scientifically justified endeavour. As Clayton and Myers (2009) show in their exploration of conservation psychology, conservation is an inherently value-laden, emotionally charged, humanistic endeavour. After all, few professional conservationists, let alone members of the public, are motivated primarily by the dry processes of sound science; most are spurred on by a passionate love for the natural world, a passion which may be informed by science but is certainly not limited to it. Many are frustrated by the professional necessity to restrict themselves to the unemotional language of scientific rationality when dealing with subject matter which inspires wonder and joy (Oates, 2008). Milton (2002) is one voice courageously calling for a new ecology which unashamedly embraces human emotions and feelings on the grounds that they are the prime motivators of thought and action. She points out the simple truth that caring is an emotional response, and she brands the conventional opposition between emotion and rationality as a myth.

If our dealings with nature were to be stripped of the normative, moral overtones of the native–alien framework and, instead, species were evaluated on the grounds of their context-specific impacts, both ecological and cultural, there could be considerable benefits. For one thing, some of the apparent anomalies and contradictions would simply evaporate. Thus, for example, the fact that the 'alien' rabbit provides many highly valued ecosystem services for nationally rare species would no longer appear anomalous, but would simply be an indication that this long-naturalized species is now so thoroughly intertwined within native ecosystems that it has become 'functionally native' as part of the 'natural' food chain (Lees and Bell, 2008). Other advantages of focusing on impacts and value rather than immigration history are that it would bypass much of the definitional tangle surrounding 'what is native?', avoid xenophobic associations, and give conservation managers more achievable and defensible objectives. It would also allow unashamed protection, where appropriate, of introduced species that deserve it (Smout, 2003). Arguably, it would also be a more participatory and democratic approach.

National pride and the desire to see the animals and plants of one's own country thriving do not spring primarily from rational scientific assessments but from emotional, historical and cultural connections. As humans, we are emotional beings, and our attachments to nature are fundamentally emotional, psychological and spiritual (Trudgill, 2008; Clayton and Myers, 2009). Why, then, should we not be 'up front' about our feelings for nature when making environmental decisions instead of cloaking them behind the language of rationality where they lose the power to move the heart and lift the spirit? Moreover, if conservation spoke more in the vernacular, it would connect more readily with the motivations of the general public. As Macdonald et al (2007) point out, people value nature 'for a mysterious jumble of aesthetic, intellectual and emotional reasons akin to those that make us value, for example, a particular piece of art'. Public discourses of nature protection, for so long exclusively rationalistic, would do well to recognize this and seek to harness rather than suppress the deep emotional connections with nature which make people so passionate about protecting 'their' wildlife and plants. Whether these flora and fauna are labelled native or alien within Scotland is far less important than that the environmental processes upon which all of human society depends are allowed to thrive, and that animals and plants themselves have the freedom to flourish. In conclusion, then, this discussion suggests that continuing to employ the increasingly discredited native–alien framework as a key criterion for deciding which species 'belong' in Scotland would be a misguided way of caring for the nation's present and future nature, and a flawed way of celebrating the country's newly enhanced sense of national identity.

Acknowledgements

I am grateful to Chris Smout for commenting on a draft of this chapter, and to Andrew Midgeley for pointing me to the paper on 'nature and nation' by Toogood (1996).

References

Aitken, G. (2004) *A New Approach to Conservation: The Importance of the Individual Through Wildlife Rehabilitation*, Ashgate, Aldershot

Ascherson, N. (1993) 'The warnings that Scotland's patient nationalism could turn nasty', *The Independent*, 21 November 1993

Baskin, Y. (2002) *A Plague of Rats and Rubbervines: the Growing Threat of Species Invasions*, Island Press, Washington, DC

Brown, N. (1997) 'Re-defining native woodland', *Forestry*, vol 70, no 3, pp191–198

Callicott, J. B. (2002) 'Choosing appropriate spatial and temporal scales for ecological restoration', *Journal of Biosciences*, vol 27, pp410–420

Clayton, S. and Myers, G. (2009) *Conservation Psychology: Understanding and Promoting Human Care for Nature*, Wiley-Blackwell, Chichester

Condor, S. and Abell, J. (2006) 'Vernacular constructions of "national identity" in post-devolution Scotland and England', in Wilson, J. and Stapleton, K. (eds) *Devolution and Identity*, Ashgate, London, pp51–75

Davis, M. A. (2009) *Invasion Biology*, Oxford University Press, Oxford

Devine, T. M. (1999) *The Scottish Nation 1700–2000,* Allen Lane, London

Devine, T. and Logue, P. (eds) (2002) *Being Scottish: Personal Reflections on Scottish Identity Today,* Polygon, Edinburgh

Fischer, A. (2009) 'Public views on biodiversity change: The role of non-native species', Presentation at Scottish Natural Heritage conference 'The Changing Nature of Scotland', Perth, 16–18 September

Fischer, A. and van der Wal, R. (2007) 'Invasive plant suppresses charismatic seabird: The construction of attitudes towards biodiversity management options', *Biological Conservation*, vol 135, no 2, pp256–267

Gröning, G. and Wolschke-Bulmahn, J. (2003) 'The native plant enthusiasm: Ecological panacea or xenophobia?', *Landscape Research*, vol 28, no 1, pp75–88

Kendle, A. D. and Rose, J. E. (2000) 'The aliens have landed! What are the justifications for "native only" policies in landscape plantings?', *Landscape and Urban Planning*, vol 47, no 1, pp19–31

Kiely, R., Bechhofer, F. and McCrone, D. (2005) 'Birth, blood and belonging: Identity claims in post-devolution Scotland', *Sociological Review*, vol 53, no 1, pp150–171

Kitchener, A. C. (1998) 'Extinctions, introductions and colonisations of Scottish mammals and birds since the last ice age', in Lambert, R. A. (ed) *Species History in Scotland: Introductions and Extinctions Since the Ice Age*, Scottish Cultural Press, Aberdeen, pp63–92

Lees, A. C. and Bell, D. J. (2008) 'Conservation paradox for the 21st century: The European wild rabbit *Oryctolagus cuniculus,* an invasive alien and an endangered native species', *Mammal Review*, vol 38, no 4, pp304–320

Leslie, A. (2005) 'The ecology and biodiversity value of sycamore (*Acer pseudoplatanus* L) with particular reference to Great Britain', *Scottish Forestry*, vol 59, no 3, pp19–26

Lodge, D. M. and Shrader-Frechette, K. (2003) 'Nonindigenous species: Ecological explanation, environmental ethics, and public policy', *Conservation Biology*, vol 17, no 1, pp31–37

Macdonald, D. W., King, C. M. and Strachan, R. (2007) 'Introduced species and the line between biodiversity conservation and naturalistic eugenics', in Macdonald, D. W. and Service, K. (eds) *Key Topics in Conservation Biology*, Blackwell, Oxford, pp186–205

Milton, K. (2002) *Loving Nature: Towards an Ecology of Emotion*, Routledge, London

Newton, A. C., Stirling, M. and Crowell, M. (2001) 'Current approaches to native woodland restoration in Scotland', *Botanical Journal of Scotland*, vol 53, no 2, pp169–195

Oates, M. (2008) 'Obfuscation and the language of nature conservation', *ECOS*, vol 29, no 1, pp10–18

O'Brien, W. (2006) 'Exotic invasions, nativism, and ecological restoration: On the persistence of a contentious debate', *Ethics, Place and Environment*, vol 9, no 1, pp63–77

Olwig, K. R. (2003) 'Natives and aliens in the national landscape', *Landscape Research*, vol 28, no 1, pp61–74

Pérez-Espona, S., Pemberton, J. M. and Putman, R. (2009) 'Red and sika deer in the British Isles, current management issues and management policy', *Mammalian Biology*, vol 74, no 4, pp247–262

Preston, C. (2009) 'The terms "native" and "alien": A biogeographical perspective', *Progress in Human Geography*, vol 33, no 5, pp702–711

Ratcliffe, P. R. (1995) 'The mongrel of the glen', *The Field*, June, pp66–69

Rotherham, I. D. (2009) 'Exotic and alien species in a changing world', *ECOS*, vol 30, no 2, pp42–49

Simberloff, D. (2003) 'Confronting introduced species: A form of xenophobia?', *Biological Invasions*, vol 5, pp179–192

Smout, T. C. (2003) 'The alien species in 20th-century Britain: Constructing a new vermin', *Landscape Research*, vol 28, no 1, pp11–20

Smout, T. C. (2009) 'History, nature and culture in British nature conservation', in Smout, T. C. (ed) *Exploring Environmental History: Selected Essays*, Edinburgh University Press, Edinburgh, pp199–215

Stewart, D. (2006) 'Scottish Biodiversity List Social Criterion: Results of a survey of the Scottish population', *Scottish Executive Social Research Findings*, no 26

Taylor, P. (1996) 'Return of the animal spirits', *Reforesting Scotland*, vol 15, pp12–15

Taylor, P. (2005) *Beyond Conservation: A wildland strategy*, Earthscan, London

Toogood, M. (1996) 'Nature and nation: Ecology and the reconstruction of the Highlands', *Scotlands*, vol 3, no 1, pp42–55

Townsend, M. (2005) 'Is the social construction of native species a threat to biodiversity?', *ECOS*, vol 26, nos 3/4, pp1–9

Trudgill, S. (2008) 'A requiem for the British flora? Emotional biogeographies and environmental change', *Area*, vol 40, no 1, pp99–107

Warren, C. R. (2002) 'Of superquarries and mountain railways: Recurring themes in Scottish environmental conflict', *Scottish Geographical Journal*, vol 118, no 2, pp101–127

Warren, C. R. (2007) 'Perspectives on the "alien" *versus* "native" species debate: A critique of concepts, language and practice', *Progress in Human Geography*, vol 31, no 4, pp427–446

Warren, C. R. (2009) *Managing Scotland's Environment*, 2nd edn, Edinburgh University Press, Edinburgh

Webb, T. J. and Raffaelli, D. (2008) 'Conversations in conservation: Revealing and dealing with language differences in environmental conflicts', *Journal of Applied Ecology*, vol 45, no 4, pp1198–1204

Welch, D., Carss, D. N., Gornall, J., Manchester, S. J., Marquiss, M., Preston, C. D., Telfer, M. G., Arnold, H. and Holbrook, J. (2001) 'An audit of alien species in Scotland', *Scottish Natural Heritage Review*, no 139

Wong, J. L. (2005) 'The "native" and "alien" issue in relation to ethnic minorities', *ECOS*, vol 26 nos 3/4, pp22–27

Woods, M. and Moriarty, P. V. (2001) 'Strangers in a strange land: The problem of exotic species', *Environmental Values*, vol 10, pp163–191

6
Who Is the Invader? Alien Species, Property Rights and the Police Power

Mark Sagoff

Introduction

Everyone has seen popular articles that lament the scourge of invasive species. A recent issue of *Newsweek* carried a typical story entitled, 'Attack of the Aliens: Migrating species may be the biggest threat to plant and animal life on the planet'.[1] Since any species may migrate and because more and more plants and animals are travelling in the wake of human activity, one may wonder if the biggest threat to plant and animal life on the planet is plant and animal life on the planet. What does all this planetary mixing mean for us?

This chapter examines the extent to which science-based laws intended to control invasive species may restrict personal liberties and property rights. It describes the legal framework – which has a long history – through which governments in the USA properly exercise the police power to control or eradicate plant and animal nuisances and pests. The chapter goes on to examine efforts by the National Invasive Species Council and analogous state agencies to develop management plans to protect 'natural ecosystems' from non-native species. I argue that these efforts are largely unjustified and thus likely to fail. They represent the attempt by a scientific community to validate or vindicate through legal enforcement its conception of the way nature ought to be – in defiance of the way nature is.

In the USA, on most accounts, the Centers for Disease Control (CDC) and the Animal and Plant Health Inspection Service (APHIS), among other organizations, deal effectively with organisms that threaten human health and agricultural and other significant economic interests. If these agencies fall short, they may need better leadership or more money, but they do not require greater legal authority. Public statutes provide all the authority these agencies need to protect human health and economic interests from threats posed by pests, pathogens and other harmful organisms.

For example, the Lacey Act, initially enacted in 1900 primarily to regulate the importation of wild birds, has been amended several times and now controls the importation of any wildlife deemed 'injurious to human beings, to the interests of agriculture', and to natural resources. Under the Lacey Act (as amended in 1981) the Department of the Interior (DOI) allows any exotic species to be imported unless designated as 'injurious'. To designate a species as 'injurious', the Fish and Wildlife Service (FWS) of DOI must complete a petition-and-review process that places the evidentiary burden on those who argue a plant or animal poses a hazard.[2] As a result the so-called FWS 'dirty list' or 'black list' comprises several dozen animal species known or shown to be injurious to some important economic interest in the USA or elsewhere.[3] The US Department of Agriculture (USDA), under the Plant Protection Act, lists about 95 species and 2 genera as 'noxious weeds' in agriculture.[4]

Ecologists and other environmental professionals urge governmental agencies to adopt a 'guilty until proven innocent' or 'clean list' approach according to which 'every proposed introduction be viewed as potentially problematic until substantial research suggests otherwise'.[5] The Environmental Law Institute (ELI) explains: 'The clean list approach ... generally presumes that all species should be prohibited unless they have been officially determined to be "clean", in that they will not pose any economic or environmental threat.'

In its 'Model State Law', ELI takes a 'clean list' approach.[6] Its model legislation makes every landowner responsible to control or eradicate all non-native species that are not permitted by the state's invasive species council.[7] In its legislative guidance, ELI recommends:

> *An effective state program affirmatively declares that all non-native invasive species are subject to regulation, thereby regulating all categories of species, including wildlife, aquatic life, plants, insects, microorganisms, and pathogens. States may also use the definition of 'invasive' to expand coverage of the laws and regulations beyond those species that impact agriculture to those that cause harm to the natural environment ...*[8]

While no state has fully adopted the ELI 'model' legislation, many states empower noxious weed or invasive species committees and councils to list as injurious non-native species thought to cause harm to the natural environment not just those that threaten human health or economic interests. In these states, official lists of 'noxious weeds' include plants considered to be threats to native species or to the ecosystem even if they have no effect on agriculture. Pennsylvania, which is typical in this respect, empowers its department of agriculture to order landowners to eradicate 'noxious weeds' on their property. The weeds must first be listed as 'noxious' after hearings before the Noxious Weed Control Committee. The Pennsylvania list comprises plants like purple loosestrife thought to be bad for the natural environment as well as plants like marijuana thought to be bad for public health or morals.[9] 'The department may issue an order requiring an individual landowner to implement control measures for noxious weeds and if a landowner fails to comply with an order, the department will do so' at the landowner's expense. 'Any landowner who fails to comply ... is guilty of a summary offense.'[10]

In Nebraska, 'It is the duty of each person who owns or controls land to effectively control noxious weeds on such land.'[11] If the landowner does not act within ten days to eradicate the weeds, 'the control authority may enter upon such property for the purpose of taking the appropriate weed control measures. Costs for the control activities of the control authority shall be at the expense of the owner of the property.'[12] None of the weeds listed by USDA as agricultural pests appears on the Nebraska Noxious Weed List. The Nebraska list includes plants thought to harm natural areas – plants such as purple loosestrife, saltcedar and Canada thistle. *Phragmites*, a common reed in wetland environments, is ubiquitous in Nebraska; one variety, hard to distinguish from the others, was found to be non-native and so designated as 'noxious'. 'Phragmites is present in Nebraska landscapes as a native plant; this designation covers non-native phragmites only.'[13]

An ordinance controlling noxious weeds makes sense if it protects public health or important economic interests. Yet in many states the 'noxious' or controlled list includes for the most part plants no one believes (even mistakenly) threaten health or agriculture, such as purple loosestrife and Japanese honeysuckle. These plants, which perfume the air with beautiful blossoms, are often enjoyed by those on whose land they grow. The Pennsylvania Noxious Weed Control List names purple loosestrife just after marijuana among plants targeted for control or eradication. A state document explains that purple loosestrife is 'invasive' and 'harms the environment'. It 'crowds out native plant species and decreases the population of animals that are dependent upon native plant species for survival'.[14]

A long legal tradition allows the use of state power to compel landowners to manage their property so that they do not harm others. The hoary principle, '*Sic utero tuo ut alienum non laedas*' (use your property so as not to harm others) creates the 'nuisance' exception to the general rule, established through the Fifth Amendment of the Constitution, that the state cannot take private property for public use without paying just compensation.[15] On what concept of harm and on what theory of nuisance does the ELI rely when it recommends that state agencies use the definition of 'invasive' to expand coverage of the laws and regulations beyond those species that cause economic injury to those that cause 'harm to the natural environment'?

What is 'harm to the natural environment'? How is it measured? Are non-native species so evil – and those who permit their presence so antisocial as to trigger the nuisance exception to the 'takings' provision of the Constitution? Many states – like Nebraska and Pennsylvania – protect the natural environment with laws that make landowners responsible for ridding their property of a laundry list of non-native species. What public interest do these laws serve?

Miller v. Schoene (1928)

To answer these questions, it is useful to recall how courts have dealt with ordinances that control plants and other organisms that threaten major agricultural crops.[16] Red cedar along with the rust it harbours existed in Virginia before the European settlement. Apple, an introduced species, was widely planted in Virginia after 1900, when the railroads and refrigeration made it the leading export crop. Cedar rust, a heteroecious fungus, requires for its life cycle both cedar and another species, preferably apple, close enough that the wind can carry spores back and forth between them. It kills the apple but is harmless to the cedar other than to produce galls.

In 1914, the Virginia House of Delegates responded to the needs of apple growers by enacting unanimously the Cedar Rust Act to allow the state entomologist – at the time of the legal action, W. J. Schoene – to order the destruction of any red cedar tree threatened by cedar rust that grew within two miles of an apple orchard. Nearly everyone in Virginia recognized that the economic value of apple orchards dwarfed that of red cedars, which grew wild and were useful primarily as firewood. Nevertheless, the 1914 Cedar Rust Act, at the suggestion of apple growers, created a fund to compensate owners of particularly valuable cedar trees. Orchard owners typically paid the costs of cutting wild red cedars on the lands of neighbours, stacked the firewood for them, and taxed themselves to support the fund used to compensate for the loss of large ornamental trees.[17] According to James Buchanan, in this kind of context compensation represents 'the only test for efficiency that can be instituted politically'.[18]

William A. Fischel, in a masterful scholarly study of *Miller* v. *Schoene*,[19] notes that, although no cedars were grown commercially within range of an apple orchard, 'the orchardists' coffers were in danger of being drained by opportunistic claims from landowners whose cedars usually had more value cut than standing'.[20] Fischel cites several documents that describe many dubious claims by landowners to compensate them for cedars they suddenly discovered to have aesthetic value. Apple growers feared that inflated claims and the transaction costs involved in settling them would drive their special orchard taxes 'to heights that would make cedar rust seem preferable'.[21] To avoid prohibitive appraisal and adjudication costs, the state settled on allowing property owners who lost large ornamental trees a flat payment of one or two hundred dollars.[22]

Dr Casper Miller, who as a member of the House of Delegates had voted for the Cedar Rust Act, later sued to have it declared unconstitutional. Miller sought to retain 200 cedar trees, which being large and ornamental added more value to his property than the $100 he would receive. 'The statute is invalid,' his counsel argued, 'in that it provides for the taking of private property, not for public use, but for the benefit of other private persons.'[23] Justice Harlan Fiske Stone, who wrote the opinion of a unanimous Supreme Court, relied on the decision of the Virginia Supreme Court to interpret the Cedar Rust Act. Justice Stone wrote: 'Neither the judgment of the court nor the statute as interpreted allows compensation for the value of the standing cedars …'[24] He upheld the statute nevertheless on the grounds that, 'the state does not exceed its constitutional powers by deciding upon the destruction of one class of property in order to save another which, in the judgment of the legislature, is of greater value to the public.'[25]

The Court rested its opinion on its finding that the Cedar Rust Act served a public purpose (the overall economy of the Commonwealth) rather than a private interest (the profits of a group of apple growers), as Miller contested it did. Justice Stone refused to 'question whether the infected cedars constitute a nuisance according to the common law, or whether they may be so declared by statute'.[26] Justice Stone paid attention instead to the tortuous legal process, which worked on the county level and required 'a request in writing of ten or more reputable free-holders', even to begin an inquiry in which the state entomologist investigated the cedar trees in a locality and determined whether they must be destroyed.[27] In view of the local focus of each inquiry, involving neighbours who had to continue to live together, and the availability of appeals and reviews, the Court refused to overturn the Cedar Rust Act on due process grounds. Even if apple-growing was essential to Virginia's economy and the Act thus served a public purpose, there was nothing antisocial about allowing cedar trees on one's land. Why, then, was Justice Stone willing to let one party (the apple growers) condemn without compensation the cedar trees of another (their neighbours) who were innocent of conniving at harm? Since there is nothing 'wrong' about cedars, which grow wild all over the state, one might ask whether (i) Miller should have been paid for the value of his ornamental trees, and (ii) apple growers should not have planted orchards within two miles of them.

As one might expect, there is a large literature on this question. It is commonplace to cite Ernst Freund, who in *The Police Power: Public Policy and Constitutional Rights* states: 'Where property is destroyed in order to save property of a greater value, a provision for indemnity is a plain dictate of justice and of the principle of equality.'[28] The background principles of common law suggest that one party cannot use the law to condemn the property of another even if its interest is so much greater that the public good is thus served; rather, compensation is required. In this light, Richard Epstein has argued that Miller should have been compensated for his loss because cedar trees are innocent or passive conduits of the rust.[29] Although sympathetic with this principle, William Fischel has shown that the orchardists did pay a tax to compensate owners of valuable cedar trees. 'The moral hazard problem,' he argues, 'undercut full compensation.' The expectation of compensation served in fact to make landowners opportunistic. 'A landowner who expects to be compensated for the cutting of her cedars might, instead of suppressing them, let them grow or even encourage them' to increase his or her compensation.[30] Justice Stone was mistaken in his belief that the Cedar Rust Act did not provide compensation for especially valuable trees.[31] The Act created a fund through which apple growers tried to compensate landowners for large ornamental cedars. The fund created an incentive, alas, for landowners seeking compensation to let otherwise worthless and infected trees grow larger. The problem of moral hazard – not the intention, wording, or history of the Cedar Rust Law – prevented Miller from receiving just compensation for his losses.

Citrus canker

When an agricultural pathogen or pest spreads within a single species the situation is different. All those who grow a plant or animal share an interest in protecting it. All the growers of a pear, apple, orange or peach, for example, have an interest in stopping diseases that affect that fruit. It is advantageous for each to bear the risk that his or her trees might be destroyed if diseased than to have no power to force others to destroy their infected orchards. Statutes that require the destruction of infected fruit trees – including healthy trees that grow in the path of a pathogen and are reasonably certain to become infected – may be justified in terms of the 'average reciprocity of advantage' of the fruit growers, since each gains (at least ex ante) more from the prospective restrictions on others than he loses by accepting them himself.[32]

When a particularly virulent strain of citrus canker appeared in Florida in the late 1990s, that state enacted a statute that required the removal of all citrus trees within 1900 feet of a tree infected with the bacterium. Florida courts have a lot of experience responding to aggrieved property owners whose trees, some of them still healthy, were cut by the state because they stood in the path of blight. Deciding a case in 1957, the Florida Supreme Court found that, while diseased trees may have no value to be compensated, those that were still healthy, even if doomed by the spread of an infection, could still have a year or two to live and thus be worth something. Referring to Freund's treatise the Court held it is 'a plain dictate of justice and of the principle of equality that compensation be made for, at least, the loss of profits sustained by the owner whose healthy trees are destroyed'.[33]

In a subsequent case, *Haire* v. *Department of Agriculture Consumer Services* (2004),[34] the Florida Supreme Court affirmed the constitutionality of a statute that required the removal of all citrus trees within 1900 feet of a tree infected with citrus canker, a disease that devastated orange groves. The state programme provided compensation at $55 or $100 for trees in residential areas depending on when they were cut, if they were still healthy but were within range of the disease. Those with residential trees could petition for more compensation after a hearing. The programme gave little compensation for trees grown commercially, even if still overtly healthy, possibly because the programme principally benefited the commercial industry. Like many householders in Florida, Patty and Jack Haire, retirees living in suburban Broward County, grew in their yard orange trees that had significant value to them. The Haires challenged a state agency determination to cut their trees. The *Haire* court, like the *Miller* court, found that the state could constitutionally require the destruction of one kind of property to protect another so important that it constituted a 'public interest'.[35] The *Haire* court found 'no basis for concluding that the eradication of citrus canker is not a legitimate use of the State's police power' in view of the importance of the citrus industry to the economy of Florida.

What about compensation? There is nothing intrinsically wrong, antisocial, vicious or even unneighbourly about having an orange or lemon tree in one's yard. In the absence of a nuisance, to repeat Freund's dictum, 'Where property is destroyed in order to save property of a greater value, a provision for indemnity is a plain dictate of justice and of the principle of equality.' Even if the principle is clear, however, the amount of indemnity is often hard to determine. On the one hand, one could argue that citrus trees in the path of the canker are doomed and so worth very little. This approach compares the destruction of healthy citrus trees within a 1900-foot radius of an infected tree with the destruction of houses in the path of a conflagration. Compensation need not be paid because the house (or the tree) would be destroyed anyway.[36] On the other hand, the trees might live a year or two and be worth something. The *Haire* court stated, 'the fact that the Legislature has determined that all citrus trees within 1900 feet of an infected tree must be destroyed does not necessarily support a finding that healthy, but exposed, residential citrus trees have no value'.[37]

The legislature in Virginia that enacted the Cedar Rust Law of 1914 and the legislature in Florida that enacted the Citrus Canker Law of 2004 understood that the property owners whose trees had to be destroyed did not act in a subnormal or antisocial way; their trees acted as passive conduits of harm to the more valuable trees owned by their neighbours but did not themselves cause a trespass. The landowners therefore were owed some compensation at least in principle because their trees were destroyed to save the trees owned by others – trees that had greater economic value, so great, indeed, they constituted a public interest. In view of the plain dictate of justice, the laws made the state responsible for removing the trees and cleaning up and, in the Virginia programme, stacking the wood. No one contended or could contend that the presence of infected cedar or citrus trees on one's property constituted a nuisance that made the owners responsible for removing the trees or for paying for their removal.

Commentators on this history of agricultural law may cite Justice Scalia's opinion for the majority in *Lucas* v. *South Carolina Coastal Council* (1992) to establish that just because a state agency or legislature declares something to be a nuisance does not automatically make it so.[38] Scalia found that when a state determines something to be a nuisance it must ground its judgment on 'background principles' of property law, for example, common expectations about how people ought to behave. Plainly, to allow cedar or citrus trees on one's property is not to create a nuisance in the sense of 'harm' that would permit the state even for the sake of a legitimate public interest to take the trees without paying compensation. If compensation is partial, the reason must be found in the circumstances, for example, the moral hazard that defeated Miller's claim or the disease that devalued Haire's trees. As Fischel points out, 'in the ordinary nuisance case there is a more or less obvious "subnormal behavior" … a condition that ordinary people, *without the aid of the law*, can look at (or smell or listen to) and say, that party is not behaving as he ought to, at least at that place and time'.[39]

If one uses one's land according to normal community standards – according to the background principles or expectations of common law one may not be forced to alter it for a public good without being paid just compensation. Is the eradication of a non-native species, such as Japanese honeysuckle or *Phragmites*, a public good sufficient to warrant legislation? Does the presence of such a plant on a person's land represent behaviour 'subnormal' or reprehensible enough to permit the state to destroy it without paying compensation and to force the landowner (as the ELI model law suggests) to bear all the costs?

Scientific reproof

Conservation biologists, ecologists and other environmental scientists argue that non-native species constitute a kind of 'pollution' that degrades, destroys and disrupts ecosystems.[40] Ordinary people – *without the aid of science* – may not be able to see this. 'To the untrained eye, Everglades National Park and nearby protected areas in Florida appear wild and natural', two ecologists have written. Yet 'foreign plant and animal species are rapidly degrading these unique ecosystems'.[41] Ecologist Daniel Simberloff explains that, while the impact of non-indigenous species 'on the biotic community can be astounding, to the causal observer of nature they may not seem to be a major threat'. As he notes, 'a plethora of introduced animals may still represent nature to the average city dweller'.[42] No matter how species-rich, beautiful and complex an ecosystem may appear to the average city dweller, the biologist will see it as degraded insofar as alien species invade it. 'The reasons why a particular invasion wreaks havoc depend on the interaction between the species and the habitat' – and this requires scientific judgment.[43] 'All non-indigenous species are potentially harmful', ecologists have stated.[44]

The belief that non-native species threaten or harm ecosystems follows logically from a prevailing ecological theory that attributes the formation of ecosystems to the co-evolution of species over millennia.[45] In an influential paper published in 1964, Paul Ehrlich and Peter Raven wrote that 'studies of co-evolution provide an excellent starting point for considering community evolution'.[46] As a prominent ecologist has recently restated: 'Large-scale patterns primarily result from, rather than drive, evolution at lower levels.'[47] According to this view, by competition and co-adaptation species over millennia partition all possible niches and thus produce closed and stable communities. The goal of ecological science is then to discover, typically by mathematical modelling, the assembly rules that structure or govern ecosystems.[48]

The theory that evolutionary processes structure ecosystems and endow them with a mathematical organization – e.g. rule-governed patterns ecologists can study – has the following implication. If invasive species enter and 'meltdown',[49] 'harm',[50] 'disrupt',[51] 'destroy' and 'degrade'[52] natural ecosystems, scientists should be able to tell by observation whether a given ecosystem is heavily invaded or remains in mint condition. Heavily invaded systems, being disrupted, will not exhibit the mathematical patterns or exemplify the orderly processes that characterize heirloom ecosystems. In the heirloom ecosystem species will play by the rules that over time fashion biotic communities. In the invaded ecosystem, in contrast, species have come from all directions and play catch-as-catch-can. The site will be poorly organized, disrupted, damaged and dissolute. These differences should be obvious to the ecologist who could then tell by inspection – not just by historical research – which places are pristine, properly functioning ecosystems and which are reprobate. Ecologists should be able to determine which organisms are native and which carpetbaggers from the biology and behaviour of those species.

In fact, once non-native species have become established, which may take only a short time, ecologists are unable by observing a system to tell whether or not a given site has been heavily invaded. Invaded and heirloom ecosystems do not differ in pattern or process, structure or function, in any general ways. Heirloom and hodgepodge ecosystems function in the same ways. Nothing about the biological characteristics or behaviour of a species, moreover, indicates that it is native or non-native (however that difference may be defined) or how long it has been at a site.[53] The field biologist who learns the history of particular places – not the armchair biologist who deduces the consequences of theoretical models – can tell which sites are invaded and which are still pristine.[54] Only by doing historical research – by determining what was there before – can the ecologist tell whether and by which species a site has been invaded.[55]

Several ecologists recognize that, 'it is important to ask whether species assemblages with novel combinations of species (including both native and exotic species) function in the same way as native assemblages, even when many of the constituent species do not have a shared evolutionary history'.[56] The inability of ecologists to tell except by doing historical research whether an ecosystem is invaded or pristine suggests an answer. Novel and native assemblages must function in the same way; otherwise ecologists who do not know the historical record could distinguish between them. If they function the same way, then non-native species do not disrupt, degrade or destroy the structure, pattern or organization of ecosystems.[57]

If the heavily invaded system is just a hodgepodge of activity, so must be the pristine system, if one cannot observe general differences in the ways they function. According to Peter Vitousek and co-authors, invading species 'do not just add players to the game, they change its rules – often to the benefit of that and other invaders'.[58] If colonizing species can change the rules, in what sense could they have been *rules* at all? Perhaps one should characterize any ecosystem as a Heraclitean flux in such constant revision that no ecologist can observe the same biological community twice.[59] Ecosystems are not orderly. 'I think that the natural world out there is more like a swirling and boiling cauldron', Mark Davis, an ecologist, has said.[60] According to prevailing ecological theory, natural ecosystems self-assemble or evolve to possess an enduring structure: they obey rules, exhibit patterns or follow principles.[61] This functional organization, if it existed, must distinguish heirloom systems from Johnny-come-lately hodgepodges of non-native species. No general difference, however, is observed. A hodgepodge appears as rule-governed as an heirloom – which is not at all. Alien species, whether or not they threaten the natural environment, *do* threaten the theory of the natural environment. Is it the ecosystem or a theory of the ecosystem biologists seek to protect?

Biodiversity

Before the 1990s, conservationists, ecologists and other scientists generally held that species co-evolved over millennia to partition niches (or allocate resources) to produce a structured and functioning community or system; this community, because its niches were filled, would resist the introduction of novel species.[62] According to the 'biotic resistance' theory, as two ecologists summarize it, in pristine ecosystems 'the biota is so saturated with plant and animal species that adding immigrating aliens causes the extinction of an equal number of native species – much like a game of musical chairs, where every player has to compete for a space in order to remain in the game'.[63] As Simberloff noted: 'Until the recent burst of interest, conservationists were often complacent about non-indigenous species, assuming that disturbed habitats and communities are those most likely to be affected by these invasions whereas pristine areas are relatively immune.'[64] Ecologists often rely on computer models that use random events or stochastic variation in relevant variables to project the likelihood of changes in a population or community. Ecologists John Stachowicz and David Tilman have written that the 'stochastic model of community assembly predicts that, within a given habitat, increasing species richness should reduce resource availability and decrease invasion success'.[65]

In the 1990s, however, conservationists warned that '[e]ven species-rich pristine habitats are threatened by non-indigenous species'.[66] To explain this phenomenon, many biologists appealed to a leading ecological theory, *r/K* selection theory, which asserts that species respond to evolutionary pressures over time by adopting either a 'generalist' strategy to occupy many empty niches (*r*-selection) or a 'specialist' strategy to survive in or partition a contended niche (*K*-selection).[67] With ecological invasion, common or 'weedy' species, on this view, will displace less common or endemic ones by using their resources; as a result biodiversity, 'the sum total of genetically based variation within and among species'[68] must decrease. 'The replacement of many losing species with a relatively small fraction of widespread winners will likely produce a much more spatially homogenized biosphere', ecologists predicted. 'This implies that ecological homogenization might also occur because many ecological specialists are replaced by the same widespread and broadly adapted ecological generalists. The ultimate degree of homogenization, if unchecked, will probably exceed even that seen in the largest past mass extinctions.'[69]

What biologists observed, however, has been entirely different.[70] Organisms that appear specialized at one site may play new roles, for example, preying on different species, at new sites.[71] 'This kind of flexibility allows well-functioning ecosystems to emerge even when the various member species do not share a long history of co-existence and mutual adaptation.'[72] Accepted theory predicted 'biotic resistance' and then mass extinction. Ecologists observed, on the contrary, that non-native species easily colonized rich ecosystems – there was little resistance – and those systems got richer. 'Thus, there is an accelerating accumulation of introduced species and effects rather than a deceleration as envisioned in the biotic resistance model.'[73]

Biologists have found that invasions by exotic species 'create almost ideal conditions for promoting evolutionary diversification'.[74] Separated from their former populations, alien species diverge from them genetically, often in many ways, forming new kinds of populations.[75] Exotic species also hybridize with natives to produce novel lineages.[76] In response to pressure from exotic species, moreover, native species may evolve, drawing on intra-specific genetic variation and sometimes mutation, thus increasing their diversity.[77] For many reasons, 'the net consequence of these invasions is generally an increase in total species richness'.[78] New species emerge; homogeneous populations diverge; biodiversity flourishes.

Exotic invaders generally increase the species richness of ecosystems – often dramatically. The Red Sea and the Mediterranean were separated for millions of years until the Suez Canal, which opened in 1869, brought them together. Researchers have found that, 'over 250 species, 34 new genera, and 13 new families have moved into the Mediterranean Sea from the Red Sea, yet there has only been one documented extinction'.[79] In a similar story, the Chagres River on the Atlantic slope and the Rio Grande on the Pacific slope of Panama were isolated before 1914 when the Panama Canal joined them. Biologists have found the species richness of both rivers – surveyed in 1912 and in 2002 – greatly increased. There were no extinctions.[80] Likewise, 'in Hawaii freshwater fish richness has increased by 800% with the introduction of 40 exotic species and the loss of none of the five native species'.[81] Researchers found in the San Francisco Estuary 'a total of 234 exotic species established in the ecosystem', and at least 125 additional 'cryptogenic' species, so called because of the absence of historical evidence of their provenance.[82] These researchers also found that 'no introduction in the Estuary has unambiguously caused the extinction of a native species'.[83] The 'evidence for invasion-caused extinction is weak or non-existent' with respect to 'marine organisms, land plants, and smaller terrestrial animals',[84] except in a few insular areas, such as a lake exposed to a predator.[85]

The situation is the same in terrestrial environments. According to two ecologists, 'Within the last few centuries following European colonization, relatively few insular endemic plant species have become extinct, whereas invading species have approximately doubled the size of island floras – from 2000 to 4000 on New Zealand; 1300 to 2300 on Hawaii; 221 to 421 on Lord Howe Island, Australia; 50 to 111 on Easter Island; and 44 to 80 on Pitcairn Island.'[86] Mark Davis has written that in the USA, which hosts thousands of non-native plants, 'there is no evidence that even a single long-term resident species has been driven to extinction ... because of competition from an introduced plant species'.[87] With few exceptions, landscapes already rich in native species 'support many more species of exotics than areas with relatively few native species'.[88] Alien species make rich ecosystems richer. 'It is apparent that there is no theoretical limit to the number of species in any community.'[89]

Executive Order 13112

More than 500 ecologists, conservation biologists and other environmental professionals in a 1997 letter to Vice President Al Gore wrote, 'A rapidly spreading invasion of exotic plants and animals ... is destroying our nation's biological diversity' and called for 'an effective national program to combat invasions by non-indigenous plants and animals'.[90] In response, President Clinton signed in 1999 Executive Order 13112, 'Invasive Species', which established the National Invasive Species Council (NISC) and instructed it to develop a Management Plan 'for preventing the introduction and spread of invasive species' and to 'provide for restoration of native species and habitat conditions' in invaded systems. The Order defines a species as 'alien' if it is 'not native' to the ecosystem in which it is found. Since no way to define or delimit ecosystems exists, 'native' is usually construed as present in the USA before the European settlement. The order defines a non-native or alien species as 'invasive' if it causes or is likely to 'cause economic or environmental harm or harm to human health'.[91] When people get sick or lose crops to disease, they know they have been harmed. However, without the aid of science, particularly theoretical models of ecosystem structure and function, people cannot know that non-native species cause 'environmental harm'. Scientists rather than ordinary people define 'environmental harm' and prescribe ways to prevent and mitigate it. To maintain political clout – and the demand for their expertise – scientists must maintain a consensus about what causes 'environmental harm' and about how great an evil it represents.

Can ecosystems be harmed or suffer harm? Consider the following syllogism: (1) An object can be harmed only if it has preferences, interests, rights, plans, purposes, or is at least aware of itself. (2) No ecological unit – certainly none beyond the scale of an individual organism – has preferences, interests, rights, etc. Therefore, (3) no ecological unit can suffer harm or be harmed.

What exactly is wrong with this syllogism? One could challenge the second premise by arguing – as did law professors Christopher Stone and Lawrence Tribe – that ecological objects such as trees, forests or estuaries should be presumed to have interests or rights or that these may be attributed to them in legal and policy contexts.[92] This strategy, which takes its cue from a dissent by Justice William O. Douglas (*Sierra Club* v. *Morton*, 405 U.S. 727, 742 (1972)), has had no significant influence on environmental policy and law, in part because the putative rights and interests of nature – given the workings of natural selection, for example – are likely to be ambiguous, in conflict and anyone's guess.

The attempt to anthropomorphize natural objects as subjects for 'liberation' by ascribing rights, interests or preferences, for example, to trees has shown little promise as an approach to the theory of environmentalism and no promise as a strategy in environmental policy and law. According to law professor Douglas Kysar, commentators have 'long critiqued notions ... that ecology can provide a unitary, stable conception of "nature" to fix environmental conservation goals'.[93] Unless one adopts a metaphysic of vitalism or pan-psychism, which has little resonance in legal thought, one may have to concede the second premise, i.e. that nature has no preferences.

Is the first premise true? Is it true that an object must be self-conscious, feel pleasure and pain, form intentions or possess preferences, to be harmed? Since ecological entities, such as forests or reefs, lack preferences or intentions, at least as far as we know, to accept the first premise would be to concede that 'harm to the environment' must ultimately be explained or 'cashed out' in terms of costs and benefits to human beings. If so, nothing can harm nature; whatever harm occurs must occur to human health, safety, welfare or some other human concern.

NISC has made its mission the prevention and mitigation of the 'environmental harm' caused by alien species.[94] In response to questions about the meaning of 'environmental harm' the Invasive Species Advisory Committee of NISC issued a White Paper which states: 'We use environmental harm to mean biologically significant decreases in native species populations.' The guidance adds: 'Environmental harm also includes significant changes in ecological processes, sometimes across entire regions, which result in conditions that native species and even entire plant and animal communities cannot tolerate.' This definition 'will apply to *all taxa of invasive species in all habitats*' and therefore to private as well as public land. In documents coordinating state and federal programmes, NISC supports the use of biological controls – 'natural enemies' including fungal infections, pathogens, and predatory beetles – to battle invasive species. The predator or pathogen intended to destroy a non-native plant, such as honeysuckle, English ivy or purple loosestrife, even if initially released on public land, may well migrate to attack plants a landowner may cultivate or otherwise value. What then? If a landowner objects, the federal or state agency must identify how the public interest is served by the eradication of what seem to be – and often are – ornamental plants that landowners enjoy and which, they may argue, do not harm but enhance the value of their property.

A state or federal agency could reasonably reply that the public interest, at least in certain places, favours native over non-native species and communities for antiquarian reasons. Just as Greece preserves the Acropolis and other ruins, Rome maintains the Colosseum, and other nations protect the remnants of their cultural heritage – which incidentally may turn out to be terrific tourist attractions – so, too, the USA has an aesthetic, ethical and historical duty and opportunity to maintain as well as it can the vestiges of its past, for example, parks like Yosemite and Yellowstone. To maintain living museums of natural history, curatorial agencies, such as the Park Service, may engage in gardening at a grand scale, fighting the forces of nature, in this context, invasive species, that constantly wear down the monuments of the past. Agencies prune the past of the present at particular sites to preserve a venerable national or local heritage.

Another aesthetic argument points to the iconic value of certain species – for example, trees such as the chestnut, ash, and elm – which may be threatened by invasive pathogens. The 'environmental harm' these pathogens cause – and the reason the APHIS has the mandate to battle them – lies in the aesthetic significance of these trees in the landscape. The aesthetic value of these historic trees may be immeasurable, but the ecosystem will go on with or without them. 'Even a mighty dominant like the American chestnut, extending over half a continent, all but disappeared without bringing the eastern deciduous forest down with it', David Ehrenfeld has written.[95]

Beside aesthetic considerations, why is native better? Scientists know that non-native species often provide an enormous stimulus to biological diversification. 'Through genetic engineering, species introduction, and environmental modification, we conceivably could manufacture a world even more biologically variable and "diverse" than the one derived through evolutionary processes', Paul Angermeier has written.[96] Nevertheless, ecologists generally rule out the possibility that alien species could add to biodiversity. 'Our definition [of biodiversity] excludes exotic organisms that have been introduced', a group of biologists stated.[97] Once the concept *native* is implied in the concept *biodiversity*, observational evidence becomes irrelevant; historical research is all that matters in comparing levels of 'biodiversity'.

To see why, suppose that an ecologist observes at one island a hundred trillion species in a fantastically intricate ecosystem and on another island only one species, say, of lichen clinging to rock. Whichever island has the most 'native' species has the most biodiversity – thus the island with the native lichen can be more diverse than the island with a hundred trillion 'introduced' species. The history – not the biology – is all that matters once 'biodiversity' is limited to 'native' species, whatever that term may mean. The thesis that exotic species diminish 'biodiversity' depends not on biological observation but on logical inference when 'native' is implied but omitted as a modifier of 'biodiversity'.[98]

NISC speaks of protecting and restoring 'the original ecosystem'[99] but it is unclear why it believes an original ecosystem is generally better than an updated one. A Creationist may answer that God designed the original ecosystem and therefore species should stay near where Noah dropped them off. Executive Order 13112 defines a species as 'native' if it occurs in an area other than by 'introduction', that is, other than with human assistance. This connects the 'original ecosystem' with Original Sin – the idea that if a species travels not by its own powers but in the wake of human activity it is alien or exotic and thus corrupt. Daniel Simberloff points out that the quality invasive species share is their association with human beings. 'The one thing for sure is that all of these species arrived with human assistance in little more than a century, and almost certainly none would have reached there in a million years on its own.'[100] If one does not share the doctrine of some Christians that humanity because of the Fall from Grace corrupts innocent Nature or the belief of some Creationists that the world came into existence only a few thousand years ago, there is no non-arbitrary way to tell what is 'original' and what is not. According to conservation biologist Michael Soulé, 'any serious attempt to define the original state of a community or ecosystem leads to a logical and scientific maze'.[101]

To defend the superiority of the 'original' ecosystem biologists do not refer to Scripture but to a mathematical and theoretical constructs. They develop stochastic models of community assembly, matrices of interaction coefficients, inter-specific trade-off curves, theories of self-organized criticality, languages of ecosystem ontogeny, hierarchical path-dependent complex dynamical computational representations with multiple basins of attraction, the logic of regime shifts, synergetic effects, state-and-transition flips, and much else to infer the pattern, process, structure or function that distinguishes an original heirloom ecosystem from an invaded and thus corrupted one. All this is accomplished *in silico* rather than *al fresco*, that is, in front of the computer rather than in the great outdoors.

Epistemic communities

'An epistemic community,' as political scientist Peter Haas has characterized it, 'is a network of professionals with recognized expertise and competence in a particular domain and an authoritative claim to policy-relevant knowledge within that domain or issue-area.'[102] As Haas and others have shown, ecologists and environmental scientists with influence in governmental agencies have played a crucial and creative role in solving environmental problems, for example, in controlling sewage and other pollutants to save the Mediterranean Sea and in banning the use of chlorofluorocarbons to protect the stratospheric ozone layer.[103] Epistemic communities, according to Haas, possess 'a shared set of normative and principled beliefs, which provide a value-based rationale for the social action of community members'. They also share causal beliefs along with 'internally defined criteria for weighing and validating knowledge in the domain of their expertise'. They have reached a consensus about science and policy and they seek to bring those with political power into that consensus.[104]

Biologists who study invasive species – invasion biologists – constitute an epistemic community. On the basis of the theory of island biogeography (the 'musical chairs' analogy discussed above), r/K selection theory, biotic resistance theory, models of community self-assembly, and other mathematical constructs, an overwhelming majority of invasion biologists have concluded that the influx non-native species (NNS) will cause biodiversity greatly to decline. The 'overwhelming consensus of ecologists and systematic specialists who study them is that NNS are a highly significant factor in endangerment and extinction – indeed second only to habitat destruction by most tallies'.[105] Power should follow knowledge. Conservation biologist Thomas Lovejoy has said that scientists 'must take on an advocacy role' for the environment because 'scientists understand best what is happening and what alternatives exist'.[106]

Invasion biologists agree that alien species threaten natural environment. They 'speak truth to power' in the name of that consensus. They have proposed regulations that would establish 'clean lists' of those non-native species that are permitted in the USA. Those who wish to introduce a species not on the 'clean' list must prove that it would not harm the environment. In spite of the scientific consensus that supports 'clean list' proposals, they are rarely mandated. These proposals 'met with strong opposition from agriculture, the pet trade, and other special interest groups,' according to Don Schmitz and Daniel Simberloff.

'Because of the political power of vested interests, federal and most state agencies ... do not demand that importers of plants and animals demonstrate that an introduction will prove innocuous.'[107] Invasion biologists and other ecologist are generally united in their belief about the threat alien species pose to the biodiversity – as the letter by 500 experts to then Vice President Gore suggests.[108] Having reached a consensus and gained influence as an epistemic community, invasion biologists have met outstanding success in obtaining public support for their research. Yet they have not seen their policy recommendations followed.[109] Why were epistemic communities of ecologists and other scientists successful in saving the Mediterranean and protecting the ozone layer, but they have not succeeded in stemming the tide of non-native species?

In supporting the Mediterranean Plan, marine ecologists pointed to sewage washing up on beaches from the Riviera to Israel to North Africa. Everyone understands that sewage stinks. Similarly, in supporting the Montreal Protocol atmospheric scientists related the thinning ozone level to the increasing incidence of skin cancer. When scientists link a condition or a practice (smoking, for example) with cancer they are likely to get the regulations they seek. The public understands what 'harm' means when it shows up as cancer, unsanitary water or unbreathable air.

Invasion biologists describe non-native species as ecological pollution.[110] Japanese honeysuckle, purple loosestrife and multiflora rose, however, do not look or smell like sewage; in fact, just the reverse. Biologists may all call these plants pollutants, but this does not make them pollutants. Invasion biologists invoke metaphors of disease, describing invasive species as overcoming 'biotic resistance' to spread aggressively like cancer. Yet invaded ecosystems do not seem to be 'dying'. The public understands that pathogens, whether native or not, harm human health; it understands that plant pests, such as the emerald ash borer, destroy iconic trees. That pathogens and pests may harm human health and economic interests, however, does not explain what scientists mean by 'harm to the natural environment'. West Nile disease has nothing to do with purple loosestrife; in battling one, why must society also detest the other?

In describing successful epistemic communities, Peter Haas has written: 'Common principles and norms gave rise to a common set of rules for pollution control.'[111] Invasion biologists demonstrate deductively that non-native species threaten biodiversity because these scientists define 'biodiversity' to include only native species. They have been unable, however, to link this narrow definition of biodiversity with any normative concept people can understand without the aid of science. Invasion biologists assume that native is better but they do not explain why. They have failed to connect their interests as scientists with the public interest.

Conclusion

'The whole aim of practical politics is to keep the populace alarmed (and hence clamorous to be led to safety) by menacing it with an endless series of hobgoblins, all of them imaginary.'[112] The thinning ozone layer was not imaginary. Effluents and emissions controlled by the Clean Air and Water Acts in the USA and by the Mediterranean Action Plan were not imaginary. People know, first, that air and water pollution cause real damage and, second, that no one has a right to inflict that kind of damage on others. Scientists in the 1970s who helped write pollution control policy did not prescribe values for society. They responded to health-and-safety considerations, which values society already had.

Environmental policy, when it depends on background principles of common law, may justify the regulation of one person's property to protect the person or property of his or her neighbour. As the statutes and cases in agricultural law discussed earlier suggest, state officials may even enter private land to destroy the trees, plants or other property of a landowner to protect an industry of such great economic importance that it constitutes a public interest. This can be justified, however, only when the expense is borne by the public and just compensation is paid.

The public interest is not to be confused with the interest of an epistemic community, however united it may be in raising an alarm and calling for power and money to answer it.

In recent years, ecologists, conservation biologists and other environmental professionals have developed as societal goals normative concepts that do not resonate in common law but originate in their own research. These normative concepts include 'ecological integrity', 'invasive species', 'biodiversity', 'ecosystem services', and 'sustainability'. How should society respond to these research-originated and science-driven norms and goals? Having no basis in common law, normative concepts such as these are malleable to the political and cultural commitments of those who make them the objects of their professional study. However, if teams of social scientists and biologists agree about what is needed to sustain ecosystems, protect biodiversity, and save the environment from harm, how can society not heed their advice?

Invasion biologists, ecological economists, conservation biologists and other scientists constitute a large and important epistemic community that argues that the current ecological crisis is so alarming and menacing that society can no longer limit the use of the police power to protect people from the environment – e.g. from hazardous substances, pollutants, pathogens and pests – but must move swiftly and effectively to protect the environment from people. That scientists can define environmental harm is the assumption that underlies recommendations of NISC and the model state laws developed by the Environmental Law Institute. On this view, scientific research determines what harms the environment; legislatures and courts may then require that society respect 'ecological integrity', protect the 'biotic community', preserve 'biodiversity', exclude and eliminate 'invasive species', and promote 'sustainability'. Society does not have leisure, however, to wait for scientists to define these concepts perfectly; rather, scientists can reach a working consensus on many goals and then direct society to achieve them.

What does one say to the model legislation ELI proposed to help states rid public and private land of invasive species? With Thomas Lovejoy, president of the Heinz Center for Science, Economics, and the Environment, one might declare that scientists 'understand best what is happening and what alternatives exist.' As long as teams of scientists are interdisciplinary they may claim to be representative. Nevertheless a nagging doubt – something in the spirit of democracy – may lead one to ask: 'Who do these people think they are?' The police power has been used for centuries to protect people from sources of injury and harm. The identification and measurement of injury and harm, moreover, have always been understood within legal processes and constitutional constraints. In the context of environmental law in the USA, the police power protects people from the environment. It protects people from toxic and hazardous substances, including pathogens and pests that move through earth, water and air. Conservation biologists, ecologists and other environmental professionals, in contrast, propose the environment itself – labelled as the 'biotic community' the 'ecosystem', or 'biodiversity' – as an object of protection. To be sure, these experts may refer to traditional goals or goods, for example to human health and to agriculture, but it is the sustainability and integrity of the native biotic community that interests them. They seek to protect the environment from people – and this has no basis in the police power or in the law of nuisance.

Environmental statutes in the USA since 1969 have rested largely on the legal foundation of the common law of nuisance. The many successes of environmental regulation have been won primarily by policies that controlled the gross emissions of industrial and municipal polluters and by policies that reduced less visible and often unquantifiable risks, for example, from small amounts of carcinogenic substances. Another wave of environmental legislation responds to aesthetic, historical, cultural and spiritual values concerning wilderness areas, iconic landscapes and rare and wonderful plants and animals. The preservation of species, the protection of wild and scenic places, and other aesthetic and aspirational goals soon came to be litigated and thus tested in terms of distributional problems, that is, problems of footing the bill. These problems, for example, in the development of habitat conservation plans for endangered species, came to be adjudicated largely on a case-by-case basis under established and familiar precedents in property, land-use, wildlife and natural resources law.[113]

Have epistemic communities the ability to forge by consensus novel conceptions of harm – for example 'environmental harm' – and therefore new categories of nuisance for the police power of the state to prevent? The Supreme Court has found that the natural environment per se lacks the kinds of rights or interests that the cold steel of the law protects. The environment – even if described in a normative way as an 'ecological community' – is not a person. It does not sustain injuries or endure harms that give it (or the scientists who represent it) standing to sue for redress in the judicial system created by Article III of the Constitution. In environmental cases, 'the relevant showing ... is not injury to the environment but injury to the plaintiff'.[114]

As traditionally interpreted, the 'takings' clause of the Fifth Amendment requires that a public authority may enter onto a person's land and destroy his or her property only if this action serves a sufficiently compelling public interest and only if (in the absence of nuisance) compensation is paid. Just because a government says something is a nuisance does not make it one; rather, the nuisance exception to compensation refers to background principles of common law. The compensation test is often described as the way to assure efficiency in the context of 'takings' jurisprudence. Even more important, an insistence that government pay all the costs, compensate a landowner, and clean up after itself is the way to assure liberty, that is, the right of every person to be left to live in peace as long as he or she respects the same right of others. The government in the name of some good it constructs – or some evil it invents may otherwise become the most virulent invader.

The courts have insisted that governments compensate any landowner whose property they invade, for example, to create a public good, such as to build a road or a school. Any governmental intervention on private property except to prevent a nuisance – a recognizable harm to others – requires the government to compensate the landowner for any loss. It is not merely a crotchet of conservative judges to insist that property owners – when they use their land without harming others – receive compensation when the government restricts their property rights. On the contrary, by denying epistemic communities the power to define harm and thus to make law, this principle is one of the most important defences democracy possesses against the tyranny implicit in scientism.

Acknowledgement

This chapter is a modified version of a paper previously published as: Mark Sagoff, 'Who Is the Invader? Alien Species, Property Rights, and the Police Power', *Social Philosophy & Policy*, vol 26, no 2 (Summer 2009), pp26–52; it is reproduced here with the permission of Cambridge University Press.

Notes

1 Mac Margolis, 'Attack of the Aliens: Migrating Species May Be the Biggest Threat to Plant and Animal Life on the Planet', *Newsweek*, 15 January 2007. Available online at: www.newsweek.com/id/56574/page/1

2 To determine that a species is 'injurious', USDA completes a review process including the following steps: Petition or Initiation of an Evaluation, Notice for Information, Proposed Rule, Economic Analysis, and Final Rule. For a flow chart annotating this process, see www.fws.gov/contaminants/ANS/pdf_files/InjuriousWildlifeEvaluationProcessFlowChart.pdf

3 A complete list can be found in the *Federal Register* at www.fws.gov/contaminants/ANS/pdf_files/50CF_16_10–05.pdf

4 Noxious Weed Regulations, 7 C.F.R. 360.200 (2008).

5 Daniel Simberloff, 'Impacts of Introduced Species in the USA', *Consequences* 2(2) (1996). Available online at www.gcrio.org/CONSEQUENCES/v012n02/article2.html. See also J. L. Ruesink, I. M. Parker, M. J. Groom, and P. M. Kareiva, 'Reducing the Risks of Nonindigenous Species Introductions: Guilty Unless Proven Innocent', *BioScience* 45 (1995): 465–477.

6 Environmental Law Institute (ELI), *Invasive Species Control: A Comprehensive Model State Law* (Washington, DC: Environmental Law Institute, 2004). See p7. In a study published May 2007, ELI reviews statutes state-by-state and finds that no state has fully adopted the 'clean list approach' although the study praises Michigan, in particular, where the law 'encourages action because it does not require listing by the agency as a prerequisite to control actions. Similarly, the automatic declaration of all pests and pest hosts as a public nuisance provides a solid base for both avoiding compensation for control actions and for requiring abatement'. Environmental Law Institute and the Nature Conservancy, 'Strategies for Effective State Early Detection/Rapid Response Programs for Plant Pests and Pathogens', published online May 2007. An electronic retrievable copy (PDF file) of this report may be obtained for no cost from the Environmental Law Institute; website: www.eli.org; click on 'ELI Publications', then search for the 'Strategies for Effective State Early Detection' report. Quotation at p50.

7 ELI, *Invasive Species Control*, p33. 'A person owning private lands, waters or wetlands, or a person occupying private lands, waters, or wetlands, or a person responsible for the maintenance of public lands shall control or eradicate all unpermitted introductions, populations or infestations of prohibited, restricted or regulated invasive species on the land.'

8 Meg Filbey, Christina Kennedy, Jessica Wilkinson, Jennifer Balch, *Halting the Invasion: State Tools for Invasive Species Management* (2002) Environmental Law Institute Washington, DC (ELI Project No. 020101, 003108), August 2002, p8. Available online at: www2.eli.org/research/invasives/pdfs/d12–06.pdf

9 The list of weeds 'it is illegal to propagate, sell, or transport' in the Commonwealth of Pennsylvania can be found at: www.agriculture.state.pa.us/agriculture/lib/agriculture/plantindustryfiles/NoxiousWeedControlList.pdf

10 State of Pennsylvania, Department of Agriculture, 'Noxious Weed Law Summary' (last modified January 18, 2007). Available online at: www.agriculture.state.pa.us/agriculture/lib/agriculture/plantindustryfiles/NoxiousWeedLawSummary.pdf

11 Title 25, Chapter 10, Nebraska Administrative Code – Noxious Weed Regulations. www.agr.state.ne.us/regulate/bpi/actbb.htm.

12 Title 25.

13 Press Release, Nebraska Department of Agriculture, 7 August 2007. www.agr.ne.gov/newsrel/august2007/phragmites.htm

14 State of Pennsylvania, Rules and Regulations, Department of Agriculture, [7 PA. CODE CH. 110] 'Noxious Weeds' [27 Pa.B. 1793]. Available online at: www.pabulletin.com/secure/data/v0127/27–15/549.html. The belief that a non-native 'generalist' or 'weedy' species must 'crowd out' native species adapted to particular environments, while a consequence of prevailing theory, has little empirical support. The difference indicates poles of a spectrum; no one has shown that biologists would agree, if tested, where on the spectrum each of a random selection of species would lie. The empirical evidence does not generally show that 'weedy' species crowd out 'specialists'; that they do seems to be a consequence of definitions. (If species *A* crowds out species *B*, then it is to that extent 'weedy'.) Suppose purple loosestrife is '*r*-selected' or 'weedy'. Substantial evidence suggests it does not crowd out but improves habitat for other species. (Loosestrife was initially introduced to support honeybee populations.) For studies that demonstrate the beneficial role of loosestrife in the natural environment see, for example, M. G. Anderson, 'Interactions between *Lythrum salicaria* and native organisms: A critical review', *Environmental Management*, 19 (1995): 225–231; and M. A. Treberg and B. C. Husband, 'Relationship Between the Abundance of *Lythrum*

salicaria (Purple Loosestrife) and Plant Species Richness along the Bar River, Canada', *Wetlands*, 19 (1999): 118–125. Two researchers have concluded that ecologists 'traced the history of purple loosestrife and its control in North America and found little scientific evidence consistent with the hypothesis that [it] has deleterious effects ... Loosestrife was initially assumed to be a problem without actually determining whether this was the case ... there is currently no scientific justification for the control of loosestrife.' H. A. Hager and K. D. McCoy, 'The implications of accepting untested hypotheses: A review of the effects of purple loosestrife (*Lythrum salicaria*) in North America', *Biodiversity and Conservation*, 7 (1998): 1069–1079. For further confirmation see, E. J. Farnsworth and D. R. Ellis, 'Is purple loosestrife (*Lythrum salicaria*) an invasive threat to freshwater wetlands? Conflicting evidence from several ecological metrics', *Wetlands* 21 (2001), 199–209; J. A. Morrison, 'Wetland vegetation before and after experimental purple loosestrife removal', *Wetlands*, 22(1) (2002): 159–169; and M. B. Whitt, H. H. Prince, and R. R. Cox, Jr., 'Avian use of purple loosestrife dominated habitat relative to other vegetation types in a Lake Huron wetland complex', *The Wilson Bulletin*, 111 (1999): 105–114.

15 For discussion, see Ellen Frankel Paul, *Property Rights and Eminent Domain* (New Brunswick: Transaction Books, 1987), especially Chapter 2, 'The genesis and development of eminent domain and police powers', pp71–184.

16 In the agricultural cases discussed here, it is assumed that a major agricultural industry in a state (e.g. apples or citrus) carries a 'public interest' not simply a private one. This assumption, which is reasonable, distinguishes these cases from *Kelo* v. *City of New London*, 545 U.S. 469 (2005), where it seems at least as plausible to suppose that only a private interest (that of certain developers) was served.

17 Resolutions, 2 Va. Fruit 159, 165 (1914) [18th VSHS (Jan. 1914)]. Sections 891 and 892 stipulate the proceeding to 'determine the amount of damages' and the method by which apple growers will be taxed to pay that amount.

18 James M. Buchanan, 'Politics, property and the law: An alternative interpretation of Miller et al. v. Schoene', *Journal of Law and Economics*, 15 (1972): 438–452; quotation at pp447–448. According to Buchanan (at p443), 'what is relevant is the necessary place of compensation in the trading process between the two parties.' Buchanan follows Justice Stone in mistakenly believing that no scheme for compensation was enacted by the state.

19 276 U.S. 272 (1928).

20 William A. Fischel, 'The law and economics of cedar-apple rust: State action and just compensation in *Miller v. Schoene*', *Review of Law & Economics*, 3(2) (2007): 133–195. Quotation at p134.

21 Fischel, 'Cedar-Apple Rust', p173.

22 The state paid also for cutting the trees, stacking the wood, and cleaning the area. For details see *Bowman* v. *Virginia State Entomologist*, 128 Va. 351 (18 Nov. 1920).

23 *Miller*, 276 U.S. 272, 273. In *Miller* v. *State Entomologist* (146 Va. 175; 135 S.E. 813; 1926 Va.), the Virginia Supreme Court rejects several other reasons Dr Casper Otto Miller alleged as invalidating the law on constitutional grounds, among which were the vagueness or indefiniteness of one of its uses of the term 'locality' and the possibility that it empowered citizens (the farmers who complained about his trees) to make law.

24 *Miller*, 276 US at 278.

25 *Miller*, 276 US at 279.

26 *Miller*, 276 US at 279. The Virginia Supreme Court had written, 'The statute, so far as it relates to damages, is not clear, and we are to gather the intention of

the legislature as best we can from a consideration of it as a whole.' *Miller* v. *State Entomologist*, 146 Va. 175, 192 (1926). The Virginia court construed compensation under the law to consist primarily in the state paying the costs of cutting, stacking the wood, and cleaning the area. 'No doubt the legislature deemed such outlays as proper damages and expenses to be paid to the owner, if the circuit court deemed them proper' (at 193–194).

27 *Miller*, 27 U.S. at 278.

28 Ernst Freund, *The Police Power: Public Policy and Constitutional Rights* (Chicago: University of Chicago Press, 1940) Section 534, p565.

29 Richard A. Epstein, *Takings: Private Property and the Power of Eminent Domain* (Cambridge: Harvard University Press, 1985), p114.

30 Fischel, 'Cedar-Apple Rust', p172. It is part of the brilliance of Fischel's analysis that he shows in terms of the details of the enactment of the law and its subsequent enforcement that the moral hazard problem defeated the otherwise constitutionally required payment.

31 The Virginia Cedar Rust Law, 1914 Va. Acts p49 et seq., explicitly creates a fund paid for by taxes on apple growers to compensate the owners of especially valuable cedar trees. The relevant sections of the statute (Sections 7 and 8) are reprinted in the Syllabus in *Bowman* v. *Virginia State Entomologist* 128 Va. 351; 105 S.E. 141; 1920 Va.

32 The doctrine of average reciprocity of advantage was first state by Justice Holmes in *Pennsylvania Coal*, 260 U.S. 393, 415 (1922).

33 *Corneal* v. *State Plant Board*, 95 So. 2d 1, 6–7 (Fla. 1957).

34 *Haire* v. *Fla. Dep't of Agricuture & Consumer Services* 870 So. 2d 774, 782 (Fla. 2004).

35 See *Haire* 870 So. 2d 774 at 781; *Miller*, 276 U.S. 1928 at 279–80.

36 *Bowditch* v. *Boston*, 101 U.S. 16, 18 (1880).

37 See *Haire* 870 So. 2d 774 at 785.

38 Lucas required that 'South Carolina ... do more than proffer the legislature's declaration that the uses Lucas desires are inconsistent with the public interest, or the conclusory assertion that they violate a common-law maxim such as *sic utere tuo ut alienum non laedas*.' 505 U.S. 1003, 1031 (1992).

39 Fischel, 'Cedar-Apple Rust', p146. Fischel cites Robert C. Ellickson, 'Alternatives to zoning: Covenants, nuisance rules, and fines as land use controls', *University of Chicago Law Review*, 40 (1973): 730.

40 For a collection of papers to this effect, see B. N. McKnight, ed, *Biological Pollution: The Control and Impact of Invasive Exotic Species* (Indianapolis, IN: Indiana Academy of Sciences, 1993).

41 Don C. Schmitz and Daniel Simberloff, 'Biological invasions: A growing threat', *Issues in Science and Technology* Online, Summer 1997. http://findarticles.com/p/articles/mi_qa3622/is_199707/ai_n8780169.

42 Daniel Simberloff, 'The Biology of Invasions', in Daniel Simberloff, Don C. Schmitz, and Tom C. Brown, eds, *Strangers in Paradise. Impact and Management of Non-Indigenous Species in Florida* (Washington, D.C. and Covelo CA: Island Press, 1997), pp3–17. Quotation at p9.

43 Daniel Simberloff, Ingrid M. Parker, and Phyllis N. Windle, 'Introduced species, policy, management, and future research needs', *Frontiers in Ecology and the Environment*, 3(1) (February 2005): 12–20. Quotation at p14. See also, Daniel Simberloff, 'Non-native species *do* threaten the natural environment', *Journal of Agricultural and Environmental Ethics*, 18 (2005): 595–607.

44 Daniel Simberloff, D. C. Schmitz, and T. C. Brown, 'Why should we care and what should we do?' in Simberloff, Schmitz, and Brown eds, *Strangers in Paradise*, pp359–367. Quotation at p364. According to Simberloff 'many scientists argue that every species should be considered a potential threat to biodiversity and sustainability if it were to be introduced ... That implies that every species proposed for deliberate introduction, whether or not it appears superficially to be innocuous, necessitates some formal risk assessment.' Daniel Simberloff, "Nonindigenous species – a global threat to biodiversity and stability',' in Peter H. Raven and T. Williams, eds, *Nature and Human Society: The Quest for a Sustainable World*, Committee for the Second Forum on Biodiversity, National Academy of Sciences (Washington, DC: National Research Council, 1997). Quotation at p329. Online at: http://books.nap.edu/books/0309065550/html/329.html.

45 The theory that species co-evolve to form ecosystems – fragile communities of highly specialized interrelated organisms – produced the metaphors of conservation biology that analogized ecological communities to delicate machines. Paul Ehrlich analogized species to 'rivets' holding up the wing on an airplane. See Paul Ehrlich and Anne Ehrlich, *Extinction: Causes and Consequences of the Extinction of Species* (New York: Random House, 1981). Writing in the same a priori tradition, Simon Levin updated the metaphor to that of a computer. According to Levin, ecosystems constitute 'complex adaptive systems assembled from sets of available components as one would assemble a new computer system'. Simon A. Levin, *Fragile Dominion: Complexity and the Commons* (Reading, MA: Perseus Books, 1999). Quotation at p101.

46 Paul R. Ehrlich and Peter H. Raven, 'Butterflies and plants: A study in conservation', *Evolution* 18 (1964): 586–608. Quotation at p605.

47 Levin, *Fragile Dominion*, p104.

48 This is the 'niche assembly perspective' extensively examined in S. P. Hubbell, *The Unified Neutral Theory of Biodiversity and Biogeography* (Princeton: Princeton University Press, 2001).

49 Daniel Simberloff and Betsy Von Holle, 'Positive interactions of nonindigenous species: Invasional meltdown? *Biological Invasions*, 1(1) (1999): 21–32.

50 David W. Ehrenfeld, 'Adulusian Bog Hounds', *Orion* (Autumn 1999): 9–11.

51 Raven and Williams, eds, *Nature and Human Society*, 1997, p325.

52 Schmitz and Simberloff, 'Biological Invasions'.

53 For discussion see for example, K. Thompson, J. G. Hodgson, and T. C. G. Rich, 'Native and alien invasive plants: More of the same?' *Ecogeography*, 18 (1995): 390–402; B. J. Goodwin, A. J. McAllister, and L. Fahrig, 'Predicting invasiveness of plant species based on biological information', *Conservation Biology*, 13 (1999): 422–426; M. Williamson, *Biological Invasions* (London: Chapman & Hall, 1996).

54 In 1985, Dan Janzen, an empirical biologist, observed that species that do not share an evolutionary history may nevertheless fit together into normal ecosystems. D. H. Janzen, 'On ecological fitting', *Oikos*, 45 (1985): 308–310. For an example of a lush rainforest ecosystem composed entirely of introduced species, see D. M. Wilkinson, 'The parable of Green Mountain: Ascension Island, ecosystem construction, and ecological fitting', *Journal of Biogeography*, 31 (2004): 1–4.

55 This point is generally conceded. See, for example, M. A. Davis and K. Thompson, 'Invasion terminology: Should ecologists define their terms differently than others? No, not if we want to be any help!', *ESA Bulletin*, 82 (2001): 206. ('In the United Kingdom, about equal numbers of native and alien plants are expanding their ranges, and an analysis of their traits shows that these two groups are effectively indistinguishable.')

56 D. F. Sax, J. J. Stachowicz, J. H. Brown, J. F. Bruno, M. N. Dawson, S. D. Gaines, R. K. Grosberg, A. Hastings, R. D. Holt, M. M. Mayfield, M. I. O'Connor, and W. R. Rice, 'Ecological and evolutionary insights from species invasions', *Trends in Ecology and Evolution*, 22(9)(July 2007): 465–471. Quotation at p468.
57 For discussion, see Richard J. Hobbs, Salvatore Arico, James Aronson, Jill S. Baron, Peter Bridgewater, Viki A. Cramer, Paul R. Epstein, John J. Ewel, Carlos A. Klink, Ariel E. Lugo, David Norton, Dennis Ojima, David M. Richardson, Eric W. Sanderson, Fernando Valladares, Montserrat Vilà, Regino Zamora, Martin Zobel, 'Novel ecosystems: Theoretical and management aspects of the new ecological world order', *Global Ecology and Biogeography*, 15(1) (2006).
58 P. M. Vitousek, L. L. Loope, and C. M. D'Antonio, 'Biological invasion as a global change', in Richard Somerville and Catherine Gautier, eds, *Elements of Change* (Aspen, CO: Aspen Global Change Institute), pp216–227.
59 For discussion, see Kurt Jax, Clive G. Jones, and Steward T. A. Pickett, 'The self-identity of ecological units', *Oikos*, 82 (1998): 253–264. The concept of the natural world as flux and ecosystems as ephemeral became prominent in the 1990s when disturbance rather than permanence became the leading metaphor. In reviewing these developments, environmental historian Donald Worster described the emerging view of nature as 'a landscape of patches ... a patchwork quilt of living things ... responding to an unceasing barrage of perturbations. The stitches in that quilt never hold for long.' Donald Worster, 'The ecology of order and chaos', in Char Miller and Hal Rothman, eds, *Out Of The Woods: Essays in Environmental History* (University of Pittsburgh Press, 1997), p10. See also Worster, *Nature's Economy* (New York: Cambridge University Press, 1974, 1994 revised).
60 Quoted in Emma Marris, 'Invasive species: Shoot to kill', *Nature*, 438 (17 November 2005): 272–273.
61 For essays seeking to bolster this assumption, see Evan Weher and Paul Keddy, eds, *Ecological Assembly Rules: Perspectives, Advances, Retreats* (Cambridge: Cambridge University Press, 1999).
62 R. H. MacArthur, 'Species-packing and competitive equilibrium for many species', *Theoretical Population Biology*, 1 (1970):1–11; R. H. MacArthur, *Geographical Ecology: Patterns in the Distribution of Species* (New York: Harper and Row, 1972).
63 James H. Brown and Dove Sax, 'Do biological invasions decrease biodiversity?' *Conservation Magazine*, 8 (2) (April 2007).
64 Simberloff, 'Biology of Invasions', p3.
65 John J. Stachowicz and David Tilman, 'Species invasions and the relationships between species diversity, community saturation, and ecosystem functioning', in Dov F. Sax, John J. Stachowicz and Steven D. Gaines, eds, *Species Invasions: Insights into Ecology, Evolution, and Biogeography* (Sunderland, MA: Sinauer, 2005), pp41–64. Quotation at p41. These authors state (p55), 'If lower resource levels lead to more intense competition, and thence to greater competitive ability, it seems plausible that a region with more species would be both harder to invade and more likely to produce successful invaders.' This a priori argument, which is representative of research in invasion biology, does not seem to match what is observed. According the Daniel Simberloff and Betsy Von Holle, the introduction of one exotic species often facilitates (rather than restricts) the introduction of others. 'There is little evidence that interference among introduced species at levels currently observed significantly impedes further invasions, and synergistic interactions among invaders may well lead to accelerated impacts on native ecosystems – an invasional "meltdown" process'. Simberloff and Von Holle, 'Positive interactions of nonindigenous species: 21–31. Quotation at p21.

66 Simberloff, 'Biology of Invasions', pp3–4.
67 The theory was developed in initially in Robert MacArthur and E. O. Wilson, *The Theory of Island Biogeography* (Princeton: Princeton University Press, 1967). It subsequently met severe criticism. See, for example, S. C. Stearns, 'Evolution of life-history traits – critique of theory and a review of data', *Annual Review of Ecology and Systematics* 8 (1977): 145–171.
68 For this definition see M. Vellend, L. J. Harmon, J. L. Lockwood, M. M. Mayfield, A. R. Hughes, J. P. Wares, and D. F. Sax, 'Effects of exotic species on evolutionary diversification', *Trends in Ecology and Evolution*, 22(9) (2007): 481–488. Quotation at p481.
69 Michael L. McKinney and Julie L. Lockwood, 'Biotic homogenization: A few winners replacing many losers in the next mass extinction', *Trends in Ecology & Evolution*, 14(11) (1999): 450–453.
70 'Thousands of species are doing quite well, thank you, in parts of the world where they did not evolve, a fact that alone provides the material for endless investigations. The editors and authors [of the book being reviewed) also note, summarizing earlier literature and contributing new information, that the general outcome of most invasions is to *increase* the overall pool of resident species.' James T. Carlton, 'Species invasions: Insights into ecology, evolution, and biogeography', *BioScience*, 56(8) (August 2006): 694–695 (reviewing Dov F. Sax, John J. Stachowicz, and Steven D. Gaines, eds *Species Invasions: Insights into Ecology, Evolution, and Biogeography* (Sunderland, MA: Sinauer, 2005)).
71 For discussion, see Geerat J. Vermeij, 'Invasion as expectation: A historical fact of life', in Sax, Stachowicz, and Gaines, eds, *Species Invasions*, pp315–339; especially pp326–331. Vermeij writes, 'species play different roles in different places', at p329.
72 Sax et al, *Species Invasions,* p329.
73 Simberloff and Von Holle, 'Positive interactions of nonindigenous species: 21–32. Quotation at p22.
74 Vellend et al, 'Effects of Exotic Species', p481.
75 This is 'Allopatric speciation: the creation of new species via genetic divergence in geographically separated populations.' Vellend et al, 'Effects of Exotic Species'.
76 '[H]ybridization between individuals from genetically divergent native populations may result in introduced populations having more genetic variation than native populations of the same species.' Fred Allendorf and Laura Lunquist, 'Introduction: Population biology, evolution, and control of invasive species', *Conservation Biology*, 17(1) (2003): 24–30. Quotation at pp24–25.
77 J. A. Lau, 'Evolutionary responses of native plants to novel community members', *Evolution*, 60 (2006): 56–63.
78 Sax et al, 'Ecological and Evolutionary Insights', p466.
79 H. A. Mooney and E. E. Cleland, 'The evolutionary impact of invasive species, *Proceedings of the National Academy of Sciences of the USA*, 98(10) (2001): 5446. Online at www.pnas.org/cgi/content/full/98/10/5446 (2001).
80 S. Smith, G. Bell and E. Bermingham, 'Cross-Cordillera Exchange Mediated by the Panama Canal Increased the Species Richness of Local Freshwater Fish Assemblages', *Proceedings of the Royal Society of London Series B, Biological Sciences*, 271 (1551) (2004): 1889–1896. Online at: www.journals.royalsoc.ac.uk.
81 Sax et al, 'Ecological and Evolutionary Insights', p467.
82 Andrew Cohen and James Carlton, 'Accelerating invasion rate in a highly invaded estuary' *Science*, 279 (5350) (23 January 1998): 555–558.

83 Andrew Cohen and James Carlton, 'Nonindigenous Aquatic Species in a USA Estuary: A Case Study of the Biological Invasions of the San Francisco Bay and Delta'. A Report for the USA Fish and Wildlife Service (Washington, DC: The National Sea Grant College Program and the Connecticut Sea Grant Program, 1995). Online at www.sgnis.org/publicat/cc1.htm

84 Vermeij, 'Invasion as Expectation', p329.

85 'Much of the evidence that introduced species cause extinction does not come from studies of introduced plants, but from those of introduced animals, generally predators, and plant diseases. Many of these studies involve animals on islands and, in particular, species of birds that have gone extinct following the introduction of a predatory species, such as the brown tree snake, *Boiga irregularis*, on Guam.' Judith H. Myers and Dawn Bazeky, *Ecology and Control of Introduced Plants* (Cambridge: Cambridge University Press, 2003), p16. The bird extinctions in Guam can hardly be said to have taken place in a 'natural' area. The island was extensively bombed by the Allies in World War II. It was then extensively developed as a gargantuan shopping mall and recreation area for Japanese and other Asian consumers and tourists. For a study of the difficulty of finding a 'natural area' on Guam, see Alan Burdick, *Out of Eden An Odyssey of Ecological Invasion* (New York: Farrar, Straus and Giroux, 2005).

86 James H. Brown and Dove Sax, 'Do biological invasions decrease biodiversity?' *Conservation Magazine*, 8(2) (April 2007).

87 M. A. Davis, 'Biotic globalization: Does competition from introduced species threaten biodiversity?' *BioScience*, 53 (2003): 481–489.

88 J. D. Fridley, J. J. Stachowicz, S. Naeem, D. F. Sax, E. W. Seabloom, M. D. Smith, T. J. Stohlgren, D. Tilman, and B. Von Holle, B. 'The invasion paradox: Reconciling pattern and process in species invasions', *Ecology*, 88 (2007): 3–17. Quotation at pp5–6.

89 Paul D. Kilburn, 'Analysis of the species-area relation' *Ecology*, 47(5) (1966): 831–843. Quotation at p842.

90 For a description of this letter, see the National Invasive Species Council, National Management Plan, 2001, *Meeting the Invasive Species Challenge*, p13. Available online at: www.invasivespecies.gov/council/mp.pdf. The text of the letter is available at: http://aquat1.ifas.ufl.edu/schlet2.html.

91 Executive Order 13112, 3 February 1999, 'Invasive Species', Federal Register: 8 February 1999 (Volume 64, Number 25). Available online at: www.invasivespeciesinfo.gov/laws/execorder.shtml

92 For a review, see Roderick Nash, *The Rights of Nature: A History of Environmental Ethics* (Wisconsin: University of Wisconsin Press, 1989).

93 Douglas Kysar, 'The Consultant's Republic', *Harvard Law Review* 121 (2008): 2041–2084.

94 The Management Plan NISC published in 2001 states its view that 'damage to natural areas is increasing in priority' relative to agriculture. National Invasive Species Council, National Management Plan, *Meeting the Invasive Species Challenge*, October 2001. Quotation at p19. Published online at www.invasivespeciesinfo.gov/docs/council/mpfinal.pdf. Of the roughly 35 invasive plants for which the Council has completed profiles only about three are listed by USDA as noxious weeds in agriculture. For discussion and citation, see Justin Pidot, 'Note: The Applicability of Nuisance Law to Invasive Plants: Can Common Law Liability Inspire Government Action?' *Virginia Environmental Law Journal*, 24 (2005); 183–230; see especially p195.

95 David Ehrenfeld, 'Why put a value on biodiversity?' in *Biodiversity,* E. O. Wilson ed. (Washington, DC: National Academy Press, 1988), pp212–216.

96 Paul Angermeier, 'Does biodiversity include artificial diversity?' *Conservation Biology*, 8(2) (1994): 600–602.

97 Osvaldo E. Sala, F. Stuart Chapin III, Juan J. Armesto, Eric Berlow, Janine Bloomfield, Rodolfo Dirzo, Elisabeth Huber-Sanwald, Laura F. Huenneke, Robert B. Jackson, Ann Kinzig, Rik Leemans, David M. Lodge, Harold A. Mooney, Martín Oesterheld, N. LeRoy Poff, Martin T. Sykes, Brian H. Walker, Marilyn Walker, and Diana H. Wall, 'Global Biodiversity Scenarios for the year 2100', *Science*, 287 (10 March 2000): 1770–1774.

98 Simberloff, 'Non-native species *do* threaten the natural environment!' at p603.

99 National Invasive Species Council, *Meeting the Invasive Species Challenge,* pp10, 12. Available online at: www.invasivespecies.gov/council/mp.pdf.

100 Simberloff, 'Non-native species *do* threaten the natural environment', at pp598–599.

101 Michael Soulé, 'The social siege of nature', in M. Soulé and G. Lease (eds), *Reinventing Nature: Responses to Postmodern Deconstruction* (Washington, DC: Island Press), pp137–170. Quotation at p143.

102 Peter M. Haas, 'Introduction: Epistemic communities and international policy coordination', *International Organisation*, 46(1) (1992): 1–36. Quotation at p3.

103 Peter M. Haas, *Saving the Mediterranean – The Politics of International Environmental Co-operation* (New York: Columbia University Press, 1990). Peter Haas, 'Banning chlorofluorocarbons: Epistemic community efforts to protect stratospheric ozone', *International Organizations*, 46(1) (1992): 187–224. See also Peter Haas, 'Obtaining international environmental protection through epistemic consensus', in Ian Rowlands and Malory Greene, eds, *Global Environmental Change and International Relations* (London: Macmillan, 1992) and 'Social constructivism and the evolution of multilateral governance', in Jeffrey Hart and Aseem Prakash, eds, *Globalization and Governance* (London: Routledge, 1999).

104 Haas, 'Epistemic communities and international policy coordination', p3.

105 Simberloff, 'Non-native species *do* harm the natural environment', p597. For a principal paper supporting this consensus, see D. S. Wilcove, D. Rothstein, J. Dubow, A. Phillips, and E. Losos, 'Quantifying threats to imperiled species in the USA', *BioScience*, 48 (1988): 607–615. For a critique of this paper, see Mark Sagoff, 'Do non-native species threaten the natural environment?' *Journal of Agricultural and Environmental Ethics*, 18 (2005): 215–236.

106 Thomas Lovejoy, 'The obligations of a biologist', *Conservation Biology*, 3(4) (1989): 329–330. Quotation at p330.

107 Schmitz and Simberloff, '*Biological Invasions: A Growing Threat.*'

108 Simberloff accurately describes those who disagree with the consensus within invasion biology as comprising primarily non-scientists and a few unreconstructed ecologists. Simberloff identifies 'a number of authors from different cultural fields, who have joined with a few ecologists in a rearguard action to convince biologists and the lay public that the ecological threat from introduced species is overblown' (parenthetical citations omitted). Daniel Simberloff, 'Invasional meltdown 6 years later: Important phenomenon, unfortunate metaphor, or both?' *Ecology Letters*, 9(8) (August 2006): 912–919. Quotation at p915.

109 Public agencies have attempted to make the monitoring and control (these are different things) of invasive species such as purple loosestrife and Eurasian watermilfoil conditions for obtaining licenses or permits. For an appreciation of the

intricacies involved, see, for example, *Rhinelander Paper Co.* v. *FERC*, 405 F.3d 1 (D.C. Cir. 2005) (upholding a ruling by the Federal Energy Regulatory Commission that required the petitioner, the operator of a hydroelectric dam, as a condition of continuing its operating license to develop and implement a plan to monitor purple loosestrife and Eurasian water-milfoil at the project).

110 B. N. McKnight, *Biological Pollution: The Control and Impact of Invasive Exotic Species* (Proceedings of a Symposium held at the University Place Conference Center, Indiana University-Purdue University at Indianapolis on October 25 & 26, 1991) (Indianapolis: Indiana Academy of Science, 1993).

111 Haas, 'Epistemic Communities and International Policy Coordination', p3.

112 H. L. Mencken, *Mencken Chrestomathy: His Own Selection of His Choicest Writing* (New York: Vintage, 1982), p29.

113 As a result, environmental law has by now largely ceased to exist as a separate form of jurisprudence. See A. Dan Tarlock, 'Is there a There There in Environmental Law?' *Journal of Land Use & Environmental Law* (Spring, 2004): 214–252. Tarlock explains, 'Environmental law, as now defined, is primarily a synthesis of pre-environmental era common law rules, principles from other areas of law, and post-environmental era statutes which are lightly influenced by the application of concepts derived from ecology and other areas of science, economics, and ethics' (p222).

114 *Friends of the Earth, Inc.* v. *Laidlaw Environmental Services (TOC), Inc.* 528 U.S. 167 (2000) at 181.

7
Whales, Whitefellas and the Ambiguity of 'Nativeness': Reflections on the Emplacement of Australian Identities

David Trigger

Introduction

In 1992, an American marine environment researcher observed a distinctive white humpback whale off the Queensland coast and named it 'Migaloo'. The nickname is a term meaning Whitefella among Aboriginal people in Queensland. It was suggested by a local Indigenous man and his aunt (the latter was said to be a 'revered aboriginal elder') after the whale researcher contacted them to express interest in finding an appropriate Aboriginal name. The website of the Pacific Whale Foundation now suggests to a world hungry for information about what has become a highly charismatic animal that Aboriginal people regard all 'albinos' – whether humans, kangaroos, crocodiles or whales – as 'special beings', 'perhaps signs or tokens from the spirit world'.[1] Thus, Migaloo the white cetacean has become an object of desire for those who celebrate whales as an inspiring species of nature; its name, deemed to be from everyday speech among Aboriginal people, may be understood to mean that Migaloo *belongs* in the marine environments of Australia.[2]

The research biologist's desire to find an Aboriginal name for the whale is understandable, and might well be commended, in that it seeks to recognize the richness of indigenous knowledge about Australian nature. Of course, many people who know about the whale (and the number of websites listed in a search for 'Migaloo' is not a small number) are very probably ignorant of the name's derivation. Aboriginal names for things are not easily remembered or embraced by Australians, whether they are recently arrived immigrants or people whose families have been here for generations. While there is plenty of evidence to support the view that names for White people in Aboriginal world-views have historically designated figures connected to the autochthonous world of disembodied human spirits,[3] ending up with such a term for a white whale seems more than slightly ambiguous – at least, with respect to any desire to name nature in a fashion commensurate with thousands of years of pre-European human occupation of Australian land and waters.

For rather than a term for whale from an east coast Aboriginal language,[4] we have a word with ambivalent meaning in Aboriginal English – as the American whale researcher himself notes, Migaloo is like the term Ha'ole in Hawaii.[5] For Polynesian Hawaiians, Ha'oles are White people, this being a name that can have either an endearing or a hostile meaning. And it implies a category of person which may be 'alien' or 'native', depending on context. As with the label for Whites in New Zealand, Pakeha,[6] terms such as Ha'ole and Migaloo are simultaneously a factual description and a name for those whose ancestors brought at least as much catastrophe as benefit.

Further ambiguity arises from the fact that whales are typically migratory through Australian seas rather than located solely within national territory.[7] While increasingly iconic of the wonder of Australian nature, whales are simultaneously animals that move across territorial boundaries. This prompts the question of whether people regard them as 'native' while also belonging elsewhere. Rather than 'nativeness' as such, is it other features of this species (size, or perhaps a popularly assumed capacity for sophisticated communication) which are made to symbolically represent the magnificence of nature[8] – albeit with a simultaneous recognition that this is a 'natural' species encountered in its 'native' or autochthonous place? More generally, if there are publicly expressed sentiments of compassion and anthropomorphism about a figure like Migaloo the white whale, is this case an instructive illustration of the nature of Australian desires for identification with the landscapes to which they or their forebears have come over the past 200 years?

Nature, nativeness and cultural belonging in a post-settler society

A desire to embrace Aboriginal names for Australian nature might be understood as part of an increasing attempt to achieve cultural embeddedness and emplacement in the landscapes and seascapes of the continent. As philosopher Linn Miller suggests, the search for what she terms 'non-Aboriginal belonging' has intensified over the past few decades.[9] Indeed, Miller points out that among adherents to both the right and left of Australian politics, this issue has at times verged on an almost 'hysterical' level of insecurity. On the one hand, Pauline Hanson's maiden speech to Parliament expressed outrage about any suggestion that her sense of belonging in the land of her birth could be morally inferior to the claims of Aboriginal people;[10] and on the left, at least some intellectuals have argued that angst and guilt about 'non-Indigenous' status in Australia can be both warranted and redemptive in moral terms.[11] These are probably extreme positions held only among minorities in the population; but they are arguably indicative of the live issue about who (and what) 'belongs' in Australia.

How do we assess the current state of an Australian sense of emplacement and relations with nature? There can be no doubt about the rich body of cultural connections to land and species in the traditions of the Aboriginal minority. Much further north than where Migaloo was first encountered in 1992, there is a small island in the Flinders group regarded as the topographic embodiment of whale (most likely the humpback);[12] cultural mapping work by anthropologists in the mid 1970s recorded names for places around the topography of the islet that translate as body parts of the animal – thus, for example, slopes are its 'ribs' (*warnkeeni*), a spur is its 'lower back' (*muyu wubu*), a narrow peninsula its 'tail' (*obuy*) and a rocky depression its 'oral cavity' (*anggul walin*).[13] The island as a whole is clearly shaped like a whale, and it is regarded in the relevant Whale Story of the Flinders Island Aboriginal people as the transformed body of the mythic Whale. This type of complex attribution to landscapes of species corporeality and associated spiritual sentience distinguishes Aboriginal relations with nature as autochthonous (literally 'from the earth') in a way unlikely to be matched by the other 97 per cent of the Australian population.

Evidence given in native title claims demonstrates that intergenerational reproduction of such knowledge among Aboriginal people is at best uneven across different regions and at worst severely threatened by forces of cultural assimilation. Nevertheless, there is the issue of whether such cultural wealth serves as a platform for non-Aboriginal learning about emplaced nature, whether orally maintained or documented in written form. If this knowledge does so serve, then in what ways does this happen? Leaving aside the risk of cultural appropriation for Aboriginal people, what is the capacity of other Australians to embrace native species and landscapes as intellectually engaging, apart from their utilitarian value?

It is now common enough for people to distance themselves from what is seen as the brutality of a purely utilitarian relationship with land and nature. Whales perhaps present the archetypal case, in that there have been 'exceptionally radical reversals that have characterized the Western relation to cetacean species over the last two centuries';[14] whereas whales were once hunted nearly to extinction, the Australian government has recently been foremost among nations that argue against such a relationship with these animals.[15] It is fine for us to eat domesticated species such as cattle and sheep, or wild fish stocks, or even a more charismatic native animal such as kangaroo (though there is no doubt more public uncertainty in that case[16]); but whales and dolphins are to be celebrated rather than killed. Indeed, this is an area of tension with Aboriginal peoples in other parts of the world. For example, despite conservationist opposition, the Inuit of northern Canada have struck a deal with government such that they can take a limited number of whales as traditional food.[17] In Australia, the parallel case concerns species such as dugong and sea

Figure 7.1 *As shown in this image from 1965, whaling in the town of Albany, southwest Australia, was at the time a routine and respectable job (Photograph courtesy National Library of Australia, nla.pic-vn3119613-v)*

turtle. While whales were not hunted traditionally here,[18] Aboriginal beliefs about many native animals include envisioning them as cuts of meat, albeit in a fashion understood as commensurate with their spiritual significance.

Figure 7.2 *Volunteers attempt to rescue beached whales (*The Advertiser, *Adelaide, South Australia, 3 June 2005) (See Peace, 2010; note 16)*

An Australian ethic of nature that might be informed by Aboriginal cultural knowledge may thus find some autochthonous elements more consistent with current environmentalist sensibilities than others.

Australian ethics regarding nature, perhaps intensifying in recent decades, fix on the value of landscapes and species as a visual spectacle. A survey of Sydney residents carried out in 2002, for example, found a majority of people partitioning 'nature' as out there in the bush, a place to visit rather than live in.[19] Indeed, we might surmise that this perception of unspoilt character linked to symbolic distance from the hubbub of city life is part of what tourists consume through such visits. Again, the case of whales is instructive.

Anthropologist Adrian Peace has spent participant observation time in Queensland waters with whale watchers and their boat skippers. He finds the major feature of this relatively new relationship with nature to be 'a richly embellished anthropomorphic discourse', whereby 'quite specific human qualities are attributed to individual whales with which skippers and guides claim some familiarity'.[20] Tour operators present the animals as complexly sentient creatures whose behaviours can be interpreted in such terms as 'taking a break', 'not in a hurry', 'hurling their bodies' and 'slapping the surface'. Peace's research reports females being described as 'caring' and 'loving' mothers towards their offspring, and a large male situated about a kilometre away from one such pair as 'riding shotgun' to protect them.[21]

While quite distinct from the Flinders Islands Aboriginal instantiation of the animal into the landscape itself, this 'non-consumptive use of cetaceans'[22] certainly propounds a different relationship with nature from their consumption solely as a utilitarian material resource. 'Whales are worth more to us alive than dead', as the saying goes; the interpretation from Peace is that whale watchers seek some kind of social relationship with the huge animals they venture to observe. Australians here may not be naming specific places using this animal's mythologized body parts, but they are imposing personhood (and sometimes, as we see from the case of Migaloo, individual names) on a species whose biological distance from humans is immense in evolutionary terms.

Peace's study makes a persuasive case that taking a whale watching trip is now 'an act of redemption for past slaughter' and an embracing of conservation ethics which seek to distance us from the brutal practices associated with commercial whaling.[23] Sustaining these enormously impressive animals is unquestionably a righteous cause, and if this means disowning what only 50 years ago was widely regarded as a 'fine and dignified profession',[24] the end possibly justifies the means. Yet we are left with the issue put well by that polymath of the humanities and sciences George Seddon: is it not counterproductive to completely destroy the faith of the young in their own culture and history? In arguing for an enlightened set of environmental practices and understandings, Seddon nevertheless seeks to elucidate what is enabling in the inherited cultural traditions of Euro-Australia, as well as what may be disabling.[25] How then are Australians to mix their environmental and cultural histories, such that each has a good chance of informing the future lives and sense of emplaced identity of young people? How should we think about whaling with the hindsight of environmental consciousness? More generally, how should we make sense of an inherited biota that now includes many species said by science experts to be invasively displacing native species, out of place and unworthy of societal acceptance, let alone celebration?

In forging a relationship with Australian nature, the population hardly draws only (or even predominantly) on the native animals, plants and places of this continent. This is a multicultural nation whose society derives from the cultural traditions and 'exotic' species of many parts of the world. Franklin's innovative research[26] on human–animal relations suggests something of an unresolved contest over the relative moral and symbolic significance of introduced species as compared with native creatures. Furthermore, some citizens, regardless of their views about what constitutes 'Australian nature' as such, may well regard *themselves* as persons who 'belong' elsewhere, as well as (or even instead of) in Australia. Nevertheless, the question of intellectual, emotional and physical entwinement of emplaced human identities with the distinctively autochthonous species and landscapes of this continent is of considerable public significance, surfacing in many ways. 'Not indigenous, merely born here', is poet Les Murray's provocative articulation of the issue, insofar as it implicates how the majority without any Aboriginal ancestry might *belong*:

Where will we hold Australia
We who have no other country?
Not Indigenous, merely born here,
Shall we be Australian in Paraguay
Again, or on a Dublin street corner?[27]

There is perhaps a pointed political agenda in Murray's question. Feigning self-deprecation, the collection in which the verse appears is titled *Subhuman Redneck Poems,* caricaturing the expected view of his work among those on the left. However, this accomplished Australian poet also puts his intellectual finger on a legitimate matter of debate. Australians without any Aboriginal ancestry often enough resent being termed 'non-Indigenous'. Perhaps especially if they and their immediate forebears were 'born here', Murray's apt statement is that this is not regarded by them as inferior on any alleged moral hierarchy of belonging. Being born here is no 'mere' aspect of Australian identity, but rather a primary determinant. Such 'natives' have not been here from 'the beginning', an 'original' time, as the term 'Aboriginal' implies, but their feelings for places and landscapes are likely to be articulated strongly if probed.[28] Prompted by Murray's poem, writer Geoff Page points this out, commenting that seeking Aboriginal names for things – not just animals like Migaloo the whale, but also suburban streets, parks and perhaps more so in earlier decades, house names – is a 'genuine, almost chthonic, impulse ... which makes it understandable why some non-Aboriginal Australians are sometimes distressed to be described as "non-Indigenous"'.[29]

The search for such names is, as the example of the white whale shows, hardly sophisticated in terms of any serious engagement with Aboriginal cultural knowledge; and, as the pockets of vigorous opposition to adopting the name 'Uluru' for Ayers Rock showed us some years ago, the politics of acknowledging Aboriginal cultural landscapes can be bitterly contested.[30] However, general discomfort with embracing the notion of a 'non-indigenous' identity, is not likely to be expressed only by 'subhuman rednecks' at all. John Morton and Nick Smith, researchers who are both long-term supporters of Aboriginal interests, ask why Australians must be constantly reminded of a status as 'intruders in the bush'.[31] Like such introduced species as the rabbit, Indian mynah bird and the plant known as 'Patterson's Curse', Australians, they say, are *already* indigenous, a part of this land's nature and 'at home' here. Furthermore, it is in fact not at all clear that Aboriginal views across the country would necessarily disagree with this position. Though some suggest that few Aboriginal people will 'concede that their fellow non-Aboriginal Australians born here are also indigenous',[32] in my view this is an open question. Moreover, many Aboriginal responses to 'exotic' plants, animals and cultural forms seek to embrace them.[33] At the least, this is a complex matter rather than any simplistic divide between natives and invaders, just as is the diversity of non-Aboriginal thinking about the constituents of identity in a globalizing world.

Conclusions

But are rabbits, Indian mynah birds and Patterson's Curse pests and weeds and in this sense ecologically out of place? And, if so, what do we do with them culturally? Do they 'belong' as much as native species (whales included)? Geoff Page comments that Australians are more comfortable distinguishing exotic plants than people (at least, when it is suggested that it is the settler-descendants from Europe who are the 'exotics'): 'No-one driving past a row of poplars or a creek lined with willows is likely to consider them native.' Perhaps, but for many Australians such certainty about which species are native or introduced is probably to be only sporadic. And, in any case, what is the emotional investment in all those plants and animals that we are told are bad for the land? How might an ecological consciousness deal with the idea that historically familiar things should now be eradicated? In the same collection, another of Les Murray's poems laments the 'present slaughter' of feral animals:

> *But so far as treetops or humans now alive know*
> *All these are indigenous beings. When didn't we have them?*
> *Each was born on this continent. Burn-off pick and dusty shade*
> *Were in their memory, not chill fall, not spiced viridian.*[34]

So if introduced plants and animals are now naturalized, i.e. adapted to Australian environments (and not to the 'chill fall' or 'spiced viridian', perhaps of other climes), what length of presence is required for them to be *counted* as 'native', as George Seddon asks – bearing in mind the literal sense that any plants that germinate from seed are native to the place where they do so (i.e. they are born there). Like others, Seddon points out for the world of plants the confusions potentially implicit in the notion of 'native' as opposed to 'colonist, or invader, or exotic'.[35] The idea of native plants is a complex issue, 'a remarkable mixture of sound biology, invalid ideas, false extensions, ethical implications, and political usages both intended and unanticipated', according to the late maestro of evolutionary biology, Stephen Jay Gould.[36] Indeed, to speak of so-called 'Australian plants' is fraught with difficulties; as Seddon explains, plants know nothing of nationality.[37] And while the nation is (unusually, in comparison to elsewhere in the world) coterminous with the continent, the Australian land mass encompasses many highly diverse environments, with plants endemic to one often failing to survive in another. Furthermore, a difficult question looms of whether relatively recent introductions that have become naturalized in biological terms should now be included among what are thought of as 'Australian plants'.

If, in the domain of plants and animals, there is ambiguity about the constituents of autochthony – as well as the risk of what Seddon describes as 'wildlife xenophobia'[38] – to what extent does this matter parallel the related issue of defining 'nativeness' and belonging to place among people? More specifically, what are the implications of Australians working through the issue of cultural belonging to the lands and seas of their continent by acknowledging, identifying with and naming nature in general, and/or charismatic species in particular? And amid such attempts at forging culturally underpinned links to the landscape and seascapes of a relatively young post-settler society, what is the significance of a desire for 'nativeness' – as this quality may be perceived to inhere (or be emergent) in certain species, physical settings and indeed persons? It is the complexity and challenges connected with these questions that are revealed in the naming of Migaloo the white whale.

Acknowledgements

Maria Connolly, Peter Dwyer, Andrea Gaynor, Lesley Head, Monica Minnegal, Jane Mulcock, Bruce Rigsby, Paul Rees and Yann Toussaint have provided stimulating discussion that has contributed to my ideas expressed in this essay. A version of this chapter was published previously in *Island* no 107, 2006, pp25–36, a journal produced in Tasmania, Australia (see: www.islandmag.com).

Notes

1 See 'Facts about Migaloo', Pacific Whale Foundation, www.migaloowhale.org. Accessed 7 May 2006.

2 On the Pacific Whale Foundation website, where a number of whale photographs entice people to 'adopt a whale', there is also a whale that has been given the name 'Uluru' – said to be 'Named after the Aboriginal word for "Sacred Rock"' (see www.pacificwhale.org/adopt/meet_whales_australia, accessed 7 May 2006) – though the term is more accurately understood as a place name for the famous site in central Australia (also known as Ayers Rock) that is sacred to the Yankunytjatjara and Pitjantjatjara Aboriginal people (R. Layton, 1986, *Uluru: An Aboriginal History of Ayers Rock*, Australian Institute of Aboriginal Studies).

3 Historian Henry Reynolds discusses many cases where first contacts with Whites led Aboriginal people to regard the new arrivals as 'ghosts' (*The Other Side of the Frontier*, Penguin, 1981, pp31, 36); at times they were reportedly considered to be the spirits of deceased kin (D. Trigger, 1992, *Whitefella Comin'*, Cambridge University Press, pp18–20). The term migaloo (*mikulu*) has been recorded as meaning spirit, ghost, corpse or white person for the Mayikudunu language of northwest Queensland (G. Breen, 1981, *The Mayi Languages of the Queensland Gulf Country*, Australian Institute of Aboriginal Studies, pp118–119), and as having a similar meaning in the coastal Queensland language of Biri (A. Terrill, 1998, *Biri*, Languages of the world, Lincom Europa, München, pp 76–77). Linguist Peter Sutton comments (personal communication 27 June 2005) that, given the prevalence of the term in coastal Aboriginal communities such as Palm Island, it was probably diffused west to the Gulf languages during the early period of colonization. Reynolds probably misses the subtleties of the word's contemporary usage when suggesting only a single meaning as 'a derisory, even contemptuous, term for white Australian' (*ibid* p.36).

4 For example, *yulingbila*, *dumgarumba* and *jalubila* are names for whale in the Yagara language of the Brisbane valley and Moreton Bay shores (T. Jeffries, 2002, *The Phonology of Yagara: The Language of the Brisbane River Valley and Southern Moreton Bay*, Honours thesis, University of Queensland).

5 Personal communication from Professor Paul Forestell, 27 June 2005. On the term 'Haole', see E. Whittaker, 1986, *The Mainland Haole: The White Experience in Hawai'i*, Columbia University Press.
6 See A. Fleras and P. Spoonley, 1999, *Recalling Aotearoa: Indigenous Politics And Ethnic Relations in New Zealand*, Oxford University Press, pp89–93.
7 M. Simmons and H. Marsh, 1986, 'Sightings of humpback whales in Great Barrier Reef waters', *Scientific Reports of the Whales Research Institute*, 37, 31–46.
8 P. Tyack, 1999, 'Communication and cognition', in *Biology of Marine Mammals*, J. Reynolds and S. Rommel (eds), Smithsonian Institution Press, p287. Similar assumptions about dolphin communication and intelligence are regarded with scepticism in scientific studies, see B. Wursig, 2002, 'Intelligence and cognition', in *Encyclopedia of Marine Mammals*, W. Perrin, B. Wursig and J. Thewissen (eds), Academic Press, p636.
9 L. Miller, 2003, 'Belonging to country: a philosophical anthropology', *Journal of Australian Studies*, 76, 215–223.
10 Pauline Hansen was the newly elected leader of the 'One Nation' political party in 1996.
11 See P. Read, 2000, 'Four historians', in *Belonging: Australians, Place and Aboriginal Ownership*, Cambridge University Press, pp172–197.
12 I am indebted to Dr Paul Memmott, University of Queensland, for initially bringing this case to my attention. On the likelihood of it being the humpback species (*Megaptera novaeangliae*) upon which the emplaced Aboriginal landscape story is focused, see Simmons and Marsh, 'Sightings', pp31–46.
13 This research was carried out by Drs Athol Chase and Peter Sutton (see P. Sutton, 1993, 'Flinders Islands and Melville National Parks land claim', For Cape York Land Council, Figure 10); *Aboriginal Land Claims to Cape Melville National Park, Flinders Group National Park, Clack Island National Park and Nearby Islands*, Report of the Queensland Land Tribunal, 1994, pp96, 112.
14 P. Armstrong, 2004, 'Moby-Dick and compassion', *Society and Animals*, 12(1), p22.
15 Indeed, the Commonwealth Government Environment Minister describes Japanese whaling as 'absurd' and 'sick', rejecting any proposition that there is a 'cultural reason' for it; though a reading of relevant anthropological research would indicate at least the likelihood of a contrary finding: 'Whales have for centuries been hunted and eaten by the Japanese, and whaling activities are intimately bound up with religious beliefs and practices' (A. Kalland, 'Whales and Japanese culture', in *The Encyclopedia of Religion and Nature*, J. Kaplan (ed), 2005, p1730; see also A. Peace, 2010, 'The whaling war: Conflicting cultural perspectives', *Anthropology Today*, 26(3), pp5–9). The Minister does appear to accept that Inuit whaling in Canada is a 'culturally sensitive' practice among an 'Indigenous population', however it is not clear how such sensitivity might be defined (see ABC News Online, 25 May 2005, available at www.abc.net.au).
16 C. Bulbeck, 2005, *Facing the Wild: Ecotourism, Conservation and Animal Encounters*, Earthscan, p185.
17 See, e.g. 'The Inuit case for whaling', http://news.bbc.co.uk/1/hi/world/asia-pacific/2005773.stm; 'Canada natives prepare controversial whale hunt', http://members.aol.com/adrcnet/marmamnews/98042701.html; 'Revival of whale hunt sparks bitter debate', www.whalewatch.co.nz/_disc3/000000b5.htm.
18 Though particularly along parts of the southern Australian coastline it seems whales beached on the shore were butchered and their bones used as frames for shelters;

and the Mirning people from the coastline of the Great Australian Bight identified Whale Dreaming (P. Memmott, 2008, 'Bone as an architectural medium', chapter in P. Memmott, *Gunyah, Goondie and Wurley: Australian Aboriginal Architecture*, University of Queensland Press, p95). Whale is regarded as edible (a form of *minh*, the term for 'meat') in the Wik language of western Cape York Peninsula, though anthropological research has not found stories about people living off beached whales in that region (Peter Sutton, personal communication, 28 July 2005).

19 Woolcott Research, 2002, *Urban Wildlife Renewal: Growing Conservation in Urban Communities*. NSW National Parks and Wildlife Service Research Project, p62.

20 A. Peace, 2005, 'Loving Leviathan: The discourse of whale-watching in Australian ecotourism', in *Animals in Person: Cultural Perspectives on Human-Animal Intimacy*, J. Knight (ed), Berg, Oxford, pp191–192.

21 Peace, 'Loving Leviathan', pp197, 199.

22 Peace, 'Loving Leviathan', p193.

23 Peace, 'Loving Leviathan', p204.

24 N. Einarson, 1993, 'All animals are equal but some are cetaceans: Conservation and culture conflict', in *Environmentalism: The View from Anthropology*, K. Milton (ed), Routledge, New York, p74. See also A. Wolfe, 2002, 'A cruel business: Whaling and the Albany community 1946–1988', in *Country: Visions of Land and People in Western Australia*, A. Gaynor, M. Trinca and A. Haebich (eds), pp86, 88, Western Australian Museum.

25 G. Seddon, 1997, *Landprints: Reflections on Place and Landscapes*, Cambridge University Press, p xiv; G. Nossal, 'Foreword', in Seddon, *Landprints,* pviii.

26 A. Franklin, 2006, *Animal Nation*, University of New South Wales Press.

27 L. Murray, 1996, *Subhuman Redneck Poems*, Duffy and Snellgrove, p47. While the reference to Dublin is obvious enough, given the substantial Irish role in the settlement of and migration to Australia, the mention of Paraguay is probably linked to attempts to establish an Australian utopian settlement there in the 1890s (see A. Whitehead, 1997, *Paradise Mislaid – in Search of the Australian Tribe of Paraguay*, University of Queensland Press, St. Lucia).

28 See, e.g. the work of P. Read, 1996, *Returning to Nothing: The Meaning of Lost Places*, Cambridge University Press.

29 G. Page, 1997, 'Not Indigenous, merely born here ...', *Quadrant,* November 1997, pp23–25.

30 Layton, *Uluru*. Other cases revealing similar politicized contests over replacing earlier English names with Aboriginal place names include the Grampians mountain range in Victoria (see T. Birch, 1996, 'A land so inviting and still without inhabitants: Erasing Koori culture from (post-) colonial landscapes', in *Text, Theory, Space,* K. Darian-Smith, L. Gunner and S. Nuttall (eds), pp173–188, Routledge).

31 J. Morton and N. Smith, 1999, 'Planting indigenous species: A subversion of Australian eco-nationalism', in *Quicksands,* K. Neumann, N. Thomas and H. Erickson (eds), pp153–161, UNSW Press.

32 Rigsby, 'Not Indigenous, merely born here'; Les Murray, 'Some thoughts on being indigenous in Australia'. Unpublished paper delivered May 2000, University of Queensland.

33 See D. Trigger, 2008, 'Indigeneity, ferality and what "belongs" in the Australian bush: Aboriginal responses to "introduced" animals and plants in a settler-descendant society', *Journal of the Royal Anthropological Institute (NS)*, 14 (4), pp628–646.

34 L. Murray, 'Some thoughts', p41.
35 G. Seddon, 2005, *The Old Country: Australian Landscapes, Plants and People*, Cambridge University Press, pp1–25.
36 S. J. Gould, 1997, 'An evolutionary perspective on strengths, fallacies and confusions in the concept of native plants', in *Nature and Ideology: Natural Garden Design in the Twentieth Century*, J. Wolschke-Bulmahn (ed), Dumbarton Oaks Research library and collection,Washington, DC, pp11–19.
37 Seddon, *The Old Country*, pp9–12.
38 Seddon, *The Old Country*, p7.

8
The Rise of Modern Invasion Biology and American Attitudes towards Introduced Species

Daniel Simberloff

Introduction

The beginning of the modern environmental movement in the USA is traditionally associated with two events: Rachel Carson's publication of *Silent Spring* in 1962 and the first Earth Day in 1970, spearheaded by Senator Gaylord Nelson. The primary focus of both the book and the event was chemical pollution – pesticides for Carson, and air and water pollution, oil spills and toxic waste dumps for the Earth Day participants. Introduced species as an environmental threat played no role in initiating this movement. Carson did discuss two introduced species, but her concern was not with their damaging impacts but with the damage caused by chemicals used to combat them. One was the South American red fire ant, imported and introduced to the southern USA. Carson pointed to deaths of non-target animals such as songbirds, raccoons, opossums and armadillos caused by chlorinated hydrocarbon pesticides used against the ant. When Carson wrote, there was no evidence of fire ant-induced environmental or conservation damage in today's terms. Rather, the ant's impacts on cattle and as a nuisance to people are what led to the misguided US Department of Agriculture campaign to eradicate it that Carson inveighed against (Buhs, 2004). Carson also discussed the European gypsy moth ravaging northeastern forests, but again her concern was deaths of species such as other insects, crabs, birds and fish from the DDT employed against the moth, whose impact and prospects for spread she minimized.

Public and scientific perceptions through the 1970s

Introduced species as an environmental concern did not achieve great traction with the US public until the 1990s. This was following widespread recognition among ecologists and conservation biologists in the 1980s of the scope and magnitude of threats that many introduced species posed to ecosystem functioning and even the continued existence of certain native species (Schmitz et al, 1997; Simberloff, 2010). The delayed perception of introduced species as a general problem, among both the public and environmental scientists, was in spite of publications and events that one might have expected to trigger such concern. Even before Carson's book, both editions (1948, 1956) of a leading plant ecology textbook by Henry J. Oosting, president in 1955 of the Ecological Society of America, warned scientists about ecological impacts of such introduced species as Japanese honeysuckle, chestnut blight, water hyacinth, the small Indian mongoose, the gypsy moth, kudzu, eucalypts, the English sparrow and the starling. Yet even among scientists working subsequently on introduced species, none recall Oosting as their inspiration. In 1956, Marston Bates, a prominent American zoologist and early environmentalist, published a wide-ranging review of impacts of introduced species entitled, 'Man as an agent in the spread of organisms', a version of a paper he presented at a heavily publicized academic symposium, 'Man's Role in Changing the Face of the Earth'. It did not crystallize great interest in the issue among scientists or the public. In 1958, the English ecologist Charles S. Elton published a book for a lay audience, *The Ecology of Invasions by Animals and Plants*, based on a series of BBC radio broadcasts. Although the book targeted the public, many biologists read it. It did not inspire an expanded interest in introduced species as a general environmental issue, but in the 1980s, when such interest exploded in the wake of a scientific programme mounted by the international Scientific Committee on Problems of the Environment (SCOPE), several participants recalled Elton's book and cited it, ensuring continued attention in the burgeoning scientific literature (Simberloff, 2010). This book was reprinted in 2000 in response to growing popular interest in invasions.

Oosting and Bates were scientists writing for scientists, so it is not too surprising that they did not inspire public concern. However, Raymond Dasmann, a wildlife biologist noted for environmental books written for a lay readership, published *No Further Retreat: The Fight to Save Florida* in 1971. This was after *Silent Spring* and Earth Day, when the environmental movement was rapidly gaining steam. Dasmann featured introduced species as a key environmental threat to Florida, citing Elton's book and enumerating a number of problems caused by introductions. His book did not engender widespread concern about introduced species, even in Florida. Rather, the American public did not see introductions as a general environmental issue until the 1990s, after conservation biologists and ecologists, beginning with the 1980s SCOPE programme, started publicizing their concerns to the populace at large and to government officials and agencies.

Some particularly harmful introductions did register on the public consciousness well before the 1990s. However, each problematic invasion aroused primarily the stakeholders who observed it first hand or whose lives were in some way affected by it. For a long time, these introductions were not conceived as parts of a larger force, the movement by humans of large numbers of species from areas in which they were native to different, often distant sites. Probably the invasion receiving the most public attention was the Asian chestnut blight that essentially eliminated American chestnut, a dominant forest tree of eastern North America, in less than 50 years beginning around 1900. This species had been such a dominant feature of Appalachian culture that its rapid disappearance there could hardly pass unnoticed; its nuts were eaten and sold as a culinary delicacy and used for medicinal purposes and livestock feed, and its wood was prized as timber (Freinkel, 2007).

However, exotic insect pests of agriculture certainly attracted attention from farmers and their representatives. For instance, the Japanese beetle was first detected in the USA at a New Jersey nursery in 1916 and had spread by the 1940s to the mid-Atlantic states, southern New England and Virginia. By the 1920s it was recognized as a major agricultural pest, and subsequent incipient invasions in the Midwest and California were met with concerted, well-funded eradication campaigns. Because the beetle attacks ornamental plants as well as crops, it became well-known and reviled well beyond the agricultural sector by the 1950s (Clair and Kramer, 1989).

The sea lamprey first reached Lake Ontario in the 1830s, probably through the Erie Canal, and in the early 1920s Lake Erie and the upper Great Lakes through the second Welland Canal. Its arrival especially in the upper Great Lakes occasioned great concern from sport and commercial fishermen as it devastated populations of prized large fish and even contributed to the extinction of three species of ciscoes (Sorensen and Bergstedt, 2010).

Kudzu, deliberately introduced from Japan to the southern USA in 1883 as an ornamental, was widely distributed by US government agencies in the early 20th century as a forage crop, and then as an aid to soil conservation and erosion prevention. It reached its peak popularity in the 1930s. By the 1950s, it was widely recognized by the public in the South as a nuisance, and it was officially listed by the US Department of Agriculture as a weed by the 1970s. At this time it was beginning its rise to the status of emblem of the South, with concurrent fame as an invader, but real public concern with its environmental impact was long restricted to affected southern states (Alderman and Alderman, 2001; Alderman, 2004). Salt cedar (*Tamarix*) from the Middle East was introduced to the USA by the early 19th century; by the mid 19th century it was being used to control bank and sandbar erosion in the Southwest and as an ornamental in the Northeast. By the 1920s, scientists and land managers were concerned that salt cedar was becoming invasive in the Southwest. By the 1950s, regional government agencies were uniting to stop salt cedar as an invader, and by 1970s, the public in the Southwest were broadly cognizant of it as an environmental scourge, primarily for its use of water to the detriment of native plants and humans (Chew, 2009).

Modern invasion biology rises, and public concern soon follows

These and similar regional or sectoral cases did not coalesce into broad-based public concern about biological invasions in general until scientists studying and managing invasions achieved this synthesis, largely inspired by the SCOPE programme in the 1980s, noted above. The SCOPE programme was inspired by the growing scientific recognition of the magnitude and number of problematic invasions and had three goals: (i) to understand why some introductions lead to invasions and others do not; (ii) to understand why some ecosystems seem easily invaded and others resistant; (iii) to use the knowledge gained in endeavours (i) and (ii) to limit and manage invasions much more effectively. The programme engaged hundreds of scientists, and Americans played highly prominent roles (Simberloff, 2010). This programme, by forcing recognition of the increasing number of problematic invasions and publicizing details of case after case, precipitated the modern science of invasion biology. It impressed upon conservation scientists the major, growing role of invasions in creating the situations they hoped to ameliorate.

Recognition by the general public soon followed. The Convention on Biodiversity, opened for signature at the Rio Earth Summit in 1992, included a strong statement on the importance of biological invasions and mandated that its signatories 'as far as possible and as appropriate ... prevent the introduction of ... those alien species which threaten ecosystems, habitats, or species.' Although the Convention was not ratified by the USA (indeed, partly because the fact that it remained unratified was widely publicized), it was highly visible to the American public and raised the profile of biodiversity issues in general and the threat of introduced species in particular. Newspaper articles proliferated, and in 1998 two new books aimed at the general public discussed introduced species, one with particular reference to the USA: Robert S. Devine's *Alien Invasion: America's Battle with Non-Native Animals and Plants* and Chris Bright's *Life out of Bounds: Bioinvasion in a Borderless World*. They both reflected and contributed to growing public anxiety over introductions. By the 2000s, Yvonne Baskin's *A Plague of Rats and Rubbervines: The Growing Threat of Species Invasions* (2002) and Alan Burdick's *Out of Eden: An Odyssey of Ecological Invasion* (2005) could be found in airport bookstores and 'interesting new additions' shelves of public libraries. Many new introductions to the USA from the 1990s onwards achieved not just regional or sectoral fame but villain status in national newspapers and television news: the zebra mussel (1988), snakehead, emerald ash borer, Asian longhorn beetle, Burmese python, silver and bighead carp, killer algae.

The reaction

Recognition that invasive introduced species are a huge environmental and conservation problem, and a greater or lesser degree of animosity towards them today, constitute the dominant mindset of both the US public and US environmental and conservation scientists. However, reactions have arisen based on several partially overlapping concerns:

Citizens who like particular non-native species or dislike control methods

Among the most vocal and visible dissenters are citizens who, for one reason or another, favour a particular introduced species that has been targeted for removal. This may be on environmental or economic grounds or because they oppose management procedures. These activists are not initially motivated by an ideological commitment to welcome, or at least not to mistreat, introduced species in general, but rather by devotion to a species they have come to love or by horror induced by how some invaders, especially vertebrates, are killed. As controversies are prolonged and become heated, the verbiage may become weighted with heavier, ideological baggage. Introduced plants that have long inhabited parts of the USA, in some cases for so long that they are not widely recognized as introduced until a controversy arises over their removal, are frequently the focus of such activity. Australian *Eucalyptus* trees of several species were first brought to California in the late 19th century. In many areas they formed extensive groves and became a characteristic landscape feature. Eucalypts are ecologically harmful to many native California species of plants, birds and insects, especially because they are more fire-adapted than native species and can propagate substantial fires (Rejmánek and Richardson, 2010). However, many ecological restoration projects entailing eucalypt removal have been delayed or stymied by eucalypt-lovers. For example, eucalypts were planted as windbreaks on Angel Island, now part of the Golden Gate National Recreation Area, between 1863 and the 1930s. As their inimical effects on native plants became known, in 1979 the US National Park Service devised a plan to remove all but 6 acres out of 86 acres of eucalypt forest, only to be fiercely opposed by a group called 'POET' (Preserve Our Eucalyptus Trees), who decried 'plant racism' and called Park Service employees 'plant Nazis'. POET delayed the restoration project for about five years, until 1990 (Williams, 2002).

A similar situation crops up repeatedly in Florida, where 'Australian pine' (which is actually several species of *Casuarina*), introduced in the 1890s, is a highly invasive ecological threat on several grounds. Its profuse leaf fall smothers native vegetation, while its interaction with microbial symbionts fixes nitrogen that fertilizes otherwise nutrient-poor soil, to the detriment of natives adapted to the natural nutrient regime. The spreading roots of Australian pine also prevent sea turtles from nesting (Schmitz et al, 1997). Nevertheless, when the state and various municipalities mount plans to remove Australian pine, they often run foul of passionate advocates who like their appearance or the shade they provide. A 1999 blog complaining about a removal in Boca Raton ended by hinting at a larger motivation than ecological restoration: 'If we follow that reasoning a little further, we should probably also get rid of the Haitians, the Cubans, the Canadians, and all the people from New Jersey –

they are all "non-native"' (Ellis, 1999). On the Big Island of Hawaii, South American strawberry guava (*Psidium cattleianum*), introduced in the early 19th century, is a highly invasive small tree that has replaced native trees over wide areas and that also harbours fruit flies that inflict substantial losses on agriculture. Yet a 2008 USDA Forest Service plan to release a host-specific leaf-galling scale insect from Brazil to attack the guava elicited loud protest from several citizens who like to eat guava and saw the plan as an attack on private property rights. An entity called 'Save Our Strawberry Guava' publicizes these views and is attempting to stop the release by enlisting state officials (Tummons, 2008). In Illinois, a 1996 plan to restore 7000 acres in west suburban Chicago to oak savannah and prairie sparked a spreading controversy that ultimately stopped several restoration projects in the Chicago area. Although many factors contributed to this cessation (including arguments about failure to inform the public about the nature and reasons for various restoration activities), the ultimate factor was that many people prefer forests, even those dominated by European plants introduced to the Midwest in the 19th century, common buckthorn (*Rhamnus cathartica*) and glossy buckthorn (*R. frangula*), to the prairies and savannahs that would reign if humans had not so greatly modified the region (Helford, 2000).

Some advocates of particular introduced species are akin to animal rights advocates, simply unable to countenance destruction of vertebrates, especially those that are charismatic. A flock of wild South American parrots (cherry-headed conures) on Telegraph Hill, San Francisco, founded in the 1990s by escapees and released pets, came to be a beloved feature of the region, fed by residents and visitors and immortalized in a documentary film in 2005 (Bittner, 2004). A plan to remove them as non-native elements that could harm native birds elicited great opposition on the grounds that they were an important cultural feature and intelligent, sentient beings. The removal was stymied, though in 2007, the San Francisco Board of Supervisors enacted a ban on feeding them. Similar objections from animal rights advocates (e.g. Welch, 1973) have met attempts in Florida, California, and other states to eradicate burgeoning populations of South American monk parakeets that cause significant economic damage by nesting on utility structures (Pruett-Jones et al, 2007).

In Hawaii, the Puerto Rican tree frog known as the coqui (*Eleutherodactylus coqui*) was introduced accidentally, probably in potted plants, before 1988. After a lag of about a decade, the population exploded over a short period in the 1990s and substantially threatened native insect populations. They are also reviled because of their loud, chirping, nocturnal calls. Calls for their eradication were countered by a small group of animal rights advocates. In addition to being moved inadvertently on nursery plants, frogs were deliberately introduced to new sites, and probably new islands, despite legal prohibitions. Opposition to actions against the frog by a similarly motivated state official delayed a government response to the invasion by about a year, with lasting impact (Kraus, 2009).

Hunters and fishers have objected to removing introduced game animals, even those implicated in substantial ecological damage. Perhaps the most remarkable such case involves an odd alliance of animal rights advocates, led by People for the Ethical Treatment of Animals (PETA), and pig hunters in the islands of Hawaii. Polynesians first introduced Asian pigs to the archipelago AD *c.*1000; these hybridized with European boar introduced in the late 18th century. Rooting by pigs in mountainous Hawaii causes enormous erosion, but the main concern of ecologists is their fostering invasions by exotic plants (such as strawberry guava) by disturbing the soil and distributing seeds. Yet the snaring that is the only feasible control method in the most mountainous areas has particularly exercised animal rights advocates, and fencing by the Nature Conservancy as well as federal and state governments has infuriated hunters, including natives whose ancestors have long hunted pigs (Burdick, 2005). The upshot is an uneasy antagonism and continued environmental damage. A similar alliance between hunters and animal rights advocates (in this case the Fund for Animals) plagued a proposal by the National Park Service to remove exotic mountain goats from Washington's Olympic National Park, ultimately limiting the operation to one core region of the park (Houston et al, 1994).

The northern pike is native to northern Europe and also to North America, but not to California. In 1994, it appeared in Lake Davis in the Sierra Mountains of the northern part of the state, probably brought by fishers from a nearby, illegally stocked lake. Conservation biologists feared these large, voracious predators would not only damage native fish populations in Lake Davis but escape to threaten imperilled salmon and steelhead populations in the Sacramento-San Joaquin River Delta. A controversial attempt in 1997 to eradicate the pike by poisoning the lake galvanized intense opposition from fishers who prized it and residents concerned about the safety of the poisoning protocol. In the event, thousands of fishes were killed and the lake was restocked with natives, but the pike reappeared, probably introduced by saboteur fishers (Elmendorf et al, 2005). A second attempt to eradicate the pike by poisoning took place in 2007, this time with much preparatory interaction with local citizens and far less opposition. However, the outcome remains in doubt, not least because of the threat of rogue reintroduction (California Department of Fish and Game, 2009).

The charge of xenophobia

A different, more general objection to attempts to control or eradicate introduced species has been persistent charges of xenophobia, nativism and racism voiced chiefly by critics from outside the community of ecologists actually working on introduced species. The historian Philip Pauly (1996) perceived early actions against introduced species as part of the nativism prominent in the USA during the Progressive Era (1880s to 1920s): 'attitudes towards foreign pests merged with ethnic prejudices: the gypsy moth and the oriental chestnut blight both took on and contributed to characteristics ascribed to their presumed human compatriots' (p54). He accused early US government officials who inveighed against introduced species, especially Charles Marlatt, of xenophobia based largely on the fact that they were active during the

Progressive Era. This was despite their having expressed legitimate, indeed, prescient, concerns about particular species. In 1910, Marlatt, acting chief of the Bureau of Entomology of the US Department of Agriculture, rejected a shipment of 2000 ornamental cherry trees from Tokyo officials to their 'sister Capital City' of Washington, DC on the grounds that they harboured several insect pests. Pauly sees the rejection as having 'subtexts of racial inequality' and 'implications of an insidious Yellow Peril hidden within beautiful packaging' (p54), though Marlatt's statement is flat, bureaucratic and noted specifically the presence of fungi, crown gall, root gall, two kinds of scale, a potentially new species of borer, and other insects. Marlatt advised the Japanese on how to cultivate and ship an uninfested group of trees, and when they did so, a second shipment was approved (Pauly, 1996; Coates, 2006).

American philosopher Mark Sagoff has been a prominent critic of invasion biology and attempts to impede invasions generally; one of his reasons (Sagoff, 1999) is at least the analogy (if not the taint) with xenophobia. He notes that the same traits that nativists attribute to human immigrants (e.g. aggressiveness, unbridled fecundity and sexuality) are often ascribed to introduced species. Anthropologist Anna Tsing (1995) and feminist scholar Banu Subramaniam (2001) make similar observations. The German garden architects Joachim Wolschke-Bulmahn and Gert Gröning (1992) have charged that the entire movement to exclude or control introduced species is a legacy of Nazis and their anti-Semitic predecessors. This charge has been repeated by many US authors, including the garden and food writer Michael Pollan (1994), California seed salesman J. L. Hudson (1998) and evolutionary biologist Steven Jay Gould (1998).

Perhaps the most extravagant charge of xenophobia is by David I. Theodoropoulos (2003), who describes himself as a 'conservation biologist who has worked in the fields of ethnobotany and plant germplasm conservation' on the jacket of his book, *Invasion Biology. Critique of a Pseudoscience*, but is in fact the seed salesman J. L. Hudson (a pseudonym). Rehashing and magnifying the claim of Wolschke-Bulmahn and Gröning, he sees two main motivations behind invasion biology, racism/nativism and greed, with scientists profiting from large grants to study invasions, government employees from jobs in the regulatory bureaucracy, and pesticide and herbicide companies from the sale of their products used on introduced species. The book is squarely in the realm of crank literature, but it has been cited without qualification by a few authors in the 'normal' literature (e.g. Gobster, 2005; Chew, 2006).

Absent from the charges of xenophobia, from Pauly through Theodoropoulos, is consideration of the real impacts of introduced species, which are extensively documented by the scientists and managers dealing with them (Simberloff, 1996, 2003, 2005). Rather, these critics have produced a social construction of a science, in this case invasion biology, in which psychological and social factors rather than objective facets of the natural world drive developments and decisions. Peter Coates (2006) has come to a similar conclusion regarding both Wolschke-Buhlman's and Gröning's claims of a Nazi motivation among current US invasion biologists and managers and suggestions that ornithologists unhappy with the spread of the English sparrow in North America a century ago were motivated by nativism. There is little doubt that some US nativists in the past lumped introduced species with human immigrants as objects of scorn; this does not mean that everyone concerned about introduced species was a xenophobe or racist.

US scientists actually working with invasions have rarely accused one another of xenophobia, although ecologist James Brown, a prominent early critic of invasion biology, came very close at the SCOPE synthesis conference:

> *There is a kind of irrational xenophobia about invading animals and plants that resembles the inherent fear and intolerance of foreign races, cultures, and religions. I detect some of this attitude at this conference. Perhaps it is understandable, given the damage caused by some alien species and the often frustrating efforts to eliminate or control them. This xenophobia needs to be replaced by a rational, scientifically justifiable view of the ecological roles of exotic species.* (Brown, 1989, p105)

The scientific opposition

Aside from concern about possible xenophobia, a contrarian minority view of invasions has arisen among US scientists working on introduced species. Its chief claim is not that the entire enterprise is infected with nativism, but rather that the problems caused by introduced species are not nearly as drastic as is bruited about, and in any event most introduced species are blameless. These critics constitute a small minority, and there is no evidence that their concerns have yet influenced public opinion, government activities or research agendas, but as respected ecologists publishing in leading scientific journals, they continue to air their arguments. Perhaps the most prominent proponent of this view today, by virtue of his recent textbook on invasion biology, is plant ecologist Mark Davis (2009). Davis's main concern, following Slobodkin (2001), Rosenzweig (2001) and Brown and Sax (2004), is a dichotomy he perceives as dominating the popular literature about invasions, and to some extent the scientific literature, between 'introduced, bad' and 'native, good'. Davis believes this dichotomy inaccurately reflects the real ecological roles of introduced species, which are much more varied and nuanced, and leads to an unwarranted visceral negative response to any introduced species, even one that may be harmless. Further, as do Chew and Laubichler (2003) and Larson (2005), he sees the martial metaphors (including 'invasion' itself) used in both scientific and especially popular literature to characterize introductions as reinforcing this dichotomous, unwarranted view of introduced species. Larson et al (2005) go so far as to declare war on the use of such metaphors to characterize biological invasions.

Davis's proposal to change this state of affairs closely follows that of US Department of Agriculture social scientist Paul Gobster (2005), who advises scientists to 'reframe' the issue of introduced species. For Gobster, who is particularly focused on removing introduced species as part of ecological restoration projects, the reframing entails focusing not on the possible evils of the introduced species but on the good of the original (native) community to be restored, and avoiding vilifying the individuals of the introduced species to be removed, who are not responsible for being there. Gobster's reframing also includes avoiding fear-inducing language, especially martial metaphors. For Davis, the reframing includes these features but is broader and requires explicit acknowledgement of the great variation among introductions and the complications and uncertainties that pervade both scientific understanding and management.

The main thread to the contrarian view is to enumerate and minimize the documented harmful impacts of introduced species. Thus, for example, Gurevitch and Padilla (2004) argued that only 6 per cent of taxa in the IUCN Red List database (2003) are threatened by introduced species, and only 2 per cent of extinctions are attributed to them. However, Clavero and García-Berthou (2005) showed this claim to be based on the fact that the Red List assigns causes to very few cases, and that a case-by-case examination gives a very different picture. For instance, of 680 extinctions in the list, a cause could strongly be inferred for 170, and for 91 of these, introduced species were a sole or contributing factor. Similar studies of fishes, mammals and birds yield similar results (references in Simberloff, 2005). Vermeij (1996) observed that introductions may often cause extinction on islands but almost never do so on continents. Sax et al (2002) suggest that, in terms of biodiversity, even this pattern is not a problem because, typically, introduced species augment diversity of plants and birds, as the number of extinctions of native plants and birds is generally less than the number of additions by introductions. Sax and Gaines (2008) show that most vertebrate extinctions associated with introduced species have been caused by introduced predators. Competition from an invader has rarely been a cause of vertebrate extinction.

The focus on extinction as the sole criterion for harm caused by introduced species is misleading. The final disappearance of a species from the face of the earth is an objective event that can be easily tallied (though often with a great delay, as humans rarely witness the death of the last individual), but many other environmental changes wrought by introduced species qualify in the public mind as harmful. Most clearly, the population sizes of many species have dwindled greatly in the face of introduced species, even though the species are not (yet) extinct. The Endangered Species Act of 1973, passed unanimously by the US Senate and by a vote of 345 to 4 in the House of Representatives, provides for the protection, including sequestering of key habitat, of any species whose population has declined to the point where its existence is threatened. Another readily observed impact of introduced species that falls short of global species extinction but is nonetheless widely perceived as harmful is the transformation of huge expanses of previous ecosystems with the concurrent decline of their native inhabitants. For instance, *c.*500,000ha of the state of Florida have been transformed from native prairies to dense forests dominated by Brazilian pepper and Australian paperbark trees. Florida citizens, acting through several state agencies, have deplored this change and funded extensive activities to redress it. Similarly, the recent invasion of the

Midwest by the emerald ash borer will not cause the extinction of ash species, just as the chestnut blight did not cause the extinction of American chestnut, but it will greatly change millions of acres of forest, just as the blight did. The great majority of citizens view the new invasion as a tragedy, just as the chestnut blight was (and is) widely deplored (Freinkel, 2007).

In sum, it is certainly true that much popular literature and many news reports reinforce an unwarranted stereotype of introduced species as all bad. However, invasion biologists have largely focused on specific impacts of particular invaders and have detected an increasing number and variety of them (Simberloff, 2005). This research shows that, far from being overblown, the impacts of introduced species as a whole are even greater than had been suspected, and that some take a very long time to manifest themselves. There is no evidence that the science of invasion biology has been pervaded by the deplored stereotyping or that scientific progress in understanding and managing invasions has been hindered by the use of military metaphors (Simberloff, 2006, 2009). Nor is there evidence that the use of such metaphors and stereotypes in the popular press has affected public attitudes towards immigrant or foreign humans.

The current atmosphere

The contrarians get into the news quite frequently, as do contrarians with respectable pedigrees in any field; global climate change is an obvious analogy. Thus, for example, the article by Sax and Gaines (2008) was quickly reported in a *New York Times* article entitled 'Friendly invaders' (Zimmer, 2008), while Davis's views are reported in a *Scientific American* feature entitled 'Alien invasions? An ecologist doubts the impact of introduced species' (Borrell, 2009). The review of Davis's book in *Nature*, by science writer Emma Marris (2009), is entitled 'The end of the invasion?'. However, whether such attention does more than add to the science writers' resumés is questionable. Both scientific and public opinion in the USA still reflects great concern with impacts of introduced species; even the scientific critics of invasion biology agree that concern is warranted. Most publications by the contrarians include some sort of disclaimer to the effect that some invasions have been extremely harmful. Davis (2009), for instance, says, 'a small proportion of non-native species are considered harmful or undesirable owing to their impacts. In some instances, the harmful impacts can be dire' (p101).

There is no evidence that US public agencies, which ultimately reflect public opinion, are lessening their vigilance and management activities. If anything, it is widely believed that border security should be enhanced (Government Accounting Office, 2006a, 2006b). Davis (2009, p150), following Slobodkin (2001), advocates an 'LTL' (Learn to Love 'Em) approach to managing many introduced species and correctly observes that this is already the public attitude towards some widespread introduced species in the USA that cause no obvious harm, pointing to a series of roadside plants in Minnesota, such as Old World Queen Anne's lace. However, roadsides are not habitats that normally house species of special concern; rather, they are by-products of road construction and the best we usually hope for them is that they will not lead to enormous erosion problems, will not be the source population for introduced invaders of habitats we care more about (as they have been in some circumstances), and will not be aesthetic horrors (most of us would rather see Queen Anne's lace than billboards). There is little evidence yet that Americans have broadly adopted the LTL approach towards species that have been perceived as problematic. For instance, there is no indication yet that federal and state agencies are worrying less about tamarisk because Stromberg and Chew (2002), Stromberg et al (2009) and Chew (2009) insist its impacts have been exaggerated.

Just as the modern field of invasion biology is young, the reaction is even more recent. Although adumbrated by Brown (1989) in the SCOPE synthesis volume, the reaction among scientists is largely a product of this first decade of the 21st century. Though the contrarians are a small minority, it is possible the movement will grow if their arguments prove cogent. I am sceptical that academic matters like the percentage of introduced species that cause problems or the frequency of species extinctions caused by invaders on islands as opposed to continents will greatly shift public or scientific opinion when many cases of great ecological, economic or aesthetic consequence stare us all in the face. It could be that pessimism about the ultimate outcome (e.g. Quammen, 1998) or the expense of combating invasions will lessen public enthusiasm for the fight, but my guess is that unless the majority of scientists become convinced that their perception of the invasion phenomenon has been faulty, public concern with introduced species will only grow.

Acknowledgements

I am grateful to Louise Robbins and Nathan Sanders for thoughtfully criticizing an early version of this manuscript.

References

Alderman, D. H. (2004) 'Channing Cope and the making of a miracle vine', *Geographical Review*, 94, 157–177

Alderman, D. H. and Alderman, D. G. (2001) 'Kudzu: A tale of two vines', *Southern Cultures*, 7, 49–64

Baskin, Y. (2002) *A Plague of Rats and Rubbervines. The Growing Threat of Species Invasions*, Island Press, Washington, DC

Bates, M. (1956) 'Man as an agent in the spread of organisms', in Thomas, W., Sauer, C. O., Bates, M. and Mumford, L. (eds), *Man's Role in Changing the Face of the Earth*, University of Chicago Press, Chicago, pp788–804

Bittner, M. (2004) *The Wild Parrots of Telegraph Hill*, Three Rivers Press, New York

Borell, B. (2009) 'Alien invasion? An ecologist doubts the impact of exotic species', *Scientific American* features, 14 August, www.scientificamerican.com/article.cfm?id=alien-invasion-ecologist-doubts-exotic, accessed 9 December 2009

Bright, C. (1998) *Life Out of Bounds: Bioinvasion in a Borderless World*, W. W. Norton, New York

Brown, J. H. (1989) 'Patterns, modes and extents of invasions by vertebrates', in Drake, J. A., Mooney, H. A., di Castri, F., Groves, R. H., Kruger, F. J., Rejmánek, M. and Williamson, M. (eds) *Biological Invasions: A Global Perspective*, pp85–109

Brown, J. H. and Sax, D. F. (2004) 'An essay on some topics concerning invasive species', *Austral Ecology*, 29, 530–536

Buhs, J. B. (2004) *The Fire Ant Wars: Nature, Science, and Public Policy in 20th-Century America*, University of Chicago Press, Chicago

Burdick, A. (2005) *Out of Eden: An Odyssey of Ecological Invasion*, Farrar, Straus and Giroux, New York

California Department of Fish and Game (2009) www.dfg.ca.gov/lakedavis/, accessed 25 December 2009

Carson, R. (1962) *Silent Spring*, Houghton Mifflin, Boston

Chew, M. K. (2006) 'Ending with Elton: Preludes to invasion biology', unpublished PhD dissertation, Arizona State University, Tempe, AZ

Chew, M. K. (2009) 'The monstering of tamarisk: How scientists made a plant into a problem', *Journal of the History of Biology*, 42, 231–266

Chew, M. K. and Laubichler, M. D. (2003) 'Natural enemies: Metaphor or misconception', *Science*, 301, 52–53

Clair, D. J. and Kramer, V. L. (1989) 'Japanese beetle', in D. L. Dahlsten and R. Garcia (eds) *Eradication of Exotic Pests*, Yale University Press, New Haven, pp89–10

Clavero, M. and García-Berthou, E. (2005) 'Invasive species are a leading cause of animal extinctions', *Trends in Ecology and Evolution*, 20, 110

Coates, P. (2006) *Strangers on the Land: American Perceptions of Immigrant and Invasive Species*, University of California Press, Berkeley

Dasmann, R. F. (1971) *No Further Retreat. The Fight to Save Florida*, Macmillan, New York

Davis, M. A. (2009) *Invasion Biology*, Oxford University Press, Oxford

Devine, R. S. (1998) *Alien Invasion. America's Battle with Non-Native Animals and Plants*, National Geographic Society, Washington, DC

Ellis, K. (1999) www.kodachrome.org/pines/, accessed 25 December 2009

Elmendorf, S., Byrnes, J., Wright, A., Olyarnik, S., Fischer, R. and Chamberlin, L. (2005) *Fear and Fishing in Lake Davis* (DVD), Flag in the Ground Productions, Davis, California

Elton, C. S. (1958) *The Ecology of Invasions by Animals and Plants*, Methuen, London (reprinted 2000, University of Chicago Press, Chicago)

Freinkel, S. (2007) *American Chestnut: The Life, Death, and Rebirth of a Perfect Tree*, University of California Press, Berkeley

Gobster, P. H. (2005) 'Invasive species as ecological threat: Is restoration an alternative to fear-based resource management?', *Ecological Restoration*, 23, 261–270

Gould, S. J. (1998) 'An evolutionary perspective on strengths, fallacies, and confusions in the concept of native plants', *Arnoldia*, 58, 11–19

Government Accounting Office (US) (2006a) *Homeland Security: Management and Coordination Problems Increase the Vulnerability of US Agriculture to Foreign Pests and Disease*, GAO-06–644

Government Accounting Office (US) (2006b) *Invasive Forest Pests: Recent Infestations and Continued Vulnerabilities at Ports of Entry Place US Forests at Risk*, GAO-06–871T

Gurevitch, J. and Padilla, D. K. (2004) 'Are invasive species a major cause of extinction?', *Trends in Ecology and Evolution*, 19, 470–474

Helford, R. M. (2000) 'Constructing nature as construction science: Expertise, activist science, and public conflict in the Chicago Wilderness', in Gobster, P. H. and Hull, R. B. (eds) *Restoring Nature: Perspectives from the Social Sciences and Humanities*, Island Press, Washington, DC, pp119–142

Houston, D. B., Schreiner, E. G. and Moorhead, B. B. (1994) *Mountain Goats in Olympic National Park: Biology and Management of an Introduced Species*, US Department of the Interior, National Park Service, Denver, Colorado

Hudson, J. L. (1998) 'J. L. Hudson, Seedman, the ethnobotanical catalog of seeds', www.jhudsonseeds.net/NativeVsExotics.htm, accessed 17 July 1998

Kraus, F. (2009) *Alien Reptiles and Amphibians*, Springer, Dordrecht, The Netherlands

Larson, B. M. H. (2005) 'The war of the roses: Demilitarizing invasion biology', *Frontiers in Ecology and the Environment*, 3, 495–500

Larson, B. M. H., Nerlich, B. and Wallis, P. (2005) 'Metaphors and biorisks: The war on infectious diseases and invasive species', *Science Communication*, 26, 243–268

Marris, E. (2009) 'The end of the invasion?', *Nature*, 459, 327–328

Oosting, H. J. (1948) *The Study of Plant Communities: An Introduction to Plant Ecology*, W. H. Freeman, San Francisco

Oosting, H. J. (1956) *The Study of Plant Communities: An Introduction to Plant Ecology*, 2nd edn, W. H. Freeman, San Francisco

Pauly, P. J. (1996) 'The beauty and menace of the Japanese cherry trees', *Isis*, 87, 51–73

Pollan, M. (1994) 'Against nativism', *New York Times Magazine* (section 6), 15 May, 52–55

Pruett-Jones, S., Newman, J. R., Newman, C. M., Avery, M. L. and Lindsay, J. R. (2007) 'Population viability analysis of monk parakeets in the United States and examination of alternative management strategies', *Human-Wildlife Conflict*, 1, 35–44

Quammen, D. (1998) 'Planet of weeds: Tallying the loss of earth's animals and plants', *Harper's*, 297, No 1781, 57–69

Rejmánek, M. and Richardson, D. M. (2010) 'Eucalypts', in Simberloff, D. and Rejmánek, M. (eds) *Encyclopedia of Biological Invasions*, University of California Press, Berkeley, pp203–209

Rosenzweig, M. L. (2001) 'The four questions: What does introduction of exotic species do to diversity?', *Evolutionary Ecology Research*, 3, 361–367

Sagoff, M. (1999) 'What's wrong with exotic species?', *Report from the Institute for Philosophy and Public Policy*, 19, (4), 16–23

Sax, D. F., and Gaines, S. D. (2008) 'Species invasions and extinction: The future of native biodiversity on islands', *Proceedings of the National Academy of Sciences (USA)*, 105, 11490–11497

Sax, D. F., Gaines, S. D. and Brown, J. H. (2002) 'Species invasions exceed extinctions on islands worldwide: A comparative study of plants and birds', *American Naturalist*, 160, 766–783

Schmitz, D. C., Simberloff, D., Hofstetter, R. H., Haller, W. and Sutton, D. (1997) 'The ecological impact of nonindigenous plants', in Simberloff, D., Schmitz, D. C. and Brown, T. C. (eds) *Strangers in Paradise. Impact and Management of Nonindigenous Species in Florida*, Island Press, Washington, DC, pp39–61

Simberloff, D. (1996) 'Letter to the editor', *Isis*, 87, 676–677

Simberloff, D. (2003) 'Confronting introduced species: a form of xenophobia?', *Biological Invasions*, 5, 179–192

Simberloff, D. (2005) 'Non-native species do threaten the natural environment!', *Journal of Agricultural and Environmental Ethics*, 18, 595–607

Simberloff, D. (2006) 'Invasional meltdown six years later – Important phenomenon, unfortunate metaphor, or both?', *Ecology Letters*, 9, 912–919

Simberloff, D. (2009) 'Invasions of plant communities – More of the same, or something very different?', *American Midland Naturalist*, 163, 219–232

Simberloff, D. (2010) 'Charles Elton – Neither founder nor siren, but prophet', in Richardson, D. M. (ed) *Fifty Years of Invasion Ecology*, Wiley, New York, pp11–24

Slobodkin, L. B. (2001) 'The good, the bad and the reified', *Evolutionary Ecology Research*, 3, 1–13

Sorensen, P. W. and Bergstedt, R. A. (2010) 'Sea lamprey', in Simberloff, D. and Rejmánek, M. (eds) *Encyclopedia of Biological Invasions*, University of California Press, Berkeley, 619–623

Stromberg, J. C. and Chew, M. K. (2002) 'Foreign visitors in riparian corridors of the American Southwest: Is xenophytophobia justified?', in Tellman, B. (ed) *Invasive Exotic Species in the Sonoran Region*, University of Arizona Press, Tucson, pp195–219

Stromberg, J. S., Blake, T. J., Sperry, J. S., Tschaplinski, T. J. and Wang, S. S. (2009) 'Changing perceptions of change: The role of scientists in tamarisk and river management', *Restoration Ecology*, 17(2), 177–186

Subramaniam, B. (2001) 'The aliens have landed! Reflections on the rhetoric of biological invasions', *Meridians: Feminism, Race, Transnationalism*, 2, no 1, 26–40

Theodoropoulos, D. I. (2003) *Invasion Biology: Critique of a Pseudoscience*, Avvar Books, Blythe, CA

Tsing, A. L. (1995) 'Empowering nature, or: Some gleanings in bee culture', in Yanagisako, S. and Delaney, C. (eds) *Naturalizing Power. Essays in Feminist Cultural Analysis*, Routledge, New York, pp113–143

Tummons, P. (2008) 'Waiwai biocontrol controversy', *Environment Hawaii*, 19 (1)

Vermeij, G. J. (1996) 'An agenda for invasion biology', *Biological Conservation*, 78, 3–9

Welch, P. G. (1973) 'Save the parakeets' (letter), *New York Times*, 26 April, p42

Williams, T. (2002) 'America's largest weed', *Audubon Magazine*, 104 (1), 24–31

Wolschke-Bulmahn, J. and Gröning, G. (1992) 'The ideology of the nature garden. Nationalistic trends in garden design in Germany during the early 20th century', *Journal of Garden History*, 12, 73–80

Zimmer, C. (2008) 'Friendly invaders', *New York Times*, 8 September, pD1

9

Anekeitaxonomy: Botany, Place and Belonging

Matthew K. Chew

Introduction

What does the local advent of a new taxon signify? Many authors have postulated that major environmental changes can precipitate social stress and even civil collapse (e.g. Cullen et al, 2000; de Menocal, 2001; Haug et al, 2003; Zhang et al, 2005; Drysdale et al, 2006). Others warn against embracing environmental determinism (Hunt, 2006). We need not look far to realize that unpredictable, uncooperative environments adversely affect individuals and societies. Sociologists describe 'defensive structuring' characterized by increasingly strong group identity symbols (e.g. Siegel, 1970); social psychology's 'life course' paradigm describes individual and group reactions to changes such as 'loss of control over desired outcomes' during 'hard times' (Elder, 1994). What about minor or incremental changes? Psychologists and sociologists have made little apparent attempt to examine reactions to the local advents of unfamiliar taxa (but conservation-related agencies and organizations occasionally query public opinion on the matter, to evaluate and enhance their influence; see e.g. Fitzgerald et al, 2007).

Cultural anthropologists have studied human reactions to biota, especially taxa with normative, religious and taboo significance; most cite Douglas (1966). Milton (2000) used Douglas as a point of departure for discussing reactions to the human-facilitated advent of (American) ruddy ducks (*Oxyura jamaicensis*) in Europe. She identified 'three culturally defined boundaries invoked by conservationists': (i) between species; (ii) between 'natives and aliens' (based on both physical and cultural geographies); and (iii) between 'human and non-human processes', i.e. culture and nature. Each is a taxonomic issue in the broadest sense of the word. Bounding species is familiar in concept, if arcane in practice. Bounding geographies is as commonplace as giving directions or closing a door. 'Western' culture traditionally defines nature as non-cultural, an intuitively appealing boundary simultaneously (perhaps ironically) anthropogenic and anthropocentric. That nature includes no human events is the long-standing criterion by which natural historians bound the scope of their investigations (Farber, 2000). It is also the core distinction between the native biota and any other. Native plants and animals have no human history.

This chapter describes the cultural history of distinguishing native from non-native biota. For reasons explained further on I call the production of such systems *anekeitaxonomy* (from ανήκει, belonging) (Chew, 2006). Examining the products of anekeitaxonomy reveals that concepts of biotic nativeness are modelled on archaic cultural conceptions of belonging. Furthermore and of greater interest to ecologists and conservation biologists, the idea that nativeness is anything but historically incidental is theoretically untenable. Natives and non-natives alike must survive and reproduce under locally prevailing conditions, and regardless of how those conditions ensue. In evolutionary terms, individuals exhibit fitness and contribute to lines of descent or they do not. If there is an objective criterion of biotic belonging tying a taxon to a place, it is fitness, evidenced by persistence (Chew and Hamilton, 2011). Other criteria are subjective and endlessly debatable.

Natives and natural history

Unlike evolutionary lines, people and cultures have self-conscious goals. Some of our goals are facilitated by evolution; others accommodate it, but many conflict with it. Paradoxically, the culturally defined goal of biodiversity conservation conflicts with the fact of evolution. Anekeitaxonomy evidences aspects of that conflict in its disparagement of human-influenced evolutionary dynamics. Again paradoxically, those judgements preceded the concepts of natural selection (by a generation) and of biodiversity (by over a century) taking form under an intellectual regime that is difficult to imagine today. The primary objective of the earliest anekeitaxonomists was defending the distinction in practice between history and natural history, even as the distinction in fact (if there had ever been one) was crumbling.

As the 19th century opened, 'native' plants and animals were, like 'native' metals and minerals, understood to be spontaneous products of nature in a given locality. Native plants were those not sown and tended. Native animals were those not husbanded. In Europe, compiling floras and faunas, both largely amateur endeavours, was approaching a critical stage of apparent completeness. Skilled practitioners could quickly recognize specimens violating expectations (i.e. predictions from experience) by occurring where they had not previously been observed (i.e. being surprising). This was and remains a more straightforward determination for botanists than zoologists, since most plants can neither hide nor flee (cf. Elton, 1927). However, at this point both camps mostly laboured under the assumptions that species represented fully discrete, distinguishable kinds (Atran, 1990) and that complete inventories (i.e. floras and faunas), could be compiled for given localities. How various kinds had arrived anywhere in particular was attributed to Providence by the uncurious. To others, it was a puzzle remaining to be solved; but frameworks for applicable hypothesizing were beginning to emerge from the work of geologists like Englishman Charles Lyell and pioneering continental phytogeographers including Alexander von Humboldt and Augustin Pyramus De Candolle (Chew, 2006).

John Henslow: The dagger and the asterisk

An early example of differentiating native plants from alien ones on the basis of belonging rather than spontaneity versus cultivation appeared in an 1835 article by John Henslow, Regius Professor of Botany at Cambridge. Henslow is most often remembered as the mentor of Charles Darwin, who recommended him as a companion to Captain Fitzroy of the HMS *Beagle*. By 1835, it had become standard practice to indicate anomalies in local floras by appending an asterisk (*) to any species record whose presence surprised the compiler. The asterisk thus appealed to a general, sometimes tacit, footnote indicating a history of human agency: plants or seeds had been moved or cultivated, or conditions had been changed by surface grading, or by dumping refuse or manure. Henslow was satisfied with applying the asterisk to the more certain cases, but not to those where questions remained. To doubtful ones Henslow proposed another footnote signified by a dagger (†). Still others he felt ought to be left out of floras completely, or signified by a degree mark (°) (Henslow, 1835). He described his categories in the wordy manner of the era, but did not really label or name them (Table 9.1).

Henslow's concern was not whether non-native plants should be tolerated in the landscape, but whether natural historians should concern themselves with accounting for them. The products of human agency masked natural

Table 9.1 *Categories of un- or uncertain nativeness proposed by John Henslow (1835)*

° (degree) 'To be rejected'	'It seems hardly correct to include in our lists, even as naturalized species, such as are only occasionally to be met with on heaps of manure, or among rubbish which has been the outcast of a garden. These plants, of acknowledged exotic origin … might be placed in an appendix, distinct from the species which are allowed to be indigenous or naturalized.'
* (asterisk) 'Naturalised'	'As it would be certainly improper to exclude those plants from our flora which have become strictly naturalised in our own country, and now form part and parcel of the wild flowers of our fields and hedges, it is perhaps the most convenient mode to register them continuously with the indigenous species, and merely to denote them by the usual mark.'
† (dagger) 'Possibly not indigenous'	'Some [may] point out other reasons for accepting or rejecting certain species; and … furnish us with … the circumstances under which … "suspected" species [are] found in their respective neighbourhoods.'

history and confounded natural historians, but Henslow did not consider the plants themselves to be impurities requiring eradication. An ordained clergyman (like all Cambridge dons of the time) and a creationist by default, he might have considered them theologically significant, but made no such statement in this context.

Watson's *Cybele Britannica*

Hewett Cottrell (H. C.) Watson was the eighth child (but eldest son and chief heir) of a successful English lawyer. Watson apprenticed at law, but did not take to it as a profession. He was an aficionado of phrenology, and studied medicine at Edinburgh (just behind Charles Darwin), hoping to find a scientific basis for taking the measure of a man by the shape of his head. Watson devoted great energy to phrenology, and was disappointed by its general failure (Egerton, 2003). But his medical studies also confirmed his taste for botany, and particularly *phytogeography*, a term Watson coined. By 1835 he had already published a guide to finding rare plants and established himself as a necessary part of the British botanical scene by facilitating specimen exchanges between far-flung enthusiasts (Allen, 1986; Egerton, 2003). He had also set out to systematically map and geographically classify British plants in a form he eventually called a *Cybele* (to distinguish it from a flora).

Watson's first significant reaction to Henslow's scheme was to add another category marked by the double dagger (‡) signifying 'such as may possibly have been introduced, being weeds of cultivated ground or inhabited places' (Watson, 1835). He also revealed that his own conception of biotic nativeness was strongly influenced by civil categories of individual nativeness, and was perhaps coloured by nationalism:

> *Species originally introduced by human agency now exist in a wild state; some … continued by unintentional sowings along with … cultivated plants; while several keep their acquired hold of the soil unaided, and often despite our efforts to dispossess them. Both these classes certainly now constitute a part of the British flora, with just as much claim as the descendants of Saxons or Normans have to be considered a part of the British nation. But there is a third class … of plants which have yet acquired a very uncertain right to be incorporated with the proper spontaneous flora of the island … species springing up occasionally from seeds or roots thrown out of gardens, and maintaining themselves a few years; and … those designedly planted for ornamental or economical purposes. Such are no more entitled to be called Britons, than are the Frenchmen or Germans who occasionally make their homes in England.*

In Watson's view plants could seemingly become more British, and perhaps fully so over time. Belonging could be earned, and possibly granted. He later made it clear that he regarded natural and cultural history to be separate realms. The self-published and routinely wordy Watson took no pains to hide or to much moderate his feelings about fellow botanists, but he never called for the eradication or suppression of non-native plant populations.

By 1840, Watson abandoned hope of a scientific phrenology and devoted himself fully to botany. He also abandoned degrees, daggers and asterisks. The first volume of his *Cybele Britannica* appeared in 1847, including a new codification of phytogeographical belonging. Like Henslow, Watson was specifically addressing a problem of disciplinary boundaries: which botanical phenomena qualified as natural history and which as cultural history. He evidently set out to succinctly codify principled distinctions. Like Henslow, he found both uncertainty and the effects of time precluded a simple dichotomy. Unlike Henslow, Watson had knowledge of legal practice and terminology to draw upon. The resulting system (Table 9.2) was 'intended to show the civil claims and local situation of the species in accordance with a scale of terms' (Watson, 1847), three of which were adapted from English Common Law regarding the status of persons (cf. Blackstone, 1922). Watson was rarely one to spare a reader his reflections, but in this case, he shared no further thoughts about the strengths, weaknesses or intended extent of the analogies that he was drawing.

Anekeitaxonomy accounted for only a fraction of Watson's goal in compiling the *Cybele*, 'to show those circumstances which are requisite for the purpose of enabling botanical geographers to make comparisons between the floral statistics of different portions of the earth' (Watson, 1847). One term from Table 9.2 was the first of two descriptive words relegated to the tenth and last 'line' of a *Cybele* species account. The second word also indicated a kind of belonging; another series of 14 terms described 'the usual situations of

Table 9.2 *H. C. Watson's (1847) anekeitaxonomic system for plants in Britain*

'*Native*'	'Apparently an aboriginal British species; there being little or no reason for supposing it to have been introduced by human agency.'
'*Denizen*'	'At present maintaining its habitats, as if a native, without the aid of man, yet liable to some suspicion of having been originally introduced.'
'Colonist'	'A weed of cultivated land, or about houses, and seldom found except in places where the ground has been adapted for its production by the operations of man; with some tendency, however, to appear also on the shores, landslips, and etc.'
'*Alien*'	'Now more or less established, but either presumed or certainly known to have been originally introduced from other countries.'
'Incognita'	'Reported as British, but requiring confirmation as such. Some ... through mistakes of [identifying] the species ... others may have been really seen [as] temporary stragglers from gardens ... others cannot now be found in the localities published for them ... some of these may yet be found again. A few may have existed for a time, and become extinct.'

Note: Italicized terms were adapted from English Common Law (see e.g. Blackstone, 1922).

the species', e.g. whether they occurred in damp meadows, pastures, moors, swamps, hedgerows, lakes, walls and rocks, fields, seashores, etc. (Watson, 1847). This series was no less artificially bounded than the first, but was more purely descriptive, and never attributed with such normative status. Watson later evidenced some displeasure at the effort and space required to fully elaborate this aspect of his system. Discursive as usual, he also revealed something of his feelings regarding lesser practitioners and his awareness of biogeographical dynamics along the way:

> *This subject of introduced plants has encroached too largely on the pages of the present work ... the Author has hitherto unsuccessfully tried to induce some other botanist to write a 'History of the British flora'; tracing out each species back to the earliest records of its occurrence in Britain ... the vast changes gradually wrought in the vegetation of Britain, by the conversion of forests into wastes, and of wastes into cultivated lands, would also find appropriate place in a work of the kind suggested. Such a history ... would require to be written under a higher impulse than that which prompts to the compiling of 'Keys' and 'Manuals' for the market, or under a worthier ambition than that of 'getting up some sort of paper, likely to make a talk among the geologists', at the next meeting of the British Association.* (Watson, 1868)

In light of these sentiments it is surprising that the philosophical, methodical Watson followed Henslow in forcing a one-dimensional categorization to account for a two-dimensional finding (including both the degree to which a species was locally established and the uncertainty of his judgement in the matter). Uncertainty would characterize anekeitaxonomies for several more decades. Meanwhile, the next substantial effort, undertaken in part as a response to Watson's, would emphasize an even more anthropocentric matter: human intention.

Alphonse De Candolle responds to Watson

In 1855, Augustin De Candolle's son Alphonse published the 1365 page *Géographie Botanique,* including an anekeitaxonomy offered explicitly as an alternative to Watson's. Candolle junior followed so closely in his father's footsteps as to inherit his position directing the Geneva Botanical Garden. Both were devoted to reconciling biblical creation accounts with empiricism, a hopeless effort that would eventually be relegated to the fringes of science by the advent of Darwin's theory of natural selection.

Candolle's system (Table 9.3) was two-tiered, making a primary cut between cultivated and spontaneous species, hewing more closely and perhaps objectively than Watson had to the history versus natural history division. From there, things took turns both familiar and unprecedented in the British work.

Table 9.3 *Alphonse De Candolle's anekeitaxonomic system for plants in Europe*

Cultivated	Voluntarily	[left undefined as if self-evident]
	Involuntarily	'Species which absolutely exist only in the fields, gardens, etc., without being in open country in a spontaneous state.'
Spontaneous	'Adventive'	'Of foreign origin, but badly established, being able to disappear from one year to another.'
	'Naturalized'	'Well established in the country, but there is positive evidence of a foreign origin.'
	'Probably foreign'	'Well established ... but according to strong indications, there are more reasons to believe them of origin foreign than primitive in the country ... the odds favouring a foreign origin are better than even.'
	'Perhaps foreign'	'Some indications of a foreign origin, though the species are long and well-established in the country. For one reason or another, one can raise some doubts about their indigenousness.'
	'Indigenous'	'Aboriginals, natives ... spontaneous species whose origins are not doubtful, which appear to have existed in the country before the influence of man, probably for geological rather than historical time.'

Source: Candolle, 1855; translation from Chew, 2006.

Candolle's emphasis on cultivation foreshadows his best-remembered book, the 1883 *Origine des Plantes Cultivées* which was anonymously translated and republished in English as *The Origin of Cultivated Plants* (1885). Otherwise, his system recalls Watson's in many respects. Like Watson and Henslow, Candolle tried to combine uncertainties of origin and degrees of establishment into a single series of classes where a cross-classification seems more intuitive in retrospect. As a matter of pure speculation, this could represent a vestige of the philosophy that produced the serial 'great chain of being' but there is little besides resemblance to support such a hypothesis. Having made his mark in 1855, Candolle left the discussion to others and concentrated on illuminating plant domestication.

Watson responds to Candolle

Charles Darwin's 1859 *On the Origin of Species by Means of Natural Selection* is understood to be a pivot on which the history of biology turns, but its most noticeable effect on anekeitaxonomy was a lack of any noticeable effect. H. C. Watson published the fourth and final volume of his *Cybele Britannica* after the *Origin* debuted, but he mentioned it only in an appended comment. Watson had been a correspondent of Darwin's for many years, and had provided both cases and critiques for Darwin's work (Egerton, 2003). However, anekeitaxonomy rested on natural history, not on evolutionary biology; otherwise, biogeographical belonging would have been overthrown by the geological, geographical and taxonomic dynamism inherent in ongoing natural selection. Watson was sympathetic to evolution and impressed by Darwin's synthesis, but doubtful that natural selection fully solved the puzzle (Egerton, 2003). If he had been more thoroughly convinced, perhaps Watson would have explicitly relegated part one of 'line ten' to a heuristic shorthand. What he actually added that year was a statement that nativeness as he defined it was 'simply negative', attributed to species when there were 'no grounds for supposing that they were first brought into Britain by human agency' (Watson, 1859). In modern terms, Watson knew that a finding of nativeness confidently denoted only an absence of evidence. He did not say so, but perhaps he suspected 'the vast changes gradually wrought in the vegetation of Britain' included indirect effects more subtle and far-reaching than those he had previously attributed to 'human agency.'

Having completed his *Cybele,* the never-satisfied Watson inaugurated the *Compendium of the Cybele Britannica* in 1868. The first volume revisited his 1847 anekeitaxonomy, making minor alterations but rejecting Candolle's problematic emphasis on intention. The only major change was replacing 'incognita' with the less uncertain 'casual':

> *chance stragglers from cultivation ... occasionally imported and sown with agricultural seeds ... introduced among wool, oil-seeds, or other merchandize* [sic]*; foreign plants found on ballast heaps deposited from ships; and generally such alien species as are most uncertain in place or persistence.* (Watson, 1868)

He restated his purpose, explaining that the system was:

> *a series of terms, drawn from our own legal and social classifications ... used to express various grades of uncertainty of belief with respect to those plants whose aboriginal nativity is more or less unsettled. [These] express a descending series from the truly wild and pre-historically established species down to ... the products of seeds accidentally imported.* (Watson, 1868)

Invoking the 'truly wild' was hardly an advance, but elsewhere in the volume the phrase underpinned another of Watson's concerns:

> *Inexperienced observers more readily believe in the true nativity of plants; while those of greater experience will frequently find grounds for doubt or distrust ... the desire of appearing as discoverers too often leads vainglorious collectors to make out the best case they can in support of the 'native claims' of a species, and the 'truly wild' character of their localities.* (Watson, 1868)

British botany was still largely an amateur, competitive affair, and (in Watson's view) suffering for it.

Neither Watson nor Candolle contributed to anekeitaxonomy after 1868. Watson died in 1881 while Candolle was still tracing developments in cultivation. Joseph Dalton Hooker and others from the growing ranks of professional botanists had extended phytogeographical studies to the farthest corners of the British Empire, in the process perhaps eclipsing homeland studies in the public imagination. That trend was not limited to Britain, nor was it confined to botany. Ecology had found a name and fielded a first generation of practitioners. Phytogeographers and zoogeographers had expanded their horizons to consider intercontinental and global phenomena, but remained sub-specialists of botany and zoology, respectively. Paul F. A. Acheson of the University of Berlin translated H. C. Watson's system into German for his 1883 *Human Impacts on Vegetation* [*Einfluss des Menschen auf die Vegetation*] (Muhlenbach, 1979). Both Watson's and Candolle's terms continued to appear in local floras, sometimes credited but often (and increasingly) as if simply 'in the air'. Meanwhile the increasing dependability, and both the global reach and scale of trade and travel, were making human agency an ever more significant influence on biotic distributions.

The Thellungian paradigm

The next and last major anekeitaxonomic innovation emerged from a suite of 'adventive' floras compiled by a new generation of German and Swiss botanists. The revision was apparently initiated by Martin Rikli, but almost immediately revised and expanded by Albert Thellung, an early contemporary of Braun-Blanquet's in the 'Zurich-Montpellier school' of botany (Muhlenbach, 1979). From 1905 to 1922, Thellung published at least eight works on adventitious plants, from short commentaries to a nearly 700-page inventory of the introduced flora of Montpellier, France. Montpellier was home to Port Juvenal on the Lez River, the site of a major processing centre for imported wool beginning in the late 17th century. Fleeces from North Africa, the Middle East and elsewhere were washed there and seeds thus liberated germinated into 'wool gardens' of foreign plants. Candolle senior and other 19th-century botanists had collected and commented on specimens from Port Juvenal; some thereby described before their geographic origins were known (Muhlenbach, 1979).

Thellung's (née Rikli's) system included both novel and derivative features (Table 9.4). New were polysyllabic category labels cobbled together from Greek roots and prefixes (e.g. *-phyte* = plant). This was apparently meant to distinguish them from the earlier reliance on colloquial metaphors, thus making the practice more scientific (and inspired the term *anekeitaxonomy*). But a metaphor in any language remains a metaphor. Thellung's ideas were no less culturally grounded than previous ones. However, the number of ultimate categories grew from four (Henslow) to five (Watson) to seven (Candolle) to nine (Thellung), mostly by increasing reference to human intentions.

Table 9.4 *Anekeitaxonomy proposed by Albert Thellung, condensed from Muhlenbach (1979)*

I. Anthropochores. Plants brought in by man.	A. By intentional activity of man: foreign cultivars and their derivatives.	1	Ergasiophytes. Cultivated plants that did not originate locally, requiring ongoing intervention to thrive.
		2	Ergasiolipophytes. Descendants of ergasiophytes that persist locally after cultivation has ceased. Typically, perennial and woody ornamentals surviving around abandoned farms, habitations and other cultural sites.
		3	Ergasiophygophytes. Descendants of ergasiophytes that have 'escaped without the knowledge of man from the place of cultivation and have settled on artificial or natural habitats.' Most die out, but 'in some cases they are able to adjust to prevalent conditions … in rare cases they may grow rampantly and drive out the native flora.'
	B. By involuntary intervention of man: foreign weeds.	4	Archaeophytes. These are field weeds inadvertently introduced in prehistoric times with crops.
		5	Neophytes. They occur relatively frequently and permanently in natural habitats and are often associated with the native vegetation. They do not depend on the continued activity of man to survive.
		6	Epecophytes. Plants that appeared in modern times are more or less numerous, but are restricted to artificial habitats. They are dependent upon man for their existence in that he maintains those habitats or is compelled to create them again.
		7	Ephemerophytes. These are plants that appear only sporadically as transients, almost exclusively on artificial habitats.
II. Apophytes. Plants originally wild on natural habitats, but later growing on artificial ones.	A. By intentional activity of man.	8	Ekiophyes. They are native cultivated plants, grown as ornamental or useful plants.
	B. Spontaneously establishing in artificial habitats	9	Spontaneous apophytes.

Like its predecessors, Thellung's system suffers from dimensional compression. A 1969 analysis by Göttingen phytogeographer F. G. Schröder concluded that intertwining several independent classification criteria was undesirable. He attempted to remedy this with a table purporting to sort them into three distinct yet relatable classifications (Table 9.5). Whether Schröder succeeded is arguable, since he did not proceed beyond reviewing to evaluation or revision. Anekeitaxonomy still categorized plants by the ways their natural histories contacted human history. As Thellung revealed when describing ergasiophygophytes (e.g. Table 5, I.A.3) being categorized as such was no predictor of how given species might respond in a given locations or environments; they in turn, remained subjective. Anekeitaxonomy was still anthropocentrically contingent. It was historical, not predictive.

Anekeitaxonomy, ecology and conservation

Evidence of large-scale anthropogenic environmental change helped unmoor anekeitaxonomy from its position as a historical sidebar to local floras. J. D. Hooker was an early adopter of anti-introduction sentiment. His 1867 article 'Insular Floras' for *Gardeners' Chronicle* included a passage deploring the introduction of European plants and animals to the south Atlantic island of St Helena. Hooker's concern was for the loss of natural history data attending the biotic overhaul of an island that might have harboured Galapagos-quality evidence of evolutionary processes.

During the succeeding century various scientists invoked Watsonian terminology while deploring human redistribution of plants and animals, most often without any apparent realization that terms like native, alien and casual were citable. The Thellungian system was hardly promoted in English-language journals, difficult to remember, less intuitively useful, and thus largely limited to appearing in continental European local floras.

Almost as soon as they were available, ecological concepts like communities, climaxes and ecosystems, and natural theology's 'balance of nature' repackaged as dynamic equilibrium inspired a more normative anekeitaxonomy. By 1917 and 1925 (respectively), Aldo Leopold and Charles Elton had published anti-alien sentiments (Chew, 2006). Stanford University bryologist Douglas H. Campbell's 1926 foray into phytogeography dismissed Midwestern American vegetation:

> *Except for the native trees, the predominant vegetation at the present time is largely exotic. None of the staple food crops are indigenous, and the same is true for most of the common fruits, although some of the latter, like grapes and berries of various kinds, are of native origin. Even the weeds are mostly foreigners and have driven out the native woodland plants which have retreated before these hardy invaders.* (Campbell, 1926)

Table 9.5 *Classification of Anthropochores by their position in flora and vegetation, after Schröder (1969)*

Vegetation grouping	Basic classification		Fixed in Place			No firm position in the vegetation	
			Original vegetation	Potentially natural, but not original vegetation	Current but not in potential natural vegetation	Found growing wild	Only in cultivation
			I. Idiochorophytes natives	II. Agriophytes immigrants	III. Epökophytes culture dependent	IV. Ephemerophytes unstable	V. Ergasiophytes cultivated
	Summary category comparison		Idiochores	Anthropochores			
			Residents			Aliens	
			Natives	Naturalized			
				Adventives			Cultivated
			Growing wild				
Grouped by time of immigration	Before human intervention		1. Idiochorophytes				
	Under human intervention	in prehistoric time		2. Archaeophytes – old adventives			
		in historic time		3. Neophytes – new adventives			
Grouped by means of immigration	Without human influence		A. Idiochorophytes				
	Under human intervention	Independent migrants		B. Akolutophytes – intruders			
		Inadvertently introduced		C. Xenophytes – introduced			
		Introduced as crop, now growing wild		D. Ergasiophygophytes – feral			

Campbell's effort was disparaged by reviewers and forgotten (Chew, 2006). Charles Elton's now legendary *The Ecology of Invasions by Animals and Plants* (1958) failed to generate a contemporaneous critique of alien unbelonging, and no such concerns impinged on the enthusiasm of invasion biologists founding a discipline on Elton's legacy in the 1980s and 1990s.

As the 21st century began, an international group of prominent invasion biologists led by David M. Richardson of Stellenbosch University cited 'confusion and misuse of existing terminology' when revisiting the native–alien distinction; despite their stated belief that 'much of the debate on terminology is essentially semantic' (Richardson et al, 2000). Based on a literature review, they recommended a new standardization of terms (Table 9.6).

It is evident that these authors are proposing an exemplary anekeitaxonomy. They classify plants according to where they belong, distinguishing them by the pre-Darwinian division between history and natural history. Like Henslow, they take nativeness for granted. Like Watson, they seem to have intended a series (in this case denoting increasingly egregious unbelonging); but they interrupted its progression by including merely 'not wanted' weeds. As with Thellung, they

Table 9.6 *Recommended terminology in plant invasion ecology after Richardson et al (2000), excluding footnotes*

Alien plants	Plant taxa in a given area whose presence there is due to intentional or accidental introduction as a result of human activity (synonyms: exotic plants, non-native plants; non-indigenous plants)
Casual alien plants	*Alien plants* that may flourish and even reproduce occasionally in an area, but which do not form self-replacing populations, and which rely on repeated introductions for their persistence (includes taxa labelled in the literature as 'waifs', 'transients', 'occasional escapes' and 'persisting after cultivation', and corresponds to De Candolle's (1855, p643) usage of the term 'adventive'.
Naturalized plants	*Alien plants* that reproduce consistently (cf. *casual alien plants*) and sustain populations over many life cycles without direct intervention by humans (or in spite of human intervention); they often recruit offspring freely, usually close to adult plants, and do not necessarily invade natural, semi-natural or human-made ecosystems.
Invasive plants	*Naturalized plants* that produce reproductive offspring, often in very large numbers, at considerable distances from parent plants (approximate scales: >100m; <50 years for taxa spreading by seeds and other propagules; >6m/3 years for taxa spreading by roots, rhizomes, stolons or creeping stems), and thus have potential to spread over a considerable area.
Weeds	Plants (not necessarily *alien*) that grow in sites where they are not wanted and which usually have detectable economic or environmental effects (synonyms: plant pests, harmful species, problem plants). 'Environmental weeds' are *alien plant* taxa that invade natural vegetation, usually adversely affecting native biodiversity and/or ecosystem functioning.
Transformers	A subset of *invasive plants* which change the character, condition, form or nature of ecosystems over a substantial area relative to the extent of that ecosystem.

eschew uncertainty; here even to the extent of specifying the acceptable rate at which populations could spread before 'naturalization' becomes 'invasion'. All the definitions refer to arbitrarily bounded places and times, including ecosystems. These are likewise anthropocentrically conceived, even the 'transformers', footnoted (not shown) with examples of moderation and stability violated, e.g. 'excessive users of resources', 'erosion promoters', 'sand [dune or bar] stabilizers' and 'salt accumulators/redistributors', (Richardson et al, 2000). The authors seem focused on preserving 'life as we know it' but hardly for its own sake. 'We', of course, are human beings, particularly dependent on 'knowing' for our continued survival. 'Life' needs to respect the natural boundaries we have striven to overcome, even if we drag it along with us.

Conclusion

Human survival is a social and cultural matter, contingent on fulfilled expectations. We cope best when things stay where we leave them, or where we put them, but we will sometimes compromise on drifting away slowly. While focusing on the predictable, we easily overlook how the 'slings and arrows of outrageous fortune' continually shape us, and how we in turn inflict them on our environments. Mass travel and bulk shipping technologies add new forces of flux to those of gravity, wind and water. These 'currents of commerce' entrain many small lives, sometimes purposefully, sometimes accidentally. Our land uses generate and undo habitats. The result is an admixture of expected and unexpected, desirable and undesirable outcomes.

Anekeitaxonomy explains little regarding the biological aspects of such outcomes, but reveals much about human inclinations in the face of change. It projects our expectations, and particularly our fears of unintended, uncontrollable change onto organisms having only one, unconscious (we suppose) agenda: surviving 'here, now'. No other place or time bears on that requirement, but much emerges from it. Imposing additional criteria of belonging casts light on efforts to distinguish humans from nature, but on little else.

References

Allen, D. E. (1986) *The Botanists: A History of the Botanical Society of the British Isles through 150 years*, St. Paul's Bibliographies, Winchester, UK

Atran, S. (1990) *Cognitive Foundations of Natural History: Towards an anthropology of science*, Cambridge University Press, Cambridge

Blackstone, W. (1922) *Commentaries on the Laws of England*, Vol1, Bisel, Philadelphia

Campbell, D. H. (1926) *An Outline of Plant Geography*, Macmillan, New York

Candolle, A. De (1855) *Géographie Botanique Raisonnée*, Masson, Paris

Candolle, A. De (1885) *Origin of Cultivated Plants*, D. Appleton, New York

Chew, M. K. (2006) 'Ending with Elton: Preludes to Invasion Biology', Unpublished PhD Thesis, Arizona State University School of Life Sciences

Chew, M. K. and A. H. Hamilton (2011) 'The rise and fall of biotic nativeness: A historical perspective.', in Richardson, D. M. (ed) *Fifty Years of Invasion Ecology: The Legacy of Charles S. Elton*. Wiley-Blackwell, London

Cullen, H. M., de Menocal, P. B., Hemming, S., Hemming, G., Brown, F. H., Guilderson, T. and Sirocko, F. (2000) 'Climate change and the collapse of the Akkadian empire: Evidence from the deep sea', *Geology*, 28, 379–382

de Menocal, P. B. (2001) 'Cultural responses to climate change during the late Holocene', *Science*, 292, 667–673

Douglas, M. (1966) *Purity and Danger: An Analysis of the Concepts of Pollution and Taboo*, Routledge, London

Drysdale, R., Zanchetta, G., Hellstrom, J., Maas, R. Fallick, A., Pickett, M., Cartwright, I. and Piccini, L. (2006) 'Late Holocene drought responsible for the collapse of Old World civilizations is recorded in an Italian cave flowstone', *Geology*, 34, 101–104

Egerton, F. N. (2003) *Hewett Cottrell Watson: Victorian Plant Ecologist and Evolutionist*, Ashgate, Aldershot

Elder, G. H. Jr (1994) 'Time, human agency, and social change: Perspectives on the life course', *Social Psychology Quarterly*, 57, 4–15

Elton, C. S. (1927) *Animal Ecology*, Sidgwick & Jackson, London

Elton, C. S. (1958) *The Ecology of Invasions by Animals and Plants*, Methuen, London

Farber, P.L. (2000) *Finding Order in Nature,* Johns Hopkins, Baltimore

Fitzgerald, G., Fitzgerald, N. and Davidson, C. (2007) *Public Attitudes Towards Invasive Animals and their Impacts*, Invasive Animals Cooperative Research Centre, Canberra

Haug, G. H., Günther, D., Peterson, L. C., Sigman, D. M., Hughen, K. A. and Aeschlimann, B. (2003) 'Climate and the collapse of Maya civilization', *Science*, 299, 1731–1735

Henslow, J. S. (1835) 'Observations concerning the indigenousness and distinctness of certain species of plants included in the British Floras', *The Magazine of Natural History*, 8, 84–88

Hooker, J. D. (1867) 'Insular Floras', *The Gardeners' Chronicle and Agricultural Gazette*, January 1867, pp6–7, 27, 50–51, 75–76

Hunt, T. (2006) 'Rethinking the fall of Easter Island', *American Scientist*, 94, 412

Milton, K. (2000) 'Ducks out of water: Nature conservation as boundary maintenance', in Knight, J. (ed) *Natural Enemies: People-Wildlife Conflicts in Anthropological Perspective,* Routledge, London, pp228–246

Muhlenbach, V. (1979) 'Contributions to the synanthropic (adventive) flora of the railroads in St. Louis, Missouri, USA', *Annals of the Missouri Botanical Garden*, 66, 1–108

Richardson, D. M., Pyšek, P., Rejmánek, M., Barbour, M. G., Panetta, F. D. and West, C. J. (2000) 'Naturalization and invasion of alien plants: Concepts and definitions', *Diversity and Distributions*, 6, 93–107

Schröder, F. G. (1969) 'Zur Klassifizierung der anthropochoren', *Vegetatio*, 16, 225–238

Siegel, B. J. (1970) 'Defensive structuring and environmental stress', *The American Journal of Sociology*, 76, 11–32

Watson, H. C. (1835) *Remarks on the Geographical Distribution of British Plants*, Longman, London

Watson, H. C. (1847) *Cybele Britannica*, Vol1, Longman, London

Watson, H. C. (1859) *Cybele Britannica*, Vol4, Longman, London

Watson, H. C. (1868) *Compendium of the Cybele Britannica*, Vol1, Privately distributed, Thames Ditton

Zhang, Q., Zhu, C. Liu, C. L. and Jiang, T. (2005) 'Environmental change and its impacts on human settlement in the Yangtze Delta, P. R. China', *Catena*, 60, 267–277

10
The Other Side of Bio-invasion: The Example of Acclimatization in Germany[1]

Iris Borowy

Introduction

Recent years have seen a virtual avalanche of publications on the migration of plants and animals and their effects on their new host environments, a phenomenon often called biological invasion, although definitions vary as to what exactly this term describes.

Its study increased dramatically in the 1980s. This was especially after the reissue of the 1958 publication *The Ecology of Invasions by Animals and Plants* by Charles Elton, now often perceived as the founding father of the discipline (Drake and Mooney, 1989; Richardson and Pyšek, 2008). Biology departments in a number of universities have opened special centres for invasion biology, establishing it as a new discipline, complete with specialist conferences and at least three specialist journals: *Diversity and Distributions* (founded 1998), *Biological Invasions* (1999) and *Neobiota* (2002). Feeding the term *biological invasion* into WorldCat produces 16,289 hits (13 November 2009). Most of the studies focus on the damage done by invasive species and on the plants and animals themselves, trying to detect shared characteristics, aiming at early identification of potentially harmful species and, as a result, at effective countermeasures. This line of research insists that at least some migrant species cause such extensive economic and ecological damage as to represent a threat to human prosperity and biodiversity (Pimentel et al, 2000, 2001; Pimentel, 2002). The popular press has eagerly adopted this concept with titles such as: *They All Ran Wild*, *Immigrant Killers*, *America's Least Wanted* or *Biological Pollution* (Rolls, 1969; King, 1985; Stein and Flack, 1996; Britton, 2004).

Meanwhile, this perspective has been challenged, as a number of researchers have pointed out inconsistencies in this view: the often questionable methods of defining and quantifying damage, the contingent nature of damage, the difficulty of defining chronological and spatial boundaries which determine the (non-)native character of species and anthropomorphic imagery at work in metaphorically framing biota as national enemies (Keulartz and van der Weele, 2009). All this has prompted some researchers to define biological invasion as a social construction rather than a biological reality and to raise accusations of an 'antiexotic bias' (Stromberg et al, 2009). In reaction, invasion biologists like Simberloff have tried to discredit these views by blaming them on a 'fringe group of philosophers, sociologists, landscape architects, and others' (Simberloff, 2008). Clearly, the issue of bioinvasion is about a lot more than migrating biota and forms a social as much as a biological issue. The balance between these two sides is as yet contested ground. Peter Coates, in his book-length analysis of the social constructions of bio-invasion and their connection with American nativism nevertheless seems to contradict his own findings when arguing that there is a 'frequently sound ecological (not to mention economic) case for promoting native species' related to 'objective realities of ecological relations' (Coates, 2006, p24).

These controversies sometimes distract from the fact that, regardless of whether biological invasion is real, socially constructed or something in between, it is certainly a social issue in the sense that it is related to human agency. The vastly intensified degree of biota mobility in recent centuries has been a function of increased human mobility (McNeill, 2000). Part of today's bio-invaders arrived in their new homes as unintended accessories of human travelling, in ballast water or as seeds in hides and crops, unwanted but also innocent partners in a development in which the movement of people and goods across oceans became easier, faster and more common (National Research Council, 1996; Carlton, 1998). Others came as a result of deliberate policy. The relative significance of intention and inadvertence in the introduction of species is difficult to assess. As general rule, Pimentel estimates that 'most plant and vertebrate animal introductions have been intentional, whereas most invertebrate animal and microbe introductions have been accidental' (Pimentel et al, 2000, p53; see also Elton, 1958, p73). Simberloff, another outspoken critic of invasive species, finds that 'half of all damaging plant invaders were deliberately introduced, not accidental hitchhikers or escapees' (Simberloff, 2008).

Thus, it seems studying human behaviour and the circumstances which made people decide to move biota around would be equally important as understanding the issue as plant and animal behaviour and the circumstances which determine their fates in new environments. Ironically, human agency in this process and its perception by humans have only attracted a fraction of the attention directed at the biological components. People have, in a way, been largely written out of a history in which they clearly take a central role. This chapter, and indeed this book, aims to contribute to an area which remains underresearched.

People, the movement of animals and plants, and acclimatization

Historically, people have found a variety of reasons to take plants or animals from one place to the other, including hunger, the pursuit of economic gain as well as homesickness and a sense of aesthetics. Until relatively recently it would not have occurred to anyone to dispute their right to move livestock, domesticated plants or any other species that promised to feed, enrich or please them or to increase their social status (Groves, 1991, p427; see also Crosby, 1986). By the 18th century the rapid emergence of modern science added a new facet as it assumed an increasingly important role in explorations and colonial expansion (MacLeod, 2000). In addition to geologists, geographers and anthropologists, self-styled botanists roamed the world as plant hunters, testing areas for potential botanical value and taking possession of the biological riches of far-away places (Hepper, 1989; Lyte 1983). Their efforts became easier when an English general practitioner, Nathaniel Ward, developed a portable glasshouse in 1830. The Wardian case much improved the chances of survival of transported plants (Tenner, 1996, p117). German explorer-scientists took an active part in this movement. Their most famous proponent, Alexander von Humboldt, alone is credited with bringing 6000 species from South America to Germany, where they had been completely unknown (Hielscher and Hücking, 2007, p9).

Inevitably these utilitarian and scientific approaches combined. If the unconsidered transfer of biota, such as maize or potatoes, could enhance the well-being of people and nations in Europe, then a systematic and experimental practice was bound to make this process faster and more effective. A new scientific discipline was born: 'acclimatization'. The term originated in 18th-century France, born from efforts to introduce Spanish merino sheep into the country, but soon came to include issues of European colonialism in tropical regions (Osborne, 2000, p137). By the mid 19th century self-proclaimed acclimatization scientists and interested laymen engaged in active experiments involving the transfer of promising species from one place to another in ways that would most benefit their native countries. More than just a passing fad, acclimatization became a movement that attracted substantial attention and followers for about 50 years between 1850 and 1900. In numerous private associations people and institutions interested in the field met and studied ways to promote their goals. The first was the *Société Zoologique d'Acclimatisation,* founded 1854 in Paris, which soon counted 2600 members. A few years later, two German societies formed: the *Akklimatisationsverein für die königlich-preußischen Staaten* [Acclimatization Society for the Royal Prussian States][2] in 1856 and the *Central-Institut für Akklimatisation in Deutschland* [Central Institute for Acclimatization in Germany] in 1859. A year later, the Acclimatization Society of the United Kingdom in London and the Acclimatization Society in Glasgow followed suit. By 1900, there were more than 50 such institutions, including in most countries in Europe (Lever, 1992, pp193–194). Clearly, acclimatization expressed some perceived need of the times.

These associations served a series of purposes, political, economic as well as cultural, all in tune with the growing nationalism in which they evolved. In the UK and France, acclimatization was driven primarily by colonial ambitions. Concepts of settler projects in North Africa, New Zealand and Australia entailed a transfer of European agrarian systems and their adaptation to non-European environments. Thus, scientific expertise was expected to serve the economic exigencies of an expansionist policy. By the same token, acclimatization associations tried to make use of the biological resources of the growing colonial empire by adopting exotic species into the domestic agricultural ecology. These efforts had a distinct propagandistic facet. As exotic plants and animals grew in the experimental fields, gardens and parks of acclimatization associations, they became tangible models of a colonial reality. Both metaphorically and physically, they rooted the living beings of the metropolitan and colonial territories into each others' soils (Osborne, 2000). The purpose was both political and practical, but its relative importance depended on circumstances. Stagnating population numbers in France and the availability of a large British Empire ready to absorb excess population in the UK meant that these countries faced far less pressure to feed growing populations than some others. A case in point was the USA, where mass immigration created the obvious question of how to produce ever-increasing amounts of food. This may explain why, in contrast to its European counterpart, the movement took on a more official character in addition to its practice by private citizens. Attached to a series of government bureaux and departments there was a small staff employed in order to identify, import and test plants deemed suitable to help satisfy the needs of the rapidly growing nation (Fairchild, 1982).

The German situation was particularly ambiguous. Politically, Germany resembled a phoenix before rising from the ashes. There were no German colonies, nor, in fact, was there a Germany. Instead, there were a multitude of German entities, grouped around the two leading states, Prussia and the Habsburg Empire. Of the two, Prussia was clearly the more dynamic. It was culturally less glamorous than France and economically less advanced than the UK, but it was growing. During the preceding century, through a mixture of diplomacy, military prowess and land reclamation projects, Prussia had evolved from a backward, swampy province into one of the main players on the European continent. Drawing on its substantial coal reserves it began belated but intensive industrialization which propelled the country into a position of economic strength and political power. This development went hand in hand with dramatic population growth, estimated to have grown by 87 per cent between 1816 and 1865 (Salmonowicz, 1995, p250). Thus, the agenda in which acclimatization had to find its place was obvious enough: defining and creating a nation, accommodating steep population growth and contributing to the economic upsurge that was expected to place Germany on an equal footing with the other large European powers.

The Acclimatization Society for the Royal Prussian States, founded in 1856, took a relatively timid approach, limiting its interest to the theoretical, scientific aspects of acclimatization, and, judging by the reference to Prussia in its title, following a politically conservative line. This hands-off attitude – and possibly the political agenda that came with it – was deemed unsatisfactory by some members, who proceeded to establish a second, more ambitious association, the Central Institute for Acclimatization in Germany in 1859, whose focus was explicitly on practical efforts (ZA, 1869, p170). Its inaugural meeting placed acclimatization in a tradition of an ancient development of biota expansion, which had enriched Europe with potatoes, wheat, barley, rye and oats. Accordingly, it was argued, the continuation of similar activities would not only strengthen German independence from foreign countries, it would also broaden the spectrum of agricultural crops and thereby lessen the danger of crop failure and generally strengthen the material and mental life of the German people (Mittheilungen, 1859, pp1–5). The reference to Germany betrayed the association's progressive impetus, tying it to the ongoing effort of forming a nation state. This political line was underscored by its justification for choosing Berlin as locale, the 'centre of social life and intelligence for Germans' (Mittheilungen, 1859, p3). This nationalist agenda coexisted with a similarly inherent internationalist outlook. The earlier society in Paris was acknowledged as a model and as serving the same laudable goals of seeking to benefit its population. Thus, economic, political and cultural ambitions melded in this overriding project of modernization.

In 1863, the two societies united into a single acclimatization association, now called *Akklimatisationsverein in Berlin* [Acclimatization Society in Berlin, hereinafter AIB]. The merger increased its membership to more than 200 but otherwise resulted in little apparent change from the earlier *Central-Institut*, which appeared to carry on its work as before (ZA, 1863, no 1). Notably, the new statutes confirmed the practical purpose of the Society whose primary tasks were to conduct experiments on acclimatization and to get others to carry out similar experiments (ZA, 1963, no 1). In this context, the AIB aimed to create a network of individuals and institutions actively involved in the field of acclimatization. The quarterly newsletter changed its name to *Zeitschrift für Akklimatisation* (ZA) but continued to document Society activities. The multifaceted agenda was reflected in the varied membership of the Society which included businessmen, academics (often university professors from both science and humanities departments), medical doctors, government officers, bankers, teachers, agronomists and diplomats. In different ways, they formed part of the new bourgeoisie driving Germany's rapid modernization process. By 1869, their number had risen to 380 individuals and to more than 95 agricultural and gardening societies (ZA, 1870, p26).

Their principal goal was to find foreign crops and livestock whose acclimatization in Germany would increase productivity and prosperity. At the centre of these efforts was a demonstration field cultivated directly by the society. This field was considered essential, not only for the experimentation it allowed under direct AIB control, but also because such a field provided a forum to present acclimatization to the public. It was therefore of substantial propaganda value, indispensable in order to create a broad following. Initially, the Society made do with a large field which was at odds with its resources.

However, in 1860 it was lucky to be offered ground almost in central Berlin next to the Moabit prison. Its soil was poor and sandy, but the location was convenient not only for its strategic position but also because it came with prisoners who provided agricultural labour free of charge (ZA, 1869, pp170–71). AIB sources do not reveal whether the prisoners had volunteered for this work or whether it was a form of forced labour. Additional cultivation experiments were in the hands of members who grew a variety of crops in their own fields and gardens. Their input was important especially since the society had no facilities for animals. Between 1859 and at least the mid-1870s the demonstration field and the fields and gardens of Society members were home to hundreds of plants of foreign origin, usually received as free gifts from people and institutions who offered their assistance from afar. Many were interested private German citizens who lived in various European and extra-European countries as businessmen, academics or diplomats. They were indispensable sources of foreign seeds and seedlings, and in addition, they often formed crucial ties to other acclimatization societies such as those in Nice, London, Brussels, Antwerp, Palermo, St Petersburg, Tiflis and Boston. Collectively, these connections created an extensive, albeit somewhat unsystematic transcontinental network.[3] In addition, the AIB actively cooperated with botanical gardens, zoos, various agrarian, breeding and trade associations and academies, museums and libraries in other German states and far beyond. They exchanged publications, reports on experience with specific crops or livestock, but also seeds or seedlings. The AIB was both on the receiving and giving end of this network. Associations such as the Royal Acclimatization Society of Moscow (ZA, 1864, p85) or the Boston Natural History Society (ZA, 1866, p72) approached it on their own initiative in order to establish regular contact. Often these connections appear to have been remarkably generous, sharing not only information but also scarce and valuable seeds. As a case in point, the director of the Tiflis botanical garden forwarded part of a package of seeds and bulbs which he had just received from Persia in 1863, including roses from the Shah's garden (ZA, 1864, p86). AIB connections were particularly close with the Royal Acclimatization Society in Paris, whose General Secretary, Dr Leopold Buvry, visited repeatedly and which the AIB joined as *Société Agrégée* in 1863 (ZA, 1863, p268). These growing ties had tangible political repercussions. The cooperation with institutions and associations in most major German provinces clearly contributed to the growing sense of German nationalism among the participating elite. However, at the same time, AIB contacts beyond German states added a pronounced international element, which seemed to coexist effortlessly with its nationalist agenda.

Within the AIB, pertinent information was communicated via a lengthy newsletter, usually as German or foreign (translated) articles on specific species or projects. This publication also announced when seeds were available for distribution and reported on the Society meetings. At its regular meetings, AIB members heard scientific presentations and discussed relevant activities, opportunities and problems of ongoing projects and future efforts. Often, members got into lively discussions on the pros and cons of specific crops or methods. Once a year, members organized an exhibition in which they sought to impress the public with over 1000 samples of their work, including birds, eggs, fruits, grain, bread, honey and various other products, trying to demonstrate the necessity of further acclimatization efforts (ZA, 1863–74).

Identifying species fit for acclimatization entailed the critical analysis of cultivation results. Theoretically, this process should satisfy rigorous scientific standards. For this purpose members received standard forms which called for the meticulous recording of a series of information, including a description of the field used, the date when seeds were sown and in what quantity, dates of sprouting, transplanting, cultivation, stages of growth, bloom, fruit and harvest, the quantity and quality of the harvest as well as a general assessment of the experiment (Mittheilungen, 1862, p75). Initially, both societies had sent seeds to all members, but it soon became clear that the fates of most samples were never heard. So, as of 1861, the Central Institute sent out lists of seeds on offer and only provided seeds to those members who explicitly asked for them, a strategy that was subsequently adopted by the AIB (ZA, 1863). Getting a scientific process established was difficult despite repeated urgent reminders that members should send detailed reports. Thus, in 1861, 124 seed varieties were offered, of these a selection was sent to 23 members, 7 of whom sent reports, of which 2 used the society form (Mittheilungen, 1862, pp68–69). In 1864, there were 30 requests for seeds which produced 20 reports, which was an improvement but, the organizers pointed out, still involved only a small percentage of members (ZA, 1864, pp51–75). But frequent reminders appeared to pay off. By 1873, the AIB the Society received over 60 reports (ZA, 1874, p31). These findings were then integrated into those of the experimental cultivations on the demonstration field and all were analysed in detail and faithfully presented in the newsletter at the end of the respective vegetation period. However, scientific ambitions ended here. There was never an attempt to distil results into generalized conclusions. Findings, therefore, remained largely empirical and descriptive, tied into agricultural practices.

Adopting new crops as a means of increasing or enhancing food output was no new idea but part of the agricultural development of the time. Like other parts of Europe, Prussia had reacted to a growing demand for food by changing the agricultural regime, including eliminating fallow ground by planting legumes or by the large-scale introduction of new or as yet unestablished crops like sugar beets and potatoes (Pounds, 1985, pp230–238). Regarding livestock, cross-breeding Spanish merino sheep with traditional Prussian races, practised since the early 19th century, served as a model (ZA, 1864, pp185–6). The selection of tested crops seemed to melt into these ongoing changes since they had a decidedly down-to-earth character. The largest groups of crops among the 254 tested in 1861 consisted of 49 types of beans, 38 types of peas, 19 types of fibre plants, 17 types of trees, 15 types of oats, 13 types of maize, 1 types each of barley, clover and legumes, in addition to cucumbers, millet, potatoes, cabbage, pumpkin, rye, sugar beets and lettuce (Mittheilungen, 1862, pp93–109). The specific types often came from distant, sometimes exotic, places, occasionally from as far away as Manchuria, Canada or Japan, but many of the species were hardly unusual on German fields. The seeds on offer for cultivation on private members' land were similarly notable for their lack of exoticism. They mainly concerned either species of which there were native types in Germany or species, such as maize or pumpkin, which, while not native, had long been established in Europe. After 1869, potatoes assumed a dominating position in the selection. Roughly 100 types in all shapes and forms, representing almost half of all used crops, were tested for their productive and culinary qualities (ZA, 1869, pp19–23).

This list of species is in stark contrast to efforts in English and French acclimatization societies, whose colonial orientation provided them with a far more spectacular selection. In July 1860, the first annual celebration dinner of the Acclimatization Society in London included birds' nest soup, tripang, nerfs de daim, kangaroo steamer, Honduras turkey, Chinese yam, seaweed jelly and other rare specialities otherwise unheard of in European kitchens (Lever, 1992, pp44–47). This dinner was eagerly reported in Berlin (Mittheilungen, 1862, pp83–84). There are signs that the AIB would have liked to offer more exciting species themselves. In 1861 its members considered getting lamas and alpacas (Mittheilungen, 1862, p3). At other points the Society appears to have considered the cultivation of yaks, donkeys, mouflons and reindeer (ZA, 1864, p18) as well as of chinchillas and guanacos (ZA, 1865, p6), but it lacked the money. Therefore, just a few Angora goats were imported from France, of which only the male goat survived and he was used for breeding (ZA, 1865, p15). In addition, some experimental cultivation of bees was attempted (ZA, 1969, p25). Thus, any sense of adventure was largely limited to the newsletter which routinely carried reports on acclimatization projects conducted elsewhere, with a long list of species: ostriches, rabbits, reindeer, donkeys, leaches, cochenille lice, reindeer, eland and other antelopes, alpaca, kangaroo, aguti or Chinese pigs and crops such as hemp, quinoa, cotton, asclepias, *Sorbus domestica*, sweet potatoes, arrachaca and poppy. Some species deemed particularly profitable commanded special interest and received frequent coverage, above all honey bees, and silk worms, mulberry trees and silk production in general, as well as wild rice.

In time, the AIB received certain governmental support. Several Prussian ministers of agrarian affairs became honorary members, as well as Crown Prince Friedrich Wilhelm, who was an important connection to the government support, and Prince Friedrich Karl of Prussia, who took an active part in acclimatization efforts on his estate at Düppel. They supplied limited and unreliable financial help together with some logistical and moral support. In 1864, the Prussian Minister of Foreign Affairs agreed to alert all foreign consulates and missions to the requirements of acclimatization (ZA, 1864, pp163–164). As a result the Society could communicate lists of coveted species to foreign consulates. These initiatives could be successful as the example of the request for American vine presented to the Consulate in Washington showed (ZA 1864, pp245–246). In March 1866 the US Department of Agriculture shipped off a package, (ZA, 1866, p75) and apparently the plants prospered and were distributed liberally among interested members (ZA, 1869, p34). More important, seed samples frequently seem to have been transported via diplomatic shipments.

This support was not unimportant but it hardly ensured the continued successful work. In 1865, the AIB bemoaned its limited financial and spatial resources, which were insufficient for furthering its goals the way acclimatization societies in other countries could. It therefore appealed to the Crown Prince to grant them land for a suitable garden, large enough to contain different types of soil and to allow the cultivation of a variety of crops, trees and livestock (ZA, 1865, pp88–89). No immediate response is recorded, and activities remained curtailed by tight finances. This situation became critical when, in 1868, the AIB lost its demonstration field when Moabit prison closed (ZA, 1869, p26). For a year, a member provisionally offered some ground on his estate (ZA, 1869, p171).

Afterwards, Prince Friedrich Karl of Prussia, a nephew of the future Kaiser Wilhelm II, allowed cultivation on his private grounds at Gut Düppel (where he had already kept the AIB Angora goat) and even included the services of a gardener (ZA, 1871, p10). This arrangement was most helpful at the time, but it was unsatisfactory as a long-term solution, since its location on the outskirts of Berlin prevented easy public exposure. So the AIB intensified its earlier efforts to get into a position for more comprehensive cultivation experiments. Things looked promising when the Prussian government granted considerable land for use by the Society. However, preparing the ground for agricultural use and continued cultivation required building fences, watering facilities, housing for a gardener and so on, estimated to cost 12,000 Taler. Raising funds soon came to command a large part of AIB attention (ZA, 1870, pp92–94).

These efforts were interrupted by the Franco-Prussian war of 1871, which distracted both AIB members and the public in general and temporarily obstructed the movement of people and goods. The long-term effects were more ambivalent. One consequence was that the war tested the sensitive balance between the national and international sides of acclimatization. The latter was amazingly strong. Even after German troops bombed Paris the French acclimatization society did not sever all ties with its German counterpart but merely saw fit to strike from its list of honorary members all those German princes who had actively taken part in hostilities. In the interests of scientific exchange all other contact was maintained (ZA, 1871, pp133–134). Meanwhile, the German Society congratulated Crown Prince Friedrich Wilhelm profusely on his victory over the 'hereditary enemy of the German nation' (*Erbfeind deutscher Nation*) (ZA, 1871, pp94–96). However, this language appears mainly to have paid lip service to what was expected of an institution that counted the Crown Prince among its members and hoped to gain further governmental support. In reality relations were less hostile. Soon after the war, the AIB newsletter carried a detailed description of the state of the Paris acclimatization garden, about whose fate its members were obviously concerned. The report was not without nationalist undertones, as it stressed that the garden had been destroyed by Federalist rather than German forces and blamed the French side for prejudiced attitudes (ZA, 1871, pp134–135). In the same issue the activities of German citizens, who sent seeds from foreign countries, were reconstructed as acts of patriotism (ZA, 1871, p160). Nevertheless, considering the patriotic fervour of the time the overall tenor remained mild. Besides, contact between the two acclimatization associations continued. As early as 1872, the German Society began a project of experimental cultivation of truffles it procured from southern France while the French society explicitly recommended the German newsletter to its members (ZA, 1873, pp26–28). Indeed, few, if any, French AIB members appear to have resigned their membership (ZA, 1874, pp8–16). In view of the fiercely nationalist climate of the time, this is a remarkable finding.

Yet, the result of the war, the creation of a German nation state, was clearly welcomed at the AIB. Optimistically, General Secretary Buvry predicted that the expansion of the country would also strengthen the AIB intellectually and materially, it being the only one of its kind in Germany (ZA, 1871, p13). Thus, it seemed a promising sign that Chancellor Bismarck joined the Society (ZA, 1872, p29). But soon these hopes were disappointed. Finding funds proved exceedingly difficult and even the land previously promised was now put to different, partly military use in 1872, and was no longer available. The field at Düppel was also given up and cultivation moved to a new plot at Charlottenburg, increasing yearly expenses and thereby turning the possibility of saving funds for a large garden into a distant vision (ZA, 1873, p174). By 1873, more or less subtle complaints about the lack of public appreciation of the work of the Society found their way into the newsletter (ZA, 1873, p28). Worse, instead of growing, the Society had begun to shrink, and by 1874 individual membership had fallen to 284 (ZA, 1874, p32). For all its illustrious members, the Society was eventually unable to raise the funds or governmental support necessary to consolidate its work. Part of this may be explained by the inherently expensive nature of acclimatization, which required land, labour and extensive transportation, all of which could be supplied by only a limited segment of German society. But, one may assume, more enthusiasm and tangible support might have been forthcoming if the AIB had had more to show for its efforts. After 15 years it had failed to produce a spectacularly new or productive crop that would revolutionize German agriculture. There simply was no equivalent of the potato in sight. As in the other countries, the movement faded, albeit with an unexpected twist.

During the 1880s, German agents became active in different parts of Africa, and Germany belatedly acquired several colonies. However, in contrast to experiences in Britain and France, in Germany discussions of acclimatization were only integrated into the colonial discourse to a very limited extent. Instead, the term 'acclimatization' became increasingly, and after some time exclusively, used to describe the adaptation of German citizens to life in the tropics (Bastian, 1889; Nocht et al, 1920; Grober, 1936). By 1905, the AIB appeared all but forgotten. An entry on acclimatization in the 1905 edition of a major German encyclopaedia (Anon., 1905) mentions the French Society and takes note of a Berlin counterpart with no more than a single word (Meyers, 1905). For several decades, lively debates in Germany focused on whether it was possible for Europeans to live and procreate in tropical areas without losing their distinctive European traits, i.e. their presumed superiority, gaining eugenic undertones (Grosse, 2000 and 2003). Interestingly, during the brief period when Nazi victories during the first half of World War II rekindled hope of a German colonial empire these considerations also came to include the possible acclimatization of European livestock in Africa (Staffe, 1944).

Meanwhile, inside Germany, nationalist efforts to increase agricultural output in order to feed and strengthen the nation were no longer concerned with the adoption of foreign species. Instead, the main focus was on optimizing existing crops with little regard to origin (Heim, 2003). Claims about National Socialist efforts to cleanse German nature from foreign species have proved unfounded (Uekötter, 2007).

Conclusions

Elsewhere, I have argued that the fate of German acclimatization demonstrates the volatile nature of European discursive needs (Borowy, 2009). But in the context of bio-invasion considerations, this point gains an additional meaning. For although they seemingly follow a contradictory agenda, the rationales of 20th-century invasion biology and 19th-century acclimatization have a lot in common. Both systems of thought share a belief in science and in the desirability of its mastery over nature. Enthusiastic acclimatizers as well as diehard enemies of non-native species would argue that their positions rest on scientific knowledge and rational thinking and that perfecting such knowledge would enable them to control nature. In both cases this claim is difficult to sustain, given the importance of ulterior motives. Ironically, both aim at the same objective: to strengthen the nation, i.e. to use science in order to stave off national danger and to increase the productivity, wealth and general well-being of its people. The difference is that one concept sees the nation best served by increasing the presence of foreign species, while the other believes in the benefit of avoiding them. In the process, both systems would rate biota according to their economic value. 'Good' biota was agriculturally productive biota either actively sought or protected against other, 'bad' invasive species. At the same time, both ideas link plants and animals with national identity, in acclimatization by appropriating biota which had not formerly been part of the national self (viewed as a process of enrichment), in invasion biology by removing biota deemed inherently foreign to the national self (viewed as a process of cleansing). Both constructions are hampered by their limited appeal. For all its efforts, acclimatization rarely convinced more than a few hundred people in any one country to become actively involved. Invasion biology is constrained by the inability of people to agree on which biota form part of their national identity, and their decision is often guided by childhood nostalgia more than by the science of presumed migration patterns of species (see Coates, 2006).

These similarities are to some extent obscured by the fact that acclimatization concerned itself with crops and livestock, domesticated nature so to speak, while invasion biology would claim a broader field of reference, including the wilderness. The differences express fundamentally different concepts of nature and man's place in it. Promoters of 19th-century acclimatization would consider nature, all nature as part of the territory for potential human usage. Nothing describes this attitude more clearly than the unapologetic title of David Fairchild's book *The World Was My Garden* (Fairchild, 1938). Similarly, acclimatization would also entail releasing non-native animals into the wild for hunting or fishing, blurring the line between wilderness and human cultivation. Invasion biology, by contrast, would draw a sharp line between areas under human cultivation, to be protected against invasive species because of their economic damage, and areas of uncultivated wilderness, to be protected against the danger of invasive species to 'natural' biodiversity. This attitude is not without its inherent contradiction and hypocrisy for several reasons: turning a piece of uncultivated land into a field would immediately change it from land whose biodiversity should be protected to land where biodiversity should be subdued in the interest of species (read crop) uniformity. More important, this approach presupposes an unrealistic distinction between culture and nature, ignoring that by the 20th century no place in nature is untouched by human influence, certainly not in Europe, and certainly not if and when it is up to people to decide which species are sufficiently 'native' to be allowed to grow there. Besides, in an age in which almost all potential arable land is under cultivation, it is increasingly difficult to maintain that these huge areas are in no way part of nature.

However, in a wider sense both approaches are united in their basic assumption: that foreign species are better or worse or at least in some way inherently different from native ones. This assumption is questionable to say the least. For all their efforts, acclimatization associations were remarkably unsuccessful. Although they managed to grow a range of foreign plants and animals for a while they never found spectacularly productive crops or livestock, such as maize, potatoes, cattle or pigs, which revolutionized people's diets in non-native continents but which had arrived long before transferring biota had become a scientific endeavour. At the same time, invasion biology has yet to make a convincing case that foreign species are inherently more damaging than native species, some of which also behave in ways that contradict people's economic and ecological expectations (Williamson, 2002, p109). At the end of the day, plants and animals may not be so different from people: there are all kinds of them everywhere, good and bad and mostly in the grey area in between, with good sides for some and bad sides for others.

Notes

1 I am grateful to John McNeill and Michael Osborne for most helpful comments on an earlier draft of the paper.
2 All translations by the author.
3 By 1865, Society members included the leading personalities of acclimatization societies of Palermo, Moscow and Paris; members of German and other consulates in Alexandria, Bordeaux, Buenos Aires, Constantinople, San Donato, Tenerife, Trieste and Turino as well as private citizens in Algiers, Athens, Barcelona, Beirut, Bellagio, Biskra, Blidah, Brazil, Charkow, Florence, Gibraltar, Hongkong, London, Madrid, Malaga, Montreal, Moscow, Novo Mirgorod, Odessa, Oporto, Parma,

Puerto Montt Cerea (Chile), Reval, Stockholm, St Petersburg and Texas (ZA, 1865, v–xvi). Others were added during subsequent years.

References

Sources

Mittheilungen des Central-Instituts für Akklimatisation in Deutschland zu Berlin (1859–63), abbreviated Mittheilungen

Zeitschrift für Akklimatisation (1863–74), abbreviated ZA

Secondary Literature

Anonymous, 'Akklimatisation', in *Meyers Großes Konversations-Lexikon* (1905), Vol 1, Leipzig, 224–227

Bastian, A. (1889) *Ueber Klima und Acclimatisation nach ethnischen Gesichtspuncten*, Ernst Siegfried Mittler und Sohn, Berlin

Borowy, I. (2009) 'Akklimatisierung: Die Umformung europäischer Landschaft als Projekt im Dienst von Wirtschaft und Wissenschaft, 1850–1900', *Themenportal Europäische Geschichte,* URL: www.europa.clio-online.de/2009/Article=386

Britton, K. (ed) (2004) *Biological Pollution: An Emerging Global Menace*, APS Press, St Paul, MN

Carlton, J. T. (1998) 'Bioinvaders in the sea: Reducing the flow of ballast water', *World Conservation*, 28(4)/29(1), 9–10

Chew, M. K. (2006) *Ending with Elton: Preludes to an Invasion Biology*, unpublished PhD thesis, Arizona State University

Coates, P. (2006) *Strangers on the Land. American Perceptions of Immigrant and Invasive Species*, University of California Press, Berkeley and Los Angeles

Crosby, A. (1986) *Ecological Imperialism*, Cambridge University Press, Cambridge

Drake, J. A. and Mooney, H. A. (eds) (1989) *Biological Invasions: A Global Perspective*, Wiley, Oxford

Elton, C. S. (1958) *The Ecology of Invasions by Animals and Plants*, Methuen and Co., London

Fairchild, D. (1938, reprint 1982) *The World Was My Garden: Travels of a Plant Explorer*, Charles Scribner's Sons, New York

Grober, J. (1936) *Die Akklimatisation*, Jena, Verlag Gustav Fischer

Grosse, P. (2000) *Kolonialismus, Eugenik und bürgerliche Gesellschaft in Deutschland 1850–1918,* Campus Verlag, Frankfurt/New York

Grosse, P. (2003) 'Turning native? Anthropology, German colonialism and the paradoxes of the "Acclimatization Questions," 1885–1914', in Penny, H. Glenn and Bunzl, Matti (eds) *Worldly Provincialism: German Anthropology in the Age of Empire*, University of Michigan Press, Ann Arbor, pp179–197

Groves, R. H. (1991) 'The biogeography of Mediterranean plant invasions', in Groves, R. H. and Di Castri, F. (eds) *Biogeography of Mediterranean Invasions*, Cambridge University Press, New York, pp427–438

Heim, S. (2003) *Kalorien, Kautschuk, Karrieren. Pflanzenzüchtung und landwirtschaftliche Forschung in Kaiser-Wilhelm-Instituten 1933–1945*, Wallstein Verlag, Göttingen

Hepper, F. N. (ed) (1989) *Plant Hunting for Kew*, Stationery Office Books, London

Hielscher, K. and Hücking, R. (eds) (2007) *Pflanzenjäger*, 3rd edn, Piper Verlag, Munich

Keulartz, J. and van der Weele, C. (2009) 'Between nativism and cosmopolitanism: Framing and reframing in invasion biology', in Drenthen, M., Keulartz, J. and

Proctor, J. (eds) (2009) *New Visions of Nature. Complexity and Authenticity*, Springer, Dordrecht, pp237–256

King, C. (1985) *Immigrant Killers*, Oxford University Press, Oxford

Lever, C. (1992) *They Dined on Eland: The Story of the Acclimatisation Societies*, Quiller Press, London

Lyte, C. (1983) *The Plant Hunters*, Orbis Books, London

MacLeod, R. (ed) (2000) 'Nature and Empire: Science and the Colonial Enterprise', *Osiris*, 15 (Special Issue)

McNeill, J. (2000) *Something New under the Sun*, W. W. Norton & Co, New York, pp252–262

National Research Council Committee on Ships' Ballast Operations (1996) *Stemming the Tide*, National Academies Press, Washington

Nocht, B., Volkens, G. and Neumann (1920) 'Akklimatisation', *Deutsches Koloniallexikon,* Vol. 1, Berlin, pp27–30

Osborne, M. A. (2000) 'Acclimatizing the world: A history of the paradigmatic colonial science', in MacLeod, R. (ed) *Nature and Empire: Science and the Colonial Enterprise*, *Osiris*, 15, pp135–151

Pimentel, D., Lach, L., Zuniga, R. and Morrison, D. (2000) 'Environmental and Economic Costs of Nonindigenous Species in the United States', *BioScience*, 50(1), 53–65

Pimentel, D., McNair, S., Janecka, J., Wightman, J., Simmonds, C., O'Connell, C., Wong, E., Russel, L., Zern, J., Aquino, T. and Tsomondo, T. (2001) 'Economic and Environmental Threats of Alien Plant, Animal, and Microbe Invasions', *Agriculture, Ecosystems and Environment*, 84, 1–20

Pimentel, D. (2002) 'Introduction: Non-native Species in the World', in Pimentel, D. (ed), *Biological Invasions. Economic and Environmental Costs of Alien Plant, Animal and Microbe Species*, CRC Press, Boca Raton, pp3–8

Pounds, N. J. G. (1985) *An Historical Geography of Europe 1800–1914*, Cambridge University Press, Cambridge

Richardson, D.M. and Pyšek, P. (2008) 'Fifty Years of Invasion Ecology – the Legacy of Charles Elton', *Diversity and Distributions*, 14 (2), 161–168

Rolls, E. (1969) *They All Ran Wild: The Animals and Plants that Plague Australia*, Angus & Robertson, Sydney

Salmonowicz, S. (1995) *Preußen: Geschichte von Staat und Gesellschaft*, Stiftung Martin-Opitz Bibliothek, Herne

Simberloff, D. (*2008*) 'Invasion Biologists and the Biofuels Boom: Cassandras or Colleagues', *Weed Science*, 56 (6), 867–872

Staffe, D. (1944) *Die Akklimatisation der Haustiere in den afrikanischen Tropen*, Verlag von E.S. Mittler & Sohn, Berlin

Stein, B. A. and Flack, S. R. (1996) *America's Least Wanted: Alien Species Invasions of U.S. Ecosystems*, The Nature Conservancy, Arlington, VA

Stromberg, J. S., Blake, T. J., Sperry, J. S., Tschaplinski, T. J. and Wang, S. S. (2009) 'Changing perceptions of change: The role of scientists in tamarisk and river management', *Restoration Ecology*, 17(2), 177–186

Tenner, E. (1996) *Why Things Bite Back*, Knopf, London

Uekötter, F. (2007) 'Nazi Plants: A Nazi Obsession?' *Landscape Research*, 32 (3), 379–383

Williamson, M. (1996) *Biological Invasions*, Chapman & Hall, London

Williamson, M. (2002) 'Alien Plants in the British Isles', in Pimentel, D. (ed) *Biological Invasions. Economic and Environmental Costs of Alien Plant, Animal and Microbe Species*, CRC Press, Boca Roca

Part III
Case Studies and Case Histories

11
Strangers in a Familiar Land: The Return of the Native 'Aliens' and the (Re)Wilding of Britain's Skies, 1850–2010

Robert A. Lambert

Introduction

This may seem, at first glance, a chapter that sits uneasily alongside many of the others in this book as it does not directly address the alien species debate as we might strictly know it. I take my title, slightly changed, from an upbeat piece in *Walk* (the magazine of the Ramblers' Association) about, 'the return of the natives' (Rowe, 2008), and from an essay on the problem of exotic species in the reader *Bio-Invaders* (Johnson, 2010). The chapter's relevance lies in a more broad and interactive nature–people relationship across time and space, and it starts with the notion that here in Britain (and indeed, in much of the rest of the Western world) we have learned to live (be it locally, regionally or nationally; be it for tens of years or hundreds of years) without many top predators. Indeed in the recent past many of these birds of prey and mammalian carnivores were as 'alien' to us, as much loaded in the sense of 'the other', culturally and socially, as many of the invasive alien species confronted in this book. Now that some have come back (either through natural colonization, reintroduction programmes or translocation schemes) they are, if you like, strangers, not in a strange land but in a familiar land. But we have rallied to their cause and started to embrace them as familiar. We lost most of the top mammal predators in Britain over the 10,000 years since the last Ice Age, for a variety of environmental, climatic or human-induced reasons. Indeed, historical biologists and species historians have charted that decline well, none better than Derek Yalden (1999; 2003) or Andrew Kitchener (1998). Birds of prey hung on for much longer in our crowded and overexploited island, but the death knell for many species came with the culturally loaded label 'vermin'. This we can trace back to the medieval period at least, and it reached its heyday in the 19th century (mid-Victorian) era when the new waves of gamekeepers on private estates launched an all-out assault on predatory animals. Environmental historians, working with primary sources such as estate papers and hugely illuminating 'gamebooks' (records of slaughter), have documented and charted what was a structured and well-organized wholesale modification of the natural world across that century. This was to protect and nurture game species that we considered profitable, or had been labelled as 'most noble' to pit one sporting talents against, be they red grouse, salmon or red deer (Smout, 2000; Lovegrove, 2007). To do that, estate owners paid others to shoot, trap and hunt down birds of prey and predatory mammals. We also shot them for sport, for mere target practice, or as sporting trophies; and in the case of some birds of prey, we took eggs for egg collections (be it scientific oology or the profitable trade for private collections) and skins for taxidermy (McGowan, 2009; see also Harvie-Brown archives). It is a salutary catalogue of loss upon loss, and here in Britain we lived with the consequences, ecologically and culturally, for much of the late 19th century and across the 20th century. The fates of individual British bird species do reflect a huge part of our socio-cultural and intellectual history in these islands (Cocker and Mabey, 2005; *Birds Britannia*, BBC, 2010). In an era of rampant globalization, we ponder and worry a great deal about our cultural separation from other human societies or groups and how we might 'rediscover' and reconnect with them, but rarely do we think these thoughts in an intraspecies context.

The return of the natives

Of course it helps enormously that returning native 'aliens' are charismatic, engaging, entrancing and exciting predatory species to watch. While we are, of course, watching our own 'native' birds of prey returned to the landscape, we are in some way casting a new 'tourist gaze', with historical precedents of a 'first' tourist gaze in the 1880s at a place like Rothiemurchus in the Scottish Highlands (Smout and Lambert, 1999), on a cultural ornithological 'alien'. We may not know it, or comprehend it fully, but the decades or centuries of alienation from this raptor species fuel its appeal, as we seek, each in our own way, to build a new functioning relationship (real and symbolic) with the creature (built on genuine delight, not blind persecution) lost from our cultural generational memory, or our visual consumption of certain cherished landscapes. Indeed, some of these birds of prey, most especially the osprey *Pandion haliaetus*, the red kite *Milvus milvus*, the white-tailed (sea) eagle *Haliaeetus albicilla* and the peregrine falcon *Falco peregrinus* have made a remarkable journey from persecution to sustainable tourism icons in just around 100 years. These birds have an intrinsic value to us now. We gain pleasure in watching them and are 'cured' of the ills of modern living through contact with them, however brief or fleeting; we also become physically and mentally active in engaging with them, or 'natural fit' as recent RSPB public health reports put it (Bird, 2004; 2007). This builds on something that the economic ornithologist Walter Collinge mused on back in 1921: 'the study of the living bird afield is rejuvenating to both mind and body. The outdoor use of eye, ear and limb ... brings one into contact with nature – out into the sunlight, where balmy airs stir the whispering pines or fresh breezes ripple the blue water' (Collinge, 1921).

The Scottish poet and nature writer Kathleen Jamie has embraced the recolonization of the returning osprey, noting how a creature once unimagined and alien to our experiences has in one generation become familiar:

> *What pleases about the ospreys is the quiet success of their return to their rightful place. A damage remedied and a change of direction in our attitudes, as the bird itself makes the turn into the prevailing wind. These were native birds ... some sites are famous; they are public spectacles with viewing places and video-linkups. There are large road signs directing us to birds' nests, and we don't find this bizarre. I like knowing these things. I like to be able to glance up from my own everyday business, to see an osprey.* (Jamie, 2005, p46)

While we might challenge Jamie's notion of a 'quiet' return for the osprey, in the full public media glare, and with up to 90,000 annual visitors thronging Loch Garten RSPB Osprey Centre (as in the exceptionally warm summer of 1976) to see what Henry Tegner called 'an avian exhibition, an annual show' (Tegner, 1971, pp45–50), her embracing of a powerful cultural sense of 'ornithological forgiveness' at play here for our human crimes is palpable. Scottish outdoor writer Jim Crumley picks up a similar thread, describing the culturally rehabilitated ospreys as reflective of 'a miracle of forgiving ornithological regeneration in the Highlands' (Crumley, 1991, pp89–101). You can also detect a breaking down of boundaries between the human and the natural, something that social anthropologists are beginning to examine in the context of wildlife management challenges for both conflict species and alien species (Milton, 2000). Writing on the nature and culture of the discipline of environmental history, John McNeill has identified a 'declensionist tendency' in the field, the narratives being relentlessly depressing accounts of destruction and loss, 'one damn decline after another', but also noting, especially in the 20th century, some cheerful developments (McNeill, 2003, p35). The species history of the osprey in Britain is very similar, and has much to teach about changing human attitudes and values over time, and the shifting parameters of cultural loss, ecological memory and subsequent rediscovery and attachment.

'The most observed birds in history': the Loch Garten ospreys

So believed the British newsreader, Robert Dougall, who also happened to be President of the RSPB from 1970 to 1975, and was thus in post at a key time in the history of the return of the osprey to Britain as a breeding bird

(Dougall, 1975, p297). Dougall played a pivotal role in the promotion of the osprey to a wider public audience; in his memoirs he records with pleasure how he orchestrated BBC Television News in London to tell the annual story of the comings and goings at this one osprey nest in a regular *Nine O'clock News* slot during the summer months. This exposure, along with RSPB-made films of ospreys viewed by about 5.7m people in Britain (Aitchison and Blackwell, 1988, p75), brought the cause of raptor conservation directly into living rooms and village halls across Britain, no doubt encouraging many in England to drive north to visit the site for themselves. This combination of celebrity, nature conservation and wildlife tourism was powerful. It was, as Dougall conceded, 'one of the most uplifting wildlife stories of our generation, a classic tale of how controlled access can safely be allowed to shy and rare breeding birds' (Dougall, 1978, pp175–182).

The osprey, a summer migrant to Britain, was persecuted from the medieval period onwards, no doubt, in part for its visibility at fishponds adjoining estates and grand houses; it was perceived as a direct competitor with us for a natural resource granted to us, not it, by providence. The fashion for managed estates in the 19th century (and direct persecution from gamekeepers and water bailiffs) ensured the bird was absent from southern England as a breeding bird by *c.*1847 (Lovegrove, 2007); lost from southern Scotland by the 1860s, but hanging on in the northern Highlands (Thom, 1986; Dennis, 1995), as it was not perceived as direct threat to the sheep which cluttered upland vistas. There was a nest at Loch Loyne on the Glengarry estate in 1913, the last 'official' British nest being there in 1916, although rumours persisted of possible breeding attempts in Inverness-shire in the 1930s and 1940s (Dennis, 2008; see RSPB archives, Watchers Committee Reports refs 01.05.01, 01.13.01 and 01.23.02). From Rothiemurchus estate near Aviemore in the Scottish Highlands we have some of the best evidence for shifting cultural attitudes towards ospreys, built upon deep attachment to the land by the estate owners, the Grants of Rothiemurchus, and growing recreational and aesthetic observations of ospreys by naturalists and holidaymakers alike (see Grants of Rothiemurchus archives, NRA(S) 102: 337). At Loch an Eilein on the estate, where the osprey nest had been robbed of eggs for profit year upon year in the mid 19th century, the Grants declared an end to pleasure boating in 1880 to protect the birds from disturbance. A network of gamekeepers were charged with recording osprey behaviour (not persecuting the birds!), and neighbouring landowners and sporting tenants were encouraged to embrace the raptor. This is all remarkable stuff, and pioneering and laudable when seen through early 21st-century eyes, and it should not be forgotten in our cultural story (Lambert, 2001, pp74–81); but this was an isolated stance. When the osprey started to recolonize the Spey Valley in Scotland in the early 1950s, two competing visions of how to manage the return emerged. The conservation establishment doctrine of the time, in the face of the mid-century curse of private egg collection, was to keep secret the nest site of any rare bird. However, George Waterston, RSPB Director in Scotland at the time, thought differently after a couple of years in the late 1950s when despite protection the birds had been robbed (Brown and Waterston, 1962). Rather, he sought to invite the public in, to tell them about the returning ospreys, to allow them access to see the birds from a centre, to be informed, to reconnect with a species with which they had no cultural connection. Around 14,000 people took him up on that offer of re-engagement in just six summer weeks of 1959 (at a time when RSPB national membership was only 10,579); in 1988, 92,800 accepted the offer. A domestic osprey-watching tourism industry (much copied in the decades since) was born under the guise of RSPB 'Operation Osprey' (Everett, 2002), percolating out across the local area; the nearby village of Boat of Garten even rebranded itself as 'The Osprey Village', truly socioculturally embracing the returning natives (Waterston, 1971; Dennis, 1991; Fowler, 2002). A cumulative total of nearly 2.3m people have now visited the Loch Garten osprey nest (Richard Thaxton, Site Manager, RSPB Loch Garten, pers. comm.; Lambert, 2008). Surely the most visited and admired nest of any bird in history!

We should see George Waterston as a visionary in the way he embraced the human dimension as part of his nature conservation thinking, and place him alongside another such 'voice in the wilderness' of that time, ecologist Frank Fraser Darling, who eventually emigrated to build a career as Vice President of the Conservation Foundation in the USA where his nature–people message (tinged with anthropomorphism and romance) was so much better received (Smout, 1993, pp22–23). Waterston understood that we had been culturally disconnected from some birds of prey for decades, and anticipated that we would in the new dawning age of ecological thought and action after 1960 rush to re-engage and reconnect with these lost birds. RSPB Secretary Philip Brown came to see the Waterston open-door policy as inspired forward-thinking: 'in the interests of bird protection in general and of the ospreys in particular ... [he has] done the right thing in opening up the site to the general public once the really hazardous period was over for the season and there were healthy chicks in the eyrie' (Brown, 1979; see also RSPB archives refs 01.02.02, 01.05.71 and 01.13.02). With over 200 breeding pairs in Scotland, and breeding pairs in England and Wales (through natural colonization or a translocation scheme begun at Rutland Water in 1997), the fiftieth anniversary of the return of the osprey was widely celebrated by conservation agencies, the media and the public in 2009 (Dennis and Rutherford, 2007; Anglian Water, 2008; Hamilton, 2009).

Other species histories of birds of prey (Lambert, 1998) map out similar stories, remarkable sea changes in human attitudes in the modern era, often prompted by new types of interaction as the birds reinvade our territorial and mental spaces. The red kite is the subject of the longest-running bird conservation effort in Britain, going back to 1903 and the establishment of a Kite Preservation Fund (Lovegrove, 1990; Carter, 2001; see also RSPB archives refs 01.05.01 and 01.15.01). Part of our urban landscape in the medieval period when it foraged for carrion in the streets of London, it was over time slowly banished to the periphery (both geographically and culturally) as land management practices changed, sanitation improved and persecution stepped up. It clung on in remote valleys of central Wales, and was brought back into our cultural consciousness with high-visibility reintroduction programmes in working farmed landscapes of lowland Britain and Scotland from 1989 onwards. Further reintroductions have seen what was once the bird of Owain Glyndŵr, epitomizing a wild, untamed Wales, now become a bird of tamed Shakespearian English landscapes. It is now perilous for birdwatchers to drive along the M40 through the Chiltern gap as red kites fill the skies over the motorway. In nearby villages, the kites, once alien and geographically and culturally distanced, have become a familiar garden bird, gobbling chicken scraps on the manicured lawns of suburbia, beneath mollycoddled blue tits *Parus caeruleus* and great tits *Parus major* hanging on peanut nets.

Few have written with such devotion and described such deep cultural attachment towards a bird of prey as J. A. Baker (2010) on the peregrines of the Essex marshlands. The bird he describes was persecuted in the 19th century by gamekeepers on estates; then (because of its penchant for carrier pigeons) was perceived to be a dastardly avian fifth-columnist during World War II, in league with Goering's Luftwaffe, and was hunted down in the national interest (Air Ministry, 1940; Anon., 1940); the peregrine then suffered regional population losses in the pesticide crisis of the 1960s as British farming turned intensively industrial and chemical (Moore, 1987). The peregrine retreated to the wilder parts of our coastline, was gone from much of southern and eastern England, and slipped out of our cultural consciousness. The fact is that around 120,000 of us per annum now visit peregrine watch points around Britain, not just in wild places like Symonds Yat in Gloucestershire (which records 50,000 visitors per annum) but also in our cityscapes, on cathedrals in Lincoln, Derby and Exeter. Peregrines now recapture our imagination as they force us to look at them afresh (Wintle, 2009), on top of the Tate Modern art gallery in London, adding a new and wild aspect to our perception of city tourism. New impromptu peregrine-watch points spring up around Britain every year, as the birds reconvene with us, on their terms, yet in our urban landscapes; we are drawn to them not just because they are magnificent raptors, but surely because in a city they are something different, unexpected and seemingly out of place. The osprey may have statues, may feature on logos and village signs, and even have a professional Welsh rugby union team named after it, but can there be any greater cultural assimilation than webcam images of nesting peregrines beamed onto a megascreen in Exchange Square in city-centre Manchester, the very same screen that shows Manchester United football games and BBC News?

The way forward: a sustainable future for Britain's raptors?

We live in more enlightened times. Yes, undoubtedly the fate of Britain's birds of prey has changed as human attitudes, values and perceptions of them have shifted quite dramatically in a very short space of time in the mid 20th century. Environmental history research has shown that the first fledgling moments in that attitudinal shift can be traced back to the last two decades of the 19th century, and the efforts of landowners like the Grants of Rothiemurchus (who were protecting and cherishing 'their' ospreys by around 1880) should not be forgotten (Smout and Lambert, 1999, pp41–44). Indeed, these pioneering efforts should be celebrated as the first historical murmurings in a wider raptor conservation story (Lambert, 2001, pp74–91). That some landowners were willing, for one species at least, to stand against the dominant ethos of the time, namely structured, funded raptor persecution in the name of late-Victorian game protection on estates, surely forces us at the start of the 21st century to similarly seek to redefine our relationship and carve out a more sustainable future for our bird of prey populations, not just because they are beautiful and we cherish them, not just because they attract us in our thousands as wildlife tourists, but because they are also key indicators of the health of upland ecosystems in the 21st century.

Roy Dennis, doyen of British raptor conservation, who cut his teeth in the field as a youthful and passionate warden at Loch Garten in the 1960s working under George Waterston, has always sought a dynamic and visionary future for birds of prey, politically, socially, culturally and geographically. Roy sees no harm in challenging established conservation doctrine, challenging public perceptions and pushing boundaries (both cultural and spatial). His travels in North America and Central and Eastern Europe (Dennis, in Lambert, 1998, pp5–7) have convinced him that people and predators (both avian and mammalian) can live well together, and as Director of the Highland Foundation for Wildlife (founded in 1995) he urges a new (re)wilding vision for the UK, based on species reintroductions, species translocations and species education programmes. At the Scottish Ornithologists' Club (SOC) Annual conference in Inverness in March 2010, Roy urged that we needed to be bold now, not to think in terms of individual bird species as we have in the past, but rather in terms of broader functioning ecosystems. Our biggest failure, he warns, 'is not that the project we tried failed, but that we never tried'. He asserts that the major constraint against all this (re)wilding ambition is social and cultural, not ecological (Dennis, in McMillan, 2010, p138). Roy is unashamedly ecologically radical in his future vision; he dislikes inactivity by conservationists or their political masters; rather he celebrates proactive wildlife management and innovative conservation research. This societal and political aspect of the nature–people relationship was picked up in a remarkable photographic essay (2007), and later touring exhibition, addressing our complex, changing and contradictory response to predators in Britain. This was thought-provoking and challenging material written by two of the UK's best known nature photographers; indeed, it was both uncomfortable and uplifting viewing. I quote from the dynamic introduction:

> *Our relationship with predators has always been fractious and they continue to court controversy as much today as at any time in Britain's history. For many people, predators are symbolic of a wildness we once knew, they are key to the ecological integrity of our countryside. They embrace notions of nobility and power. They make us feel good. For others, predators*

> *represent competition for our game interests and pose a threat to our domesticated animals. They are an inconvenient drain on resources and compromise our leisure pursuits – they have no place in our orderly lives where Man has dominion over nature.* (Cairns and Hamblin, 2007, p7)

This book, *Tooth and Claw*, was in essence a snapshot survey of national attitudes to a certain part of our wildlife in the run-up to 2010. It presented a positive spin, but then we are, in general, a nation obsessed by animals. Environmental historians (Sheail, 1976; Evans, 1997; Smout, 2000), historians of conservation organizations (Samstag, 1989; Rothschild and Marren, 1997; Sheail, 1998) and of shifting human attitudes (Brown, 1966; Allen, 1976; Thomas, 1984; Coates, 1998) have long mapped out the way that we have rallied to protect and offer better welfare for animals; first, those most economically useful to us (livestock) at the start of the 19th century (Moss, 1961; Harrison, 1973; Brown, 1974), then game species, then pets who share our lives and homes, and finally, in the last three decades of the 19th century, wild animals for their own intrinsic value. A recent audience research survey conducted by the BBC, reported in *Natural World* (the magazine of the county Wildlife Trusts), identified the cumulative might of the domestic wildlife constituency to be in the order of 30m people. It broke it down as follows: 2m were devoted amateur naturalists and concerned campaigners: birdwatchers, lepidopterists, mammal watchers, botanists; the other 28m were a great swathe of the public interested in, fascinated by or engaged with nature and wildlife. By any stretch of the imagination these are huge figures to comprehend; 'half the population of the UK loves wildlife', proclaimed the survey (Rollins, 2006, p31). It is these people, the *Springwatch* and *Autumnwatch* television millions, who have fuelled the rise of a domestic British wildlife tourism industry, which, in the absence of big mammalian predators from the landscape (with all the frisson of awe and danger that they excite in us), has turned to birds of prey (once gone but now returned home geographically and culturally) as iconic predators drawing our gaze. Observing, interacting and identifying with these native 'aliens' has become entrenched in the recent burgeoning of popular, intimate and reflective new British ornithological writing of the early 2000s (Mabey, 1997; Oddie, 2000; Lister-Kaye, 2003; Moss, 2003; Barnes, 2004; Riley, 2004; Hume, 2005; Moss, 2006; Cocker, 2008; Moss 2008; Armitage and Dee, 2009; Dee, 2009; Elder, 2009; Horne, 2009; Allen and Ellis, 2010). One of its leading exponents, the environment and sports journalist Simon Barnes, refers to our cultural 'aliens' the osprey, red kite and peregrine falcon as 'pilgrimage birds', birds that we have lost and 'found' again, and asks where that act of cultural (re)discovery might ultimately lead us (Barnes, 2005, pp237–277). Of the peregrine, he says they are:

> *perhaps the ultimate pilgrimage bird, and they are an inspiration. An inspiration to carry on, to see more birds, to enjoy birds more, to enjoy life more. If the fastest bird that has ever*

lived can be seen, and by you, then how much more is within your scope?. (Barnes, 2005, p277)

Welcoming back the native 'aliens'

For some years now the RSPB Economics Department has commissioned research to quantify just how large this wildlife-watching sector is in the rural and urban economic scene, and just which species are proving to have the broadest public appeal. The reports sometimes fail to grasp the depth of our earlier cultural severance from these birds of prey and how 'alien' and 'of the other' first visual encounters with them must be for many. That said, we are rallying in our thousands to embrace these returning native 'aliens', a tradition firmly built on the template for welcoming back these birds to our cultural working landscapes laid down by George Waterston in the late 1950s at RSPB Loch Garten. Ospreys have undoubtedly become an iconic species, a major public and conservation success story, a cultural and economic asset (Harley and Hanley, 1989). In the summer of 2006, *c*.290,000 people went osprey-watching in Britain, contributing £3.5m to the areas around these sites, helping to support local incomes and employment, and generating widespread educational and conservation side benefits (RSPB, 2006, pp15–21). Many of these are RSPB-organized projects: the Welsh Osprey Project opened in 2004 and hosts about 73,000 visitors per annum; the Lake District Osprey Project opened in 2001 and hosts about 70,500 visitors per annum. The Scottish Wildlife Trust's (SWT) Loch of the Lowes Osprey Centre in highland Perthshire opened in 1969, on the back of public success up the A9 at Loch Garten RSPB, and by 2007 had welcomed 1.25m visitors to the nest site. The return of the white-tailed (sea) eagle to breed on the west coast of Scotland in reintroduction schemes centred round the island of Rhum from 1975 onwards (Love, 1983) has prompted the diversification of the local tourism industry on nearby islands like Mull, an industry once founded on genealogical and heritage tourism. RSPB figures suggest that of the £38m spent by tourists on Mull each year, around £1.6m is attributable to the public draw of sea eagles, once absent and forgotten (alienated) from the landscape, now becoming more and more accepted, indeed, familiar (RSPB, 2006, pp11–14). Along with the children's television series *Balamory*, the sea eagles have become emblematic of Mull today. They had been extinct as a breeding bird in Britain since 1918 (Holloway, 1996, p106; Kitchener, 1998, p77).

Will this familiarity breed contempt? Will the returned native 'aliens' become so abundant, so commonplace, so much a part of our cherished urban and rural landscapes, that we will no longer care enough to glance upwards, or queue at a telescope at a long-established public viewing point? Some might see that as positive, as the final destination of a returning 'journey of acceptance' for these once 'alien' birds back into our sociocultural landscapes. Others fear a loss of income if these birds become all too restored, all too accessible, all too uninspiring, because currently returning native 'aliens' bring in big money and crowds. Will that mean less money for conservation projects, less money for eNGO coffers and less money pumped into local economies? Increasingly conservation organizations like the RSPB have to justify the existence of nature reserves (designated sites) to government policy makers and other stakeholders by talking openly of the economic benefits they bring to local communities or regions through employment and tourism. It may be tempting to see this as a very recent new agenda for the conservation management movement (the language of 'this species is worth this much in £s') but, as this chapter argues, it does have historical roots going back to the mid 20th century at least. The returning native 'aliens' support economic development in a number of ways: direct employment; spending by NGO employees; spending by volunteers; spending by tourists (accommodation, food, transport); direct expenditure by reserves on goods and services provided by local or regional suppliers; grazing lets and agricultural tenancies. Nature conservation itself, as a practice and a land use, helps to support a significant proportion of the UK rural tourism market (Murray and Simcox, 2003), worth around £14 billion per year in England alone. This figure is supported by spending of £19m per year by the RSPB and visitors to its reserves, with tourism expenditure because of the reserves and birds calculated at around £12m per year (RSPB, 1997; 2001; 2002). Birdwatching is one of the fastest-growing leisure pursuits in the western world, and tourism based on birds and birdwatching is now one of the fastest-growing segments of the whole nature tourism sector. The grand old man of British ornithology, Ian Wallace, has estimated the total British marketplace for bird-inspired holiday travel at between £9.5m and £11m per annum (Wallace, 2004). The annual British Birdwatching Fair (BBWF) ('the birders' Glastonbury') attracts over 22,000 visitors to Rutland Water in the East Midlands over a three-day weekend in August. Started in 1989, it has now raised over £2.5m for global bird and habitat conservation work (Tim Appleton, BBWF Organizer, pers. comm.). Britain has led the way in the development of a domestic and international bird tourism and ecotourism market (Cater and Lowman, 1994; Diamantis, 2004; Fennell, 2008), although some of the best examples of mature sustainable tourism wildlife projects exist in this country (Holden, 2008; Page and Dowling, 2002). As strange as they may seem to outside observers (and surely deserving of rich psychological study), the legions of British birdwatchers can make real contributions to nature conservation work, local economies and community-led tourism development; in other words, they can, in some ways, contribute to the further emergence of sustainable tourism practices. It would be wrong to believe that birdwatching en masse is just a peculiarly British obsession, passionate as UK domestic devotees are (Ogilvie and Winter, 1989; Cocker, 2001; Moss, 2004) it is a hugely popular form of mass outdoor recreation in countries such as Sweden, The Netherlands and the USA (Kauffman, 1997; Snetsinger, 2003; Obmascik, 2004; Koeppel, 2005), and an emerging pastime in Australasia (Dooley, 2005; 2007; Braunias, 2007). A US Fish and Wildlife Service Report suggested that 1 in 5 Americans were bird-lovers (47.693m people to be precise), and as a hobby it contributed $36 billion to the US economy in 2006 (Carver, 2009). Estimates for the value of the British birdwatching industry range from £187m in 1998

(Murray and Simcox, 2003) to an estimated £300m by 2004 (Wallace, 2004, p219). The RSPB (in August 2010) had a national membership of 1,076,112, more than all the political parties in Britain put together (Ian Dawson, RSPB Librarian, pers. comm.).

If established conservation NGOs and government agencies appear tentative, private sector action may be a way forward. Of course, it helps if you are independently wealthy and own land! Paul Lister, ecophilanthropist and heir to the MFI fortune, owns Glen Alladale Estate in the Scottish Highlands. Inspired by efforts in Yellowstone National Park, USA, and ecolodges in South Africa, he aims to (re)wild his own landholding by reintroducing as much as he can on his 23,000 acres, behind a high fence. Ambitious, bold and visionary (and ultimately dependent on the full support of his neighbours), his private initiatives have provoked extreme public responses both for and against, but surely it all culturally stinks a wee bit of a return to mid 19th-century landlord/estate-owner arrogance. He is not shy of publicity, believing that if people know, and come and see, they will support his dream (*The Real Monarch of the Glen*, BBC, 2008; O'Connell, 2009). Some would convincingly argue that in the face of government (official) inertia on the issue, we might need private enterprise to show that it can be done – that we can reconnect people with animals, red in tooth and claw, from which they have been temporally and culturally separated for decades or longer; animals that they still perceive as 'alien' to their current existence, but which they could, in time, embrace as familiar.

There is a vision of Britain's raptorial landscapes that embraces the nature–people story, one that blends on a local or regional scale, what the global environmental historian John McNeill (2000, pxxii) has called mixing the planet's history with the people's history. We should be brave enough, and hopeful enough, to anticipate a bright future for some raptor species, as top ecological predators, sustainable tourism icons, educational and public relations tools, and powerful symbols of active conservation management in the busy working countryside. This would be a hugely symbolic step for us, as we embrace and cherished the 'other', the 'alien', once outside our normal comprehension, understanding and experiences, but now a part of our daily lives or our domestic wildlife tourism experiences (Newsome et al, 2005). It is a spiritual and ecological journey (and an economic one), and embraces an accommodation that some of the most visionary and bold environmental thinkers of the past, Aldo Leopold, Frank Fraser Darling and Peter Scott, would surely approve of. In understanding and confronting our changing relationships with birds of prey over time, we learn a lot, not just about nature, but about ourselves.

Archival sources

The archives of the Grants of Rothiemurchus are privately held at the Doune of Rothiemurchus, Inverness-shire, Scotland, and are catalogued in the National Archives of Scotland, Edinburgh, at NRA(S) 102.

The archives of the Royal Society for the Protection of Birds (RSPB) are privately held at The Lodge, Sandy, Bedfordshire, England as RSPB. See Philippa Bassett (1980) *A List of the Historical Records of the Royal Society for the Protection of Birds*, Centre for Urban and Regional Studies at University of Birmingham and Institute of Agricultural History at University of Reading.

The archives of the Scottish naturalist J. A. Harvie-Brown are held in the National Museums Scotland Library, Edinburgh. See Joy Pitman (1983) *Manuscripts in the Royal Scottish Museum, Edinburgh, Part 3: J. A. Harvie-Brown Papers*. Royal Scottish Museum, Information Series.

References

Air Ministry (1940) *Destruction of Peregrine Falcons Order*, Air Ministry, London

Aitchison, J. and Blackwell, A. (1988) *The History of the Use of Film and Video, RSPB Film Unit*, RSPB, Sandy, Beds, UK

Allen, D. A. (1976) *The Naturalist in Britain: A Social History*, Allen Lane, London

Allen, M. and Ellis, S. P. (eds) (2010) *Nature Tales: Encounters with British Wildlife*, Elliot and Thompson/The Wildlife Trusts, London

Anglian Water (2008) *The Return of the Osprey: Your Reference Guide*, Anglian Water Services, Huntingdon, UK

Anon. (1940) 'Control of Wild Life', *Nature*, 146(2 November), 586

Armitage, S. and Dee, T. (eds) (2009) *The Poetry of Birds*, Viking, London

Baker, J. A. (2010) *The Peregrine*, Collins, London [first published in 1967]

Barnes, S. (2004) *How to Be a Bad Birdwatcher*, Short Books, London

Barnes, S. (2005) *A Bad Birdwatcher's Companion*, Short Books, London

Bird, W. (2004) *Natural Fit: Can Green Space and Biodiversity Increase Levels of Physical Activity*, Report to RSPB, Sandy, Beds, UK, endorsed by Faculty of Public Health

Bird, W. (2007) *Natural Thinking: Investigating the Links Between the Natural Environment, Biodiversity and Mental Health*, Report to RSPB, Sandy, Beds, UK

Braunias, S. (2007) *How to Watch a Bird*, AWA Press, Wellington, NZ

Brown, A. (1974) *Who Cares for Animals? 150 Years of the RSPCA*, Heinemann, London

Brown, P. (1966) *Birds in the Balance*, Andre Deutsch, London

Brown, P. (1979) *The Scottish Ospreys: From Extinction to Survival*, Heinemann, London

Brown, P. and Waterston, G. (1962) *The Return of the Osprey*, Collins, London

Cairns, P. and Hamblin, M. (2007) *Tooth and Claw: Living Alongside Britain's Predators*, Whittles Publishing, Dunbeath, Scotland

Carter, I. (2001) *The Red Kite*, Arlequin Press, Chelmsford, UK

Carver, E. (2009) *Birding in the United States: A Demographic and Economic Analysis: Addendum to the 2006 National Survey of Fishing, Hunting and Wildlife-Associated Recreation*, Report 2006–4, USFWS, Arlington, VA

Cater, E. and Lowman, G. (eds) (1994) *Ecotourism: A Sustainable Option?*, John Wiley, Chichester, UK

Coates, P. (1998) *Nature: Western Attitudes Since Ancient Times*, Polity Press, Cambridge

Cocker, M. (2001) *Birders: Tales of a Tribe*, Jonathan Cape, London

Cocker, M. (2008) *Crow Country: A Meditation on Birds, Landscape and Nature*, Vintage Books, London

Cocker, M. and Mabey, R. (2005) *Birds Britannica*, Chatto & Windus, London
Collinge, W. E. (1921) 'The necessity of state action for the protection of wild birds', in *The Smithsonian Report for 1919*, US Government Printing Office, Washington, DC
Crumley, J. (1991) *A High and Lonely Place: The Sanctuary and Plight of the Cairngorms*, Jonathan Cape, London
Dee, T. (2009) *The Running Sky: A Birdwatching Life*, Jonathan Cape, London
Dennis, R. (1991) *Ospreys*, Colin Baxter Photography, Grantown-on-Spey, Scotland
Dennis, R. (1995) *The Birds of Badenoch and Strathspey*, Colin Baxter Photography, Grantown-on-Spey, Scotland
Dennis, R. (1998) 'The re-introduction of birds and mammals to Scotland', in Lambert, R. A. (ed) *Species History in Scotland: Introductions and Extinctions Since the Ice Age*, Chapter 1, Scottish Cultural Press, Edinburgh, pp5–7
Dennis, R. (2008) *A Life of Ospreys*, Whittles Publishing, Dunbeath, Scotland
Dennis, R. and Rutherford, I. (2007) 'Wing and a Prayer', *Scotsman Magazine*, Saturday 4 August, pp6–9
Diamantis, D. (2004) *Ecotourism*, Thomson, London
Dooley, S. (2005) *The Big Twitch*, Allen & Unwin, Crows Nest, NSW, Australia
Dooley, S. (2007) *Anoraks to Zitting Cisticola: A Whole Lot of Stuff About Birdwatching*, Allen & Unwin, Crows Nest, NSW, Australia
Dougall, R. (1975) *In and Out of the Box: An Autobiography*, Fontana/Collins, London
Dougall, R. (1978) *A Celebration of Birds*, Collins & Harvill Press, London
Elder, C. (2009) *While Flocks Last*, Corgi Books, London
Evans, D. (1997) *A History of Nature Conservation in Britain*, Routledge, London
Everett, M. (2002) 'Ospreys in the UK', *Birds*, Winter, 19(4), 32–38
Fennell, D. (2008) *Ecotourism*, 3rd edn, Routledge, London
Fowler, J. (2002) *Landscapes and Lives: The Scottish Forest Through the Ages*, Canongate, Edinburgh
Hamilton, F. (2009) 'Osprey anniversary', *Scottish Bird News*, 91, March, 3
Harley, D. C. and Hanley, N. D. (1989) *Economic Benefit Estimates for Nature Reserves: Methods and Results*, Discussion Paper in Economics 89/6, Department of Economics, University of Stirling, Stirling, Scotland
Harrison, B. (1973) 'Animals and the state in nineteenth-century England', *English Historical Review*, LXXXVIII (CCCXLIX), 786–820
Holden, A. (2008), *Environment and Tourism*, 2nd edn, Routledge, Abingdon, UK
Holloway, S. (1996) *The Historical Atlas of Breeding Birds in Britain and Ireland, 1875–1900*, T & AD Poyser, London
Horne, A. (2009) *Birdwatching*, Virgin Books, London
Hume, R. (2005) *Life with Birds*, David & Charles, Newton Abbot, Devon, UK
Jamie, K. (2005) *Findings*, Sort of Books, London
Johnson, S. (ed) (2010) *Bio-Invaders: Themes in Environmental History*, White Horse Press, Cambridge
Kauffman, K. (1997) *Kingbird Highway*, Houghton Mifflin, Boston, MA
Kitchener, A. C. (1998) 'Extinctions, introductions and colonisations of Scottish mammals and birds since the last ice age', in Lambert, R. A. (ed) *Species History in Scotland: Introductions and Extinctions since the Ice Age*, Chapter 6, pp63–92
Koeppel, D. (2005) *To See Every Bird on Earth*, Michael Joseph/Penguin, London
Lambert, R. A. (ed) (1998) *Species History in Scotland: Introductions and Extinctions Since the Ice Age*, Scottish Cultural Press, Edinburgh

Lambert, R. A. (2001) *Contested Mountains: Nature, Development and Environment in the Cairngorms Region of Scotland, 1880–1980*, White Horse Press, Cambridge
Lambert, R. A. (2008) '"Therapy of the Green Leaf": public responses to the provision of forest and woodland recreation in twentieth-century Britain', *Journal of Sustainable Tourism*, 16(4), 408–427
Lister-Kaye, J. (2003) *Song of the Rolling Earth*, Time Warner, London
Love, J. A. (1983) *The Return of the Sea Eagle*, CUP, Cambridge
Lovegrove, R. (1990) *The Kite's Tale: The Story of the Red Kite in Wales*, RSPB, Sandy, Beds, UK
Lovegrove, R. (2007) *Silent Fields: The Long Decline of a Nation's Wildlife*, OUP, Oxford
Mabey, R. (1997) *The Book of Nightingales*, Sinclair-Stevenson, London
McGowan, B. (2009) 'The decline of the Scottish ospreys – who was to blame?', *Scottish Birds*, 29(1), 55–58
McMillan, B. (2010) 'Report on the SOC Conference, 20th March, Inverness', *Scottish Birds*, 30(2), 138
McNeill, J. (2000) *Something New Under the Sun: An Environmental History of the Twentieth Century*, Allen Lane, London
McNeill, J. (2003) 'Observations on the nature and culture of environmental history', *History and Theory*, Theme Issue 42, December, 5–43
Millington, R. (1981) *A Twitcher's Diary*, Blandford Press, Poole, UK
Milton, K. (2000) 'Ducks out of water: nature conservation as boundary maintenance', in Knight, J. (ed) *Natural Enemies: People-Wildlife Conflicts in Anthropological Perspective*, Routledge, London. Chapter 11, pp229–244
Moore, N. W. (1987) *The Bird of Time: The Science and Politics of Nature Conservation*, CUP, Cambridge
Moss, A. W. (1961) *Valiant Crusade: The History of the RSPCA*, Cassell, London
Moss, S. (ed) (2003) *Blokes and Birds*, New Holland, London
Moss, S. (2004) *A Bird in the Bush: A Social History of Birdwatching*, Aurum Press, London
Moss, S. (2006) *This Birding Life*, The Guardian/Aurum, London
Moss, S. (2008) *A Sky Full of Starlings: A Diary of the Birding Year*, Aurum, London
Murray, M. and Simcox, H. (2003) *Use of Wild Living Resources in the United Kingdom – A Review*, IUCN-UK, London
Newsome, D., Dowling, R. and Moore, S. (2005) *Wildlife Tourism*, Channel View Publications, Clevedon, UK
Obmascik, M. (2004) *The Big Year*, Doubleday/Transworld Publishers, London
O'Connell, S. (2009) 'Born to be wild (again)', *The Daily Telegraph*, 17 March, 27
Oddie, B. (2000) *Gripping Yarns: Tales of Birds and Birding*, Christopher Helm, London
Ogilvie, M. and Winter, S. (eds) (1989) *Best Days with British Birds*, British Birds Ltd, Blunham, Beds, UK
Page, S. J. and Dowling, R. K. (2002) *Ecotourism*, Pearson Education, Harlow, UK
Riley, A. (2004) *Arrivals and Rivals: A Birding Oddity*, Brambleby Books, Harpenden, UK
Rollins, J. (2006) 'It's official: half the UK loves wildlife', *Natural World*, Spring, 31
Rothschild, M. and Marren, P. (1997) *Rothschild's Reserves: Time and Fragile Nature*, Balaban/Harley Books, Colchester, UK
Rowe, M. (2008) 'Return of the Natives', *Walk: The Magazine of the Ramblers' Association*, Winter, 21, 37–45

RSPB (1997) *Working with Nature in Britain: Case Studies of Nature Conservation, Employment and Local Economies*, RSPB, Sandy, Beds, UK

RSPB (2001) *Conservation Works for Local Economies in the UK*, RSPB, Sandy, Beds, UK

RSPB (2002) *RSPB Reserves and Local Economies*, RSPB, Sandy, Beds, UK

RSPB. (2006) *Watched Like Never Before: The Local Economic Benefits of Spectacular Bird Species*, RSPB, Sandy, Beds, UK

Samstag, T. (1989) *For Love of Birds: The Story of the RSPB,* RSPB, Sandy, Beds, UK

Sheail, J. (1976) *Nature in Trust – the History of Nature Conservation in Britain*, Blackie & Son, London

Sheail, J. (1998) *Nature Conservation in Britain: The Formative Years*, HMSO, London

Smout, T. C. (1993) *The Highlands And the Roots of Green Consciousness, 1750–1990*. Occasional Paper No 1, Scottish National Heritage, Battleby, Scotland

Smout, T. C. (2000) *Nature Contested: Environmental History in Scotland and Northern England Since 1600*, Edinburgh University Press, Edinburgh

Smout, T. C. and Lambert, R. A. (eds) (1999) *Rothiemurchus: Nature and People on a Highland Estate, 1500–2000*, Scottish Cultural Press, Dalkeith, Scotland

Snetsinger, P. (2003) *Birding on Borrowed Time*, ABA, Colorado Springs, CO

Tegner, H. (1971) *A Naturalist on Speyside*, Geoffrey Bles, London

Thom, V. M. (1986) *Birds in Scotland*, T & AD Poyser, Calton, Glasgow

Thomas, K. (1984) *Man and the Natural World: Changing Attitudes in England, 1500–1800*, Penguin Books, London

Wallace, I. (2004) *Beguiled by Birds: Ian Wallace on British Birdwatching*, Christopher Helm, London

Waterston, G. (1971) *Ospreys in Speyside*, RSPB, Sandy, Beds, UK

Wintle, A. (2009) 'On a wing and a prayer', *Telegraph Weekend*, Saturday 21 February, 17

Yalden, D. (1999) *The History of British Mammals*, T & AD Poyser, London

Yalden, D. (2003) 'Mammals in Britain – a historical perspective', *British Wildlife*, April 2003, 243–251

Television

Birds Britannia, BBC Natural History Unit, BBC4, 2010. 4 x 60 minute programmes. Produced by Stephen Moss

The Real Monarch of the Glen, BBC Scotland, 2008. 6 x 30 minute episodes. Produced by Mike Birkhead

12
Public Perception of Invasive Alien Species in Mediterranean Europe

Francesca Gherardi

I never lose an opportunity of saying exactly what I think about this particularly odious representative ... this eyesore ... this reptile of a growth with which a pack of misguided enthusiasts have disfigured the entire Mediterranean basin ... A single eucalyptus will ruin the fairest landscape.

Norman Douglas (1915)

Abstract

Public perception and attitude towards biodiversity issues are still poorly understood. Such an understanding is however critical to effectively tackle the many threats to biodiversity as those produced by invasive alien species (IAS). Perception of the negative impacts exerted by IAS differs greatly among stakeholder groups and the attitude towards their introduction and eradication varies accordingly. Here I will discuss this disparity of views and its effects with a focus on the European countries of the Mediterranean. Particularly in this region, the lack of cohesion between scientific researchers, the general public, the economic sector and policy makers may be at the root of the failure to develop and implement effective management measures. On the contrary, a direct participation by an informed and motivated public will play a critical role in finding adaptive solutions to the problems generated by IAS.

Introduction

Today there is increasing awareness that biological invasions are as much a social issue as a scientific one (Reaser, 2001). Alien species (AS, species intentionally or accidentally introduced by humans beyond their native ranges) may have in fact a large impact on our socioeconomic systems. A fraction of them, the 'invasive' alien species (IAS), inflicts economic losses (from millions to billions of dollars per year; Pimentel et al, 2005), pose threats to human health and to the health of domesticated organisms (Cox, 2004), and change the composition and functioning of ecosystems (Lloret et al, 2004). Other AS are, on the contrary, harmless or even affect human production systems that underpin economies in a positive way: they may be used for food, shelter, medicine, ecosystem services, aesthetic enjoyment and cultural identity, and provide employment opportunities in diversified economic sectors (Ewel et al, 1999).

As is widely acknowledged, an understanding by the public of the threats posed by IAS is typically associated with a general support of the decisions made on their management and prevention (Bremner and Park, 2007). However, because of the dual nature of AS as 'angels and demons', different stakeholder groups may perceive the impact of these species in a discordant way and, as a consequence, may exhibit an opposing attitude towards their management (e.g. García-Llorente et al, 2008). Here I will discuss this disparity of views and its effects with a focus on the European countries of the Mediterranean region.

The Mediterranean has a long history of species introductions (Abulafia, 1994; Baskin, 2002): for millennia its ports and islands have been centres of trade (Gritti et al, 2006) – and invasion 'hubs' as well. In recent decades, free transport and trade across EU boundaries, together with increased tourism, have significantly intensified the flux of AS into the region (Stoate et al, 2001). A result of the many pathways of species introduction into the Mediterranean is the large number (over 570) of alien marine metazoans recorded so far (Galil, 2009), a number which is significantly larger than the lists of 200 and 62 AS of the Atlantic coast of Europe and the Baltic Sea, respectively (Galil et al, 2008) (Figure 12.1). The increasing role of the Mediterranean as a hub of international commercial shipping, the surge in development of marine aquaculture farming, the continuous enlargement of the Suez Canal and global warming have all contributed to a drastic increase in AS since the 1950s (Galil, 2009).

Such a bombardment of species poses severe threats to the exceptionally rich biodiversity hosted by the Mediterranean region, one of 25 biodiversity hotspots in the world (Myers et al, 2000) with *c.*13,000 endemic plants (4.3 per cent of global plants) and 235 endemic vertebrates (out of a total of 770 species) (Kark et al, 2009). It is one of the most populated regions in the world, reaching a human density of *c.*110km^{-2} in 1995. This is mostly above the world average (42km^{-2}), with a growth rate exceeding the average for developed countries (0.3 per cent yr^{-1}) (Cincotta et al, 2000). If on the one hand such a high population density is the reason for the many threats to biodiversity in this region, on the other, the services offered by healthy ecosystems are essential prerequisites to the welfare of its people (Vilà et al, 2009).

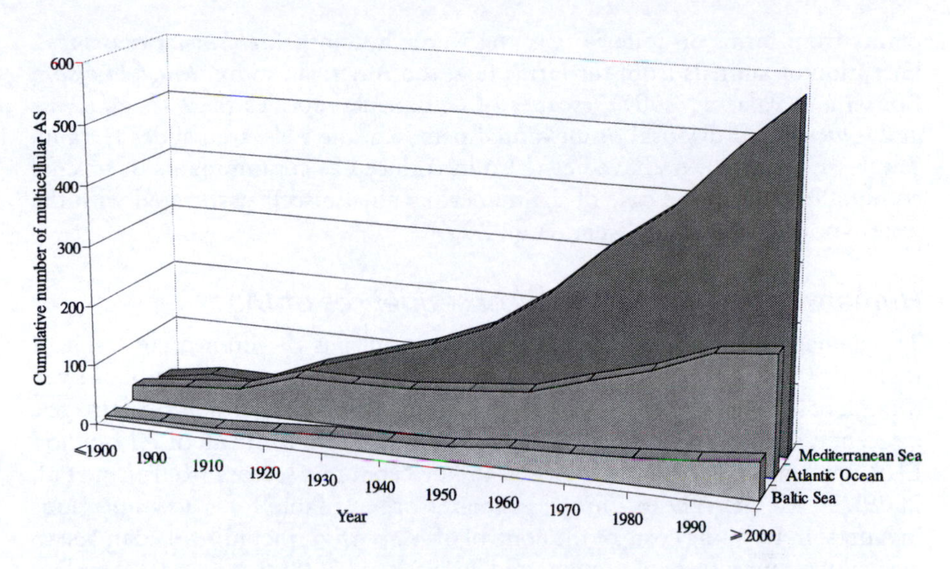

Figure 12.1 *Cumulative number of multicellular alien species recorded in European seas for each decade over the past century (after Galil et al, 2008, modified)*

The human dimension of biological invasions

Humans are involved in the entire invasive process: they serve as vectors for deliberate or accidental introductions, suffer the consequences of IAS, and possess the capacity to act and make decisions for managing these species and for preventing further introductions. The role they play in biological invasions is critical, particularly in the context of Mediterranean Europe, as shown below.

Humans are directly responsible for the establishment of IAS

Across the different biomes of the Mediterranean region, alien animals have been intentionally released as game animals (the majority of alien fishes), to improve local fauna (e.g. the red swamp crayfish *Procambarus clarkii*), as alien plants for cover (*Carpobrotus edulis*), as aquatic invertebrates as bait (the crustacean *Artemia francescana*), and several organisms as biocontrol agents (e.g. the mosquitofish *Gambusia holbrooki*). In Italy, for instance, over 30 per cent of the 112 freshwater animal AS so far recorded (and over 60 per cent of alien vertebrates) were introduced deliberately, most often in association with stock enhancement (48 per cent) and aquaculture (37 per cent) (Gherardi et al, 2008). Similarly, mariculture is responsible for the release into the Mediterranean of 12 per cent of AS on average, including the commercially important shellfish *Crassostrea gigas* and *Ruditapes philippinarum*, with a peak of 42 per cent in the western Mediterranean (Galil, 2009). The escape of alien plants and animals from managed environments is also frequent: they include feral crops, livestock and farmed animals (e.g. *Myocastor coypus*), plants from farms or aquaria (e.g. the water hyacinth *Eichhornia crassipes*), liberation of animals from fur farms (e.g. the American mink, *Mustela vison*; Bonesi and Palazon, 2007), escapes of ornamental species (e.g. *Threskiornis aethiopicus*) and disposal of unwanted pets (e.g. the red-eared slider terrapin *Trachemys scripta*). AS have been also introduced as contaminants of specific commodities, as in the case of the numerous alien insects associated with the international plant trade (Kenis et al, 2007).

Humans suffer the negative consequences of IAS

The damage that IAS inflict on human economies is the immediate result of their negative impact, and easier to understand by the general public. Even though based on a small portion of species and excluding non-market damages, a recent estimate indicates in Europe a cost associated with IAS of *c.*12 billion EUR per year with about half of that in key economic sectors (Kettunen et al, 2009). In Mediterranean Europe estimates are available for a few notorious invaders. In Italy, the costs of the control of *M. coypus*, including its damage to agriculture and river banks, amounted in 1995 and 2000 to 2.6 and 3.8 million EUR per year, respectively, but, due to the rapid expansion of the species, they are expected to reach 12 million EUR per year (Panzacchi et al, 2007). The cost of the eradication of rodents in Malta reached about 5000 EUR in two years, whereas the money spent on the control of *Eucalyptus* spp., *Pennisetum setaceum*, *C. edulis* and *E. crassipes* in different regions of Spain amounts to 1.6, 0.6, 0.3, and 3.7 million EUR per year, respectively (Shine et al, 2008).

Generally the costs of IAS are assessed in a different way according to the perception and interest of each stakeholder group: conflicts often arise when the same species is viewed as invasive by one group and beneficial by another. For instance, the European catfish (*Silurus glanis*) in the Ebro River (Spain) is regarded as a problem by environmentalist groups but it provides economic benefits to municipalities, tourist operators and fishermen (Binimelis et al, 2007).

Humans are the key factors in managing IAS

Although the case for eradication is well supported by scientists (Simberloff, 2009), very few eradication programmes have been undertaken in Europe and less in Mediterranean Europe (Genovesi, 2005). In this region, eradication attempts have been mainly directed to mammals as target species, and in particular to the ship rat *Rattus rattus* in islands (Ruffino et al, 2009). An exception is the eradication of the common carp *Cyprinus carpio* in the wetland Laguna de Zonar (Andalusia, Spain): the application of a rotenone-based piscicide led in one year to increased water transparency, recolonization by macrophytes and invertebrates, and the desired recovery of the threatened population of diving ducks, the white-headed duck (*Oxyura leucocephala*) and the common pochard (*Aythya ferina*) (Britton et al, 2008).

The scarcity of eradication projects seems to be caused by various motives, including political and public ignorance about the threats posed by

IAS (Bertolino and Genovesi, 2003), the view that eradication is an impossible goal (Bomford and O'Brien, 1995) and the lack of enthusiasm among conservationists for an action that most people consider distasteful (Temple, 1990). Lack of awareness and pessimism are the reasons for delayed interventions against invaders, as exemplified by the story of the invasion of Pacific algae *Caulerpa taxifolia* in the Mediterranean. A small infestation of the algae first appeared in 1984 just offshore of the Oceanographic Museum, Monaco. The French and Monacan government agencies argued first about who was responsible for the infestation, then about whether it would become invasive and finally, as it became apparent that an invasion was under way, about how to deal with it. The result of this debate is that C. *taxifolia* now infests over 10,000ha off the coasts of Spain, France, Monaco, Italy, Croatia and Tunisia (Simberloff, 2009).

Once an eradication or control project has been undertaken, its success or failure depends again on humans. The methods used and the species involved likely influence the level of public support. In general, poisoning and chemical control, such as the use of herbicides, are the least supported methods and the same applies to lethal injection or sterilization (Bremner and Park, 2007). People also appear to be biased towards the target species: 'hated' invaders, such as rats, are universally disliked and therefore more likely to be subject to control (Veitch and Clout, 2001), while invaders such as rhododendron or some mammals that are attractive for aesthetic or emotional reasons are less likely to be controlled. This is well illustrated by the attempt to eradicate the grey squirrel (*Sciurus carolinensis*) in Italy. In 1997, the National Wildlife Institute in cooperation with the University of Turin produced an action plan to eradicate the American grey squirrel (Bertolino and Genovesi, 2003). The first step, a trial eradication of a small population of grey squirrels at Racconigi (Turin), started in May 1997. Preliminary results showed that eradication was feasible, but the project was opposed by radical animal rights groups which took the National Wildlife Institute to court in 1997. This legal action caused a suspension of the project and led to a lengthy judicial enquiry that ended in 2000 with the acquittal of the Institute. Nevertheless, the three-year suspension of all actions led to a significant expansion of the grey squirrel's range, which made eradication no longer practical.

Humans are the key-factors for preventing introductions

In Mediterranean Europe and in other parts of Europe as well, the political will to regulate importing practices is still weak and any understanding of the invasive species problem by the industries importing and selling AS is limited. Except for aquaculture (EC Regulation 708/2007), there is no explicit legal requirement in EU member states to screen entering commodities for invasiveness risks to biodiversity (Shine et al, 2008). The European countries of the Mediterranean are all signatories to non-binding international instruments, developed in diverse sector contexts, including global (e.g. the Convention on Biological Diversity and the Code of Conduct for Responsible Fisheries) and regional (e.g. the International Council for the Exploration of the Sea, ICES, Code of Practice) instruments for biodiversity conservation and instruments that relate to sanitary and phytosanitary measures (e.g. the International Plant Protection Convention, IPPC) or to international transport (the International Civil Aviation Organization, ICAO) (De Poorter, 2009). However, any restriction or prohibition in international trade needs to be consistent with the rules and disciplines adopted by the World Trade Organization (WTO), which may hamper preventative actions. Besides, the capacity to address unintentional introductions remains low across all Europe: only France and Spain, for instance, have so far ratified the IMO Ballast Water Management Convention (Shine et al, 2008). Existing bans on imports also suffer a wide degree of heterogeneity. As an example, the sale of water hyacinth is banned in Valencia (Spain) but no equivalent measure is in place in adjacent regions. Similarly, trade in grey squirrels, prohibited in France and Switzerland, is on the contrary authorized in Italy.

Key stakeholders: scientists and wildlife managers

Scientific studies focused on AS have increased exponentially since the 1990s (Gherardi, 2007) indicating the rise in popularity that invasion biology has gained as an appealing field of research (Kolar and Lodge, 2001). However, as shown by the number of articles published between 1999 and 2009 in the specialized journal *Biological Invasions*, the large majority of studies are from North America, whereas the recorded growth of publications is slow in Europe and even slower in the Mediterranean (Figure 12.2).

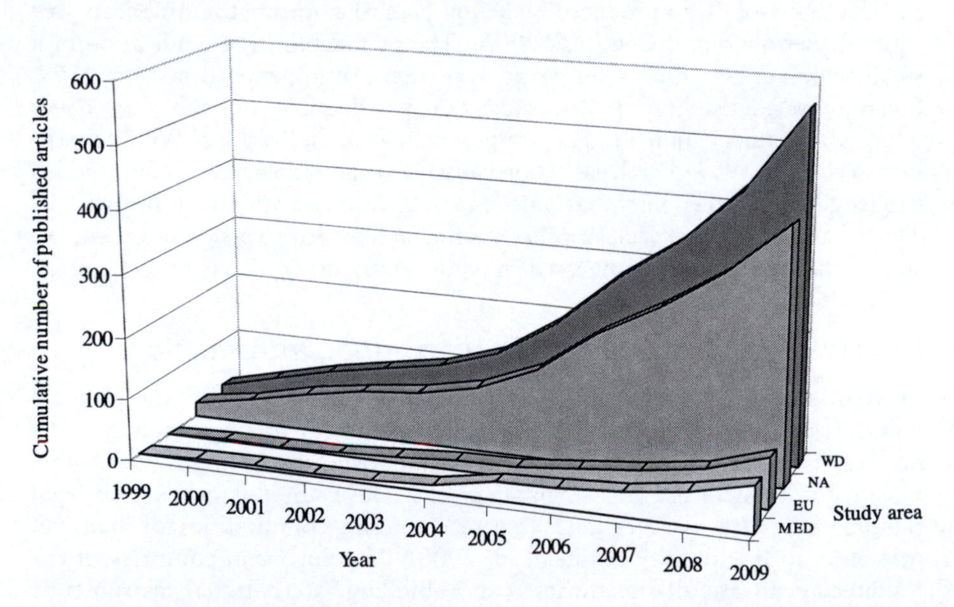

Figure 12.2 *Cumulative number of articles focused on alien species published in the journal* Biological Invasions *since 1999 compared among study areas. MED = Mediterranean Europe; EU = other European countries; NA = North America; WD = the world*

Few published articles having the Mediterranean region as the study area (17 out of a total of 63) focus on the 100 most invasive AS for Europe (DAISIE, 2008). Study species include cosmopolitan invaders (e.g. the crayfish *P. clarkii* and the dicotyledons *Ailanthus altissima* and *Oxalis pes-caprae*), invaders that exert direct impacts on human economy and health (e.g. the zebra mussel *Dreissena polymorpha*, the tiger mosquito *Aedes albopictus* and the Colorado potato beetle, *Leptinotarsa decemlineata*) and invaders that have generated discussion inside and outside the scientific community (e.g. the algae, *C. taxifolia* and *C. racemosa*, and the grey squirrel *S. carolinensis*). On the contrary, relatively little interest has been directed towards potential new invaders and few articles analyse management options or discuss the methods that might be applied to predict the risks of introductions.

The dearth of studies on AS, on one hand, and the little attention paid to actual or potential invaders and to their prevention or management, on the other, inevitably bring negative fallouts into the 'real world'. Scientists have often seemed to offer little in the way of contributions to increasing general knowledge of biological invasions and exert little influence on decisions about their management. This is also perceived by the wildlife managers interviewed by Andreu et al (2009) in Spain. Limited economic resources and negligible legislation to support action were seen by them as the main reasons that prevented necessary interventions. But the list of perceived constraints also included insufficient coordination, paucity of research on management strategies and few guidelines for prioritization. As a result, at least in the protected areas of Spain, about 80 per cent of alien plant species with a documented low impact are subject to control measures and over 95 per cent of the total expenditure on management (amounting to 50,492,437 EUR in 10 years) was targeted at five species only (*Azolla filiculoides*, *Carpobrotus* spp., *E. crassipes*, *Eucalyptus* spp. and *P. setaceum*) (Andreu et al, 2009). Without scientific coordination, wildlife managers may handle invasions differently: on average, they prioritize direct control over prevention, education and outreach; with legislation being perceived as the least relevant and effective (Andreu et al, 2009). Mechanical methods of control are promoted because they are considered to be the least harmful for the environment. Control is most often followed by annual monitoring, but only seldom has the longer-term success of actions been supervised or have standardized indicators been used to monitor any possible success. Restoration of habitats previously invaded was undertaken in just a very few cases (Andreu et al, 2009).

Notwithstanding their limited scientific understanding of the phenomenon, wildlife managers in Mediterranean Europe appear to be aware of biological invasions and national/regional agencies seem to be willing to pay to tackle the issue. This is evident from the large involvement by Mediterranean countries in projects aimed at controlling or eradicating IAS within the LIFE programme (the main EU's financial instrument aimed at promoting actions for nature conservation). From 1992 to 2006, Italian and Spanish agencies co-financed 81 projects of this type (out of a total of 187) (Figure 12.3) (Scalera, 2008). However, as lamented by Scalera (2008), these projects resulted in 'occasional'

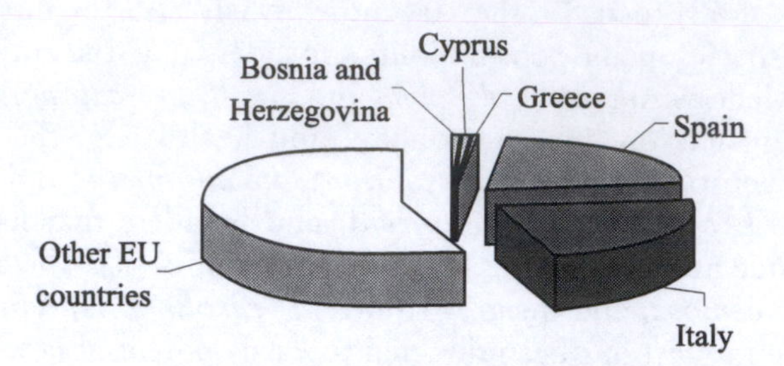

Figure 12.3 *Percentage of projects aimed at controlling or eradicating invasive alien species co-financed within the LIFE programme (1992–2006) (after Scalera, 2008, modified)*

or even 'casual' initiatives towards a handful of species (mostly *Rattus* spp. and the coypu among animals; *Populus* hybrids, *Eucalyptus* spp. and *A. altissima* among plants), without the support of either any comprehensive strategy or dedicated long-term financial programmes.

Public surveys

Compared to Australia and New Zealand, where public surveys have often been undertaken (Fraser, 2006), very little such research has been carried out in Europe, and even less in Mediterranean Europe. Although surveying techniques have limitations due to non-response bias, non-representative samples, sampling errors and influence of political and social contexts, the information they offer reflect awareness and concerns of the public and are generally a valid predictor of people's behaviour, denoting its change over time and differences among groups (Bloom, 1995).

In Mediterranean Europe, questionnaire-based studies have been focused on alien plant species on three Mediterranean islands (Mallorca in Spain, Sardinia in Italy and Crete in Greece) (Bardsley and Edwards-Jones, 2007) and in continental Spain (Andreu et al, 2009), and on a group of AS in Doñana National Park in Spain (García-Llorente et al, 2008). Qualitative techniques were also used to analyse the stakeholders' attitude towards the invasion of the zebra mussel (*D. polymorpha*) in the Ebro region (Spain) (Binimelis et al, 2007). Although this effort should be expanded to other contexts and areas, a number of useful generalizations emerge from the data gathered by the above surveys and are discussed below.

Biological invasions are viewed as an intermediate threat to biodiversity

Greater concern by wildlife managers in Spain is expressed about habitat loss and urbanization, followed by habitat fragmentation and land-use changes;

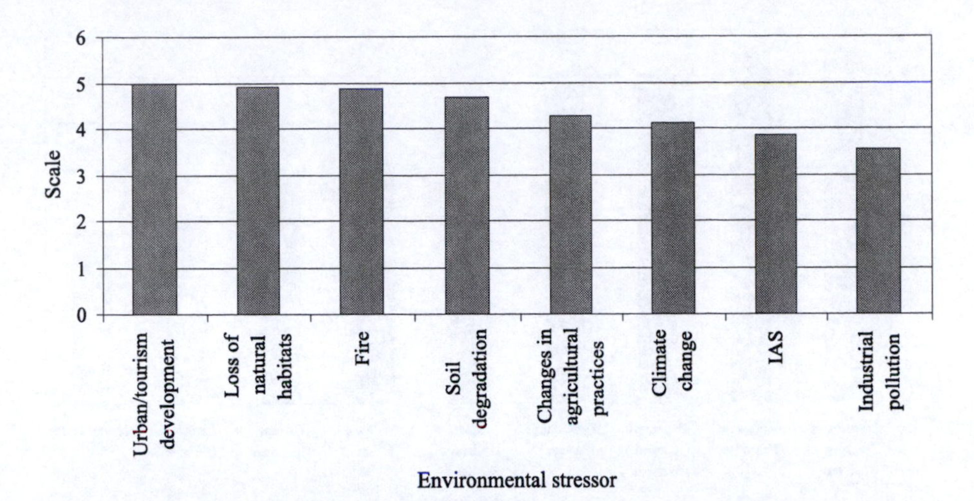

Figure 12.4 *Mean evaluation of different environmental stressors by key stakeholders in three Mediterranean islands (Crete, Mallorca and Sardinia) (after Bardsley and Edwards-Jones, 2007, modified). Scale is 1–6, with 6 = high. Total number of respondents was 142*

invasive species are ranked lower at about the same level as wildfires, pollution and climate change (Andreu et al, 2009). Similarly, the negative impact of IAS is not viewed as the principal environmental issue by key stakeholders (e.g. ecologists, government officials, wildlife managers, etc.) in Mediterranean islands (Bardsley and Edwards-Jones, 2007): urban/tourism development, loss of natural habitats, fire and soil degradation are perceived to be most important (Figure 12.4). IAS show a low visibility in other European countries as well: from a survey carried out for the EU Biodiversity Communication Campaign 2008–2010 it appears that only 2 per cent of European citizens think that IAS are an important threat to biodiversity; on the contrary, pollution (27 per cent), human-made disasters (27 per cent), climate change (19 per cent), intensive agriculture (13 per cent) and land use/development (8 per cent) generate more concern (Gellis Communications, 2008).

The perceived negative impact of IAS is mainly ecological

Alien plants are most often perceived to affect indigenous species and communities due to their ability to compete for space and soil resources with indigenous species, to lead them to extinction and to change the integrity and stability of ecosystems (Andreu et al, 2009). Similarly, key stakeholders rank the invasion of natural habitats and ecosystem disruption as the highest negative impact by IAS in Mediterranean islands, followed by economic reasons, i.e. expensive control methods and agricultural weeds; interestingly, the impact on human health is perceived as the least important (Figure 12.5) (Bardsley and Edwards-Jones, 2007).

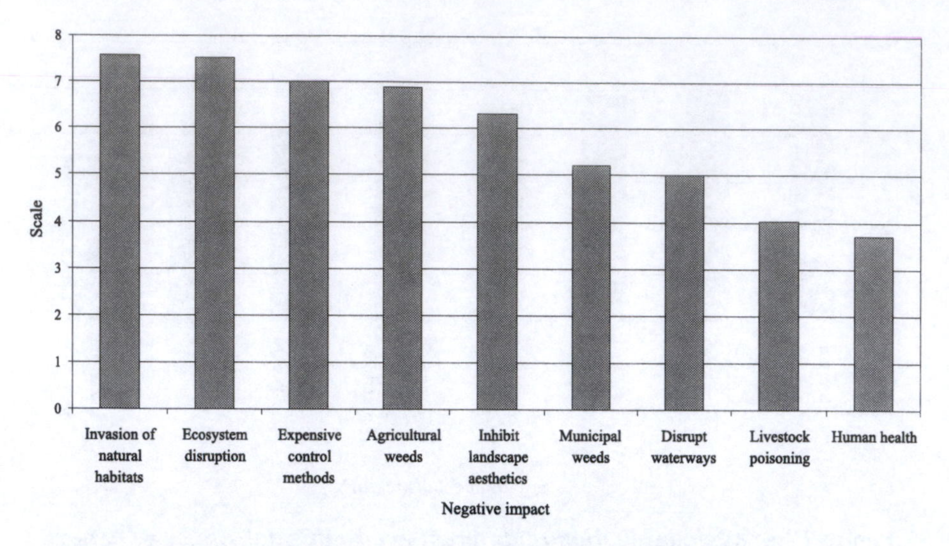

Figure 12.5 *The negative impacts of invasive alien plants perceived by key stakeholders in three Mediterranean islands (Crete, Mallorca and Sardinia) (after Bardsley and Edwards-Jones, 2007, modified). Scale is 1–11, with 11 = high. Total number of respondents was 142*

People recognize some positive impacts of IAS

The perceived positive impacts of IAS include their use as ornament and for erosion control, followed by medical and industrial uses, food and herbal uses, forestry uses and historical and cultural values (Bardsley and Edwards-Jones, 2007). People regard as beneficial even the most notorious plant invaders of Mediterranean islands, such as *Ailanthus* (which provides shade in urban areas and grows in dry climates and poor soil) and *Carpobrotus* (which serves for dune stabilization and for landscape gardening). Local people make frequent use of invasive plants: for instance, *Oxalis pes-caprae* is used as a green feed by some farmers on Mallorca, as a cover plant to reduce the impact of erosion in Sardinia and as a ground cover under olive trees and in vineyards of Crete (Bardsley and Edwards-Jones, 2007).

Even the same stakeholder group may have different opinions regarding IAS

In Doñana, some conservationists appear to be more willing to consider the different ecological and social factors involved in the process of invasion and acknowledge that the management of IAS should incorporate human practices, attitudes and perceptions (García-Llorente et al, 2008). Other conservationists, mostly members of environmental NGOs, give more relevance to the intrinsic value of biodiversity, observing that the ecological impact caused by IAS is itself a reason for their management. As shown in a similar study (Martín-López et al, 2007), environmental professionals seem to be more motivated to conserve biodiversity for scientific reasons, such as distributional uniqueness, ecological role on ecosystem functioning or endangered status, whereas anthropomorphism is the main motive for 'nature users' such as ecotourists.

Alien species is a dynamic concept

The more recently the species has been introduced, the more recognizable that species is as alien. As shown by García-Llorente et al (2008), species introduced into Doñana in the distant past, such as the common carp *C. carpio* (introduced in the 17th century during the Habsburg dynasty), were recognized as alien by few respondents, whereas species of recent introduction, such as the crayfish *P. clarkii* (introduced in 1974) and *Eucalyptus* spp. (first recorded in 1980), were recognized as alien by most respondents (García-Llorente et al, 2008). Similarly, the large majority of the wildlife managers interviewed by Andreu et al (2009) mentioned neophytes (i.e. plants introduced after 1500) more than archaeophytes as noxious species.

People are much more aware of the invasiveness of species with which they are familiar. So in Doñana, respondents were familiar with a subset of species that have been abundantly publicized, such as the ice plant *C. edulis* and the red-eared slider terrapin *T. scripta*. Similarly, widely diffused invaders (e.g. *A. altissima*, *C. edulis* and *O. pes-caprae*), are viewed as the most problematic IAS by key stakeholders in Mediterranean islands (Bardsley and Edwards-Jones, 2007).

Stakeholders accept eradication but are scarcely willing to pay for it

In Doñana, most respondents agreed that eradication of some IAS is necessary because of their deleterious impact. The meaning they assign to 'deleterious impact', however, highly differs, ranging from the damage that IAS inflict on the local economy to the threats posed to indigenous species and the right that vulnerable species have to exist (García-Llorente et al, 2008). In any case, about 60 per cent of those did not express any willingness to pay for eradication projects, mostly claiming that responsibility for solving IAS problems lay with governments and not with citizens. The majority of the other respondents appeared to be more willing to pay for eradicating species that produce acute impacts on the ecosystem structure and function than for those that threaten indigenous species. That is, freshwater plants (red waterfern *A. filiculoides* and water lettuce *Pistia stratiotes*), some fish (the mosquito fish *G. holbrooki* and *C. carpio*), and *Eucalyptus* spp. scored higher than *T. scripta*, mummichog *Fundulus heteroclitus*, the pumpkin seed *Lepomis gibbosus* and the large-mouthed bass *Micropterus salmoides*. However, a difference was found among stakeholder groups: local users were more willing to pay for eradicating species that affect them directly, such as the Argentine ant *Linepithema humile*, whereas some conservation professionals were more concerned about species without any socio-economic use but which have a far-reaching

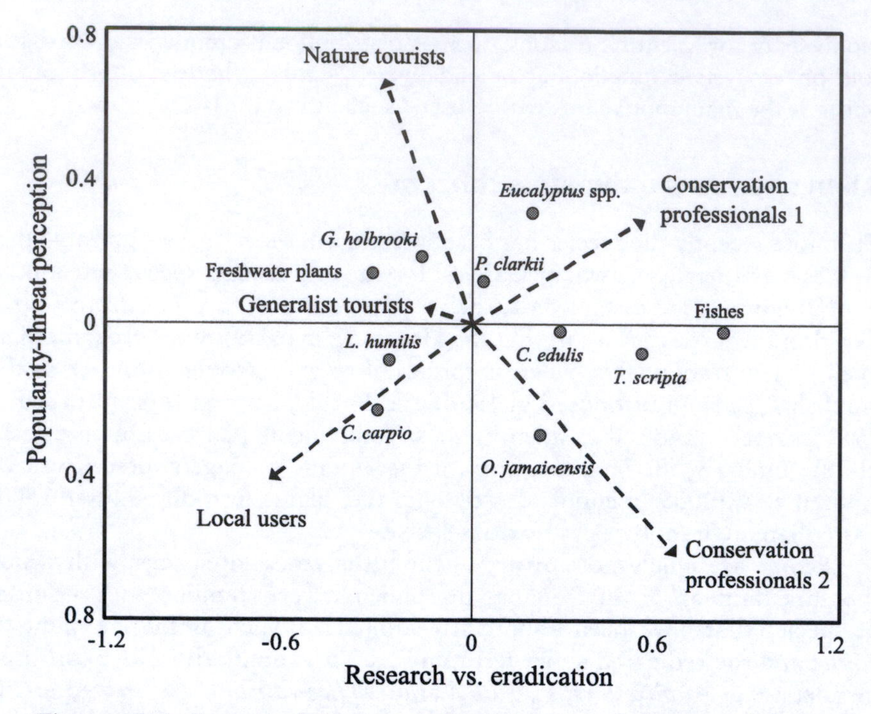

Figure 12.6 *Canonical Correspondence Analysis ordination diagram showing the relationship between stakeholder groups and their willingness to pay for eradication of selected species in Doñana National Park (Spain) (after García-Llorente et al, 2008, modified)*

impact, including hybridization with indigenous species, such as the ruddy duck *Oxyura jamaicensis* (García-Llorente et al, 2008) (Figure 12.6).

Sources of information

A recent survey has shown that for a European citizen, the main sources of knowledge regarding environmental issues are those that transmit the highest volume of information in general, i.e. the mass media (European Commission, 2008). This is particularly true in Mediterranean Europe, where, independently of the level of education of the respondents, television news reach a peak in Greece (83 per cent), followed by Cyprus (80 per cent), Spain (65 per cent), Italy (63 per cent) and Malta (62 per cent), whereas newspapers rank second. Notwithstanding this, citizens have least confidence in the mass media. When asked which information providers they trust the most, respondents from Mediterranean Europe ranked environmental protection associations and scientists far above television. Educational institutions (schools or universities), on the contrary, achieve a very low ranking (only 5 per cent of respondents named them).

Although the question remains that indeed if media coverage is a cause or a consequence of public opinion, then the role that the mass media play in disseminating information can be problematic. Environmental issues may be concealed. Since 1993, biological invasions have been the subject of an average of only 0.5 articles per year published in a total of eight Italian national newspapers. When environmental problems are examined, the media tend to sensationalize news items (Goulding and Roper, 2002), thus magnifying misunderstandings among readers (Lodge and Shrader-Frechette, 2003). Disasters instead of successes are usually reported, which contributes to developing a pessimistic attitude among the readership towards any form of mitigation of IAS and their consensus towards the 'business-as-usual' option. So, the invasion of the Mediterranean by the killer algae *C. taxifolia* generated headlines throughout Europe, featuring failed attempts to control the invader; its arrival in Californian waters was publicized with alarm and early failures to eradicate it were highlighted. On the contrary, its successful eradication in 2006 was never reported in Europe and only cited on an inside page of the local press in California (Simberloff, 2009).

Conclusions

For the citizens of Mediterranean Europe environmental protection apparently comes before the economy. When asked to indicate whether environmental protection must be given priority over the competitiveness of the economy or not, a large majority of citizens in Greece (86 per cent), Italy (65 per cent) and Spain (56 per cent) said yes (Gellis Communications, 2008). Notwithstanding this, their knowledge of the phenomenon of biological invasions is limited and the dangers of IAS are underestimated with respect to other environmental problems, such as pollution and climate change. The low relevance assigned to IAS is probably associated with the limited coverage of the problem by the mass media and with the mixed messages they broadcast. Scientists seem to direct only limited interest to biological invasions and schools certainly do not contribute to the required awareness raising. The result is inadequate scientific support to wildlife managers and limited pressure from the public to make proactive political decisions about management and prevention. Indeed, only Spain and Malta in the Mediterranean have adopted dedicated IAS strategies or action plans (Shine et al, 2008). Although the EU countries of the Mediterranean have binding and non-binding instruments that might regulate species introductions and promote control or eradication of IAS, the scope of these measures remains most often uneven for species coverage and economic sectors (Shine et al, 2008). Administrative roles and responsibilities are often unclear and, particularly in Italy and Spain, they are fragmented at subnational levels (Shine et al, 2008).

In the near future we expect accelerated rates of species introductions to be intensified by the movement of people and goods around the world and increased risks of invasions as exacerbated by climate change. To face these phenomena, significant policy measures to manage and prevent IAS must be prioritized, and the involvement of the general public in the matter is viewed as a prerequisite to make such measures effective (Boudjelas, 2009). Participation by an informed and motivated public in decision making is in fact a central tenet of any democratic country, as also recognized in 1998 by the Aarhus Convention, but it is also an effective tool that may help set priorities of intervention against IAS, provide support for their successful management and form a low-cost and wide network for monitoring and surveillance of introductions (Boudjelas, 2009). Proactive people have the potential to play a role in finding adaptive solutions to the damage inflicted by IAS. This role will be particularly crucial in the context of highly biodiverse and fragile areas, such as the Mediterranean region.

References

Abulafia, D. (1994) *A Mediterranean Emporium: The Catalan Kingdom of Mallorca*, Cambridge University Press, Cambridge

Andreu, J., Vilà, M. and Hulme, P. E. (2009) 'An assessment of stakeholder perceptions and management of noxious alien plants in Spain', *Environmental Management*, 43, 1244–1255

Bardsley, D. K. and Edwards-Jones, G. (2007) 'Invasive species policy and climate change: Social perceptions of environmental change in the Mediterranean', *Environmental Science and Policy*, 10, 230–242

Baskin, Y. (2002) *A Plague of Rats and Rubbervines: The Growing Threat of Species Invasions*, Island Press, Washington, DC

Bertolino, S. and Genovesi, P. (2003) 'Spread and attempted eradication of the grey squirrel (*Sciurus carolinensis*) in Italy, and consequences for the red squirrel (*Sciurus vulgaris*) in Eurasia', *Biological Conservation*, 109, 351–358

Binimelis, R., Monterroso, I. and Rodríguez-Labajos, B. (2007) 'A social analysis of the bioinvasions of *Dreissena polymorpha* in Spain and *Hydrilla verticillata* in Guatemala', *Environmental Management*, 40, 555–566

Bloom, D. E. (1995) 'International public opinion on the environment', *Science*, 269, 355–358

Bomford, M. and O'Brien, P. (1995) 'Eradication or control for vertebrate pests?' *Wildlife Society Bulletin*, 23, 249–255

Bonesi, L. and Palazon, S. (2007) 'The American mink in Europe: Status, impacts, and control', *Biological Conservation*, 134, 470–483

Boudjelas, S. (2009) 'Public participation in invasive species management', in Clout, M. N. and Williams, P. A. (eds) *Invasive Species Management*, Oxford University Press, Oxford and New York, pp93–107

Bremner, A. and Park, K. (2007) 'Public attitudes to the management of invasive non-native species in Scotland', *Biological Conservation*, 139, 306–314

Britton, R., Midtlyng, P., Persson, G., Joly, J.-P., Gherardi, F., Hickley, P. and Cowx, I. G. (2008) 'Assessment of mitigation and remediation procedures, and of contingency plans', Project no 044142, IMPASSE: Environmental impacts of alien species in aquaculture, 6th Framework Programme for Research (2002–2006)

Cincotta, R. P., Wisnewski, J. and Engelman, R. (2000) 'Human population in the biodiversity hotspots', *Nature*, 404, 990–992

Cox, G. W. (2004) *Alien Species and Evolution*, Island Press, Washington, Covelo and London

DAISIE (2008) *The Handbook of Alien Species in Europe*, Springer, Dordrecht, The Netherlands

De Poorter, M. (2009) 'International legal instruments and frameworks for invasive species', in Clout, M. N. and Williams, P. A. (eds), *Invasive Species Management*, Oxford University Press, Oxford and New York, pp108–125

Douglas, N. (1915) *Old Calabria*, Secker and Warburg, London

European Commission (2008) 'Attitudes of European citizens towards the environment', Special Eurobarometer, European Commission, Brussels, 295

Ewel, J. J., O'Dowd, D. J., Bergelson, J., Daehler, C. C., D'Antonio, C. M., Gomez, D., Gordon, D. R., Hobbs, R. J., Holt, A., Hopper, K. R., Hughes, C. E., Lahart, M., Leakey, R. R. B., Lee, W. G., Loope, L. L., Lorence, D. H., Louda, S. M., Lugo, A. E., Mcevoy, P. B., Richardson, D. M.and Vitousek, P. M. (1999) 'Deliberate introductions of species: Research needs', *BioScience*, 49, 619–630

Fraser, A. (2006) *Public Attitudes to Pest Control: A Literature Review*, Department of Conservation, Wellington, New Zealand

Galil, B. S. (2009) 'Taking stock: Inventory of alien species in the Mediterranean Sea', *Biological Invasions*, 11, 359–372

Galil, B.S., Gollasch, S., Minchin, D. and Olenin, S. (2008) 'Alien marine biota of Europe', in DAISIE, *The Handbook of Alien Species in Europe*, Springer, Dordrecht, The Netherlands, pp93–104

García-Llorente, M., Martín-López, B., González, J. A., Alcorlo, P. and Montes, C. (2008) 'Social perceptions of the impacts and benefits of invasive alien species: Implications for management', *Biological Conservation*, 141, 2969–2983

Gellis Communications (2008) 'Scoping study for an EU wide communications campaign on biodiversity and nature', Final report to the Commission/DG ENV, European Commission, Brussels

Genovesi, P. (2005) 'Eradications of invasive alien species in Europe: A review', *Biological Invasions*, 7, 127–133

Gherardi, F. (2007) 'Biological invasions in inland waters: An overview', in Gherardi, F. (ed), *Biological Invaders in Inland Waters: Profiles, Distribution, and Threats*, Springer, Dordrecht, The Netherlands, pp3–25

Gherardi, F., Bertolino, S., Bodon, M., Casellato, S., Cianfanelli, S., Ferraguti, M., Lori, E., Mura, G., Nocita, A., Riccardi, N., Rossetti, G., Rota, E., Scalera, R., Zerunian, S. and Tricarico, E. (2008) 'Animal xenodiversity in Italian inland waters: Distribution, modes of arrival, and pathways', *Biological Invasions*, 10, 435–454

Goulding, M. J. and Roper, T. J. (2002) 'Press responses to the presence of free-living Wild Boar (*Sus scrofa*) in southern England', *Mammal Review*, 32, 272–282

Gritti, E. S., Smith, B. and Sykes, M. T. (2006) 'Vulnerability of Mediterranean Basin ecosystems to climate change and invasion by exotic plant species', *Journal of Biogeography*, 33, 145–157

Kark, S., Levin, N., Grantham, H. S. and Possingham, H. P. (2009) 'Between-country collaboration and consideration of costs increase conservation planning efficiency in the Mediterranean Basin', *Proceedings of the National Academy of Science*, 106, 15360–15365

Kenis, M., Rabitsch, W., Auger-Rozenberg, M. A. and Roques, A. (2007) 'How can alien species inventories and interception data help us prevent insect invasions?', *Bulletin of Entomological Research*, 97, 489–502

Kettunen, M., Genovesi, P., Gollasch, S., Pagad, S., Starfinger, U., ten Brink, P. and Shine, C. (2009) 'Technical support to EU strategy on invasive species (IAS) – Assessment of the impacts of IAS in Europe and the EU', Final report to the European Commission, Institute for European Environmental Policy (IEEP), Brussels, Belgium

Kolar, C. S. and Lodge, D M. (2001) 'Progress in invasion biology: predicting invaders', *Trends in Ecology & Evolution*,16 (4), 199–204

Lloret, F., Médail, F., Brundu, G. and Hulme, P. E. (2004) 'Local and regional abundance of exotic plant species on Mediterranean islands: are species traits important?', *Global Ecology and Biogeography*, 13, 37–45

Lodge, D. M. and Shrader-Frechette, K. (2003) 'Nonindigenous species: Ecological explanation, environmental ethics, and public policy', *Conservation Biology*, 17, 31–37

Martín-López, B., Montes, C. and Benayas, J. (2007) 'The non-economic motives behind the willingness to pay for biodiversity conservation', *Biological Conservation*, 139, 67–82

Myers, N., Mittermeier, R. A., Mittermeier, C. G., da Fonseca G. A. B.and Kent, J. (2000) 'Biodiversity hotspots for conservation priorities', *Nature*, 403, 853–858

Panzacchi, M., Bertolino, S., Cocchi, R. and Genovesi, P. (2007) 'Population control of coypu *Myocastor coypus* in Italy compared to eradication in UK: A cost–benefit analysis', *Wildlife Biology*, 13, 159–171

Pimentel, D., Zuniga, R. and Morrison, D. (2005) 'Update on the environmental and economic costs associated with alien-invasive species in the United States', *Ecological Economics*, 52, 273–288

Reaser, J. K. (2001) 'Invasive alien species prevention and control: The art and science of managing people', in McNeely, J. A. (ed), *The Great Reshuffling: Human Dimensions of Invasive Alien Species*, IUCN, Gland, Switzerland and Cambridge, UK, pp89–104

Ruffino, L., Bourgeois, K., Vidal, E., Duhem, C., Paracuellos, M., Escribano, F., Sposimo, P., Baccetti, N., Pascal, M. and Oro, D. (2009) 'Invasive rats and seabirds after 2000 years of an unwanted coexistence on Mediterranean islands', *Biological Invasions*, 11, 1631–1651

Scalera, R. (2008) 'EU funding for management and research of invasive alien species in Europe', Final report for the project 'Streamlining European 2010 Biodiversity Indicators (SEBI2010)', European Commission, Brussels

Shine, C., Kettunen, M., Genovesi, P., Gollasch, S., Pagad, S. and Starfinger, U. (2008) 'Technical support to EU strategy on invasive species (IAS) – Policy options to control the negative impacts of IAS on biodiversity in Europe and the EU', Final report to the European Commission, Institute for European Environmental Policy (IEEP), Brussels, Belgium

Simberloff, D. (2009) 'We can eliminate invasions or live with them: Successful management projects', *Biological Invasions*, 11, 149–157

Stoate, C., Boatman, N. D., Borralho, R. J., Rio Carvalho, C., de Snoo, G. R. and Eden, P. (2001) 'Ecological impacts of arable intensification in Europe', *Journal of Environmental Management*, 63, 337–365

Temple, S. A. (1990) 'The nasty necessity: Eradicating exotics', *Conservation Biology*, 4, 113–115

Veitch, C. R. and Clout, M. N.(2001) 'Human dimensions in the management of invasive species in New Zealand', in McNeely, J. A. (ed) *The Great Reshuffling: Human Dimensions of Invasive Alien Species*, IUCN, Gland, Switzerland and Cambridge, UK, pp63–71

Vilà, M., Basnou, C., Pysek, P., Josefsson, M., Genovesi, P., Gollasch, S., Nentwig, W., Olenin, S., Roques, A., Roy, D. and Hulme, P. (2009) 'How well do we understand the impacts of alien species on ecosystem services? A pan-European cross-taxa assessment', *Frontiers in Ecology and the Environment*. doi: 10.1890/080083

13
The Human Dimensions of Invasive Plants in Tropical Africa

Pierre Binggeli

Introduction

The human dimensions of invasive plants have received some attention (McNeely, 2001), but the focus of the research has been on developed countries. Although most instances of invasive plant species have been reported from and investigated in the temperate zones and a limited number of publications focusing on the tropics have been produced, tropical systems are far from immune from invasive organisms (Binggeli, 1996). This review focuses solely on plants in tropical Africa and surrounding islands (for a global review of invasive woody plants, see Binggeli, 2001a) and aims to highlight how populations with a low standard of living and restricted access to the modern world perceive introduced and invasive species. The impacts of invasives on human daily activities in relation to socio-economic factors, occasionally leading to conflicts of interest between different stakeholder groups, are reviewed. Scientific activities in relation to invasive species are also investigated. In view of the dearth of publications on the topic and the difficulty in tracing information, this paper can only hope to bring a preliminary account on the way in which introduced and invasive plants are viewed, used and managed in tropical Africa.

An historical perspective of plant introductions

Plants have undoubtedly been transported by humans for millennia and were an essential component of early agricultural societies. Indeed, many regions of the tropics, and oceanic islands in particular, would not have been colonized without the introduction of various food crops. Tropical Africa, when compared to the tropical Americas and Asia, has few important native food crops and 86 introduced crop species have been documented (Alpern, 2008).

The significance of introduced crop species to African agriculture and diet is not generally appreciated. For instance, a group of mostly graduate African and European students attending a Tropical Biology Association field course identified most commonly used staple foods as of African origin. While many non-native plant species have been highly beneficial, if not essential, to the survival and development of humanity, an increasing number of species have become detrimental to the well-being of humans and their daily activities.

The history of plant introductions is closely linked with that of transportation (improved sailing technology) and European exploration of the planet (16th–19th centuries) (Grove, 1995; Mack, 2004). Oceanic islands (e.g. Mauritius) and some coastal areas were the focus of attention until the colonization of the African interior in the late 19th century. During the early period most introductions were aimed at food production to provide sailors with subsistence or species providing essential raw materials. Each colonial power established botanical gardens and experimental stations on tropical islands (e.g. Mauritius), later on mainland coastal areas and finally in more inland locations. By the 20th century, the purpose of introductions shifted from food plants to timber and other species yielding non-agricultural products. Introductions of other species, deliberate, mostly ornamentals, or accidental (e.g. seed contaminants) increased dramatically during the latter part of the 20th century. In tropical Africa four main phases of plant introductions may be recognized (after Binggeli et al, 1998):

- Early exploration and slave trade when a few fruit tree species and vegetable varieties were planted around forts at a number of coastal locations. Some species associated with religious beliefs were brought from South America.
- During the early colonial period (late 19th century) a number of experimental gardens were established by missionaries and private individuals. These were followed by several major botanic gardens established by the respective colonial powers in the 1890s and after. These gardens specialized in testing the economic value exotic plants with commercial potential but ornamentals were also introduced.
- Colonial exploitation, when large-scale forestry plantations were established after World War II using a number of introduced timber tree species.
- Post-colonial development with an increase in the number of tree species and provenances as well as the number of planting locations.

Although accidental introductions (e.g. seed contaminants) have taken place throughout these four periods, many of the most noxious weedy species were introduced during the 20th century. Secondary introductions of plants within countries are poorly reported but must have been commonplace. Two East African examples and one from Malagasy illustrate this practice. *Prosopis* spp was introduced to Kenya on more than one occasion by private individuals (seed sources are reported to have been from Hawaii and Brazil). In one instance it was done in order to increase farm productivity. Then, secondary introductions were made by the Nightingale family of Njoro and Mr de

Haller (coastal Bamburi Cement quarry) to a number of locations along the Mombasa–Nairobi railway line, and to Mrs Roberts at Lake Baringo (Nightingale, 1981). Lake Baringo is now heavily infested with the tree and has received much media attention in recent years. The Kenyan Government and FAO have been blamed for the introduction in 1980, resulting in a court case (e.g. Mawathe, 2006), suggesting that the species was planted at least on two occasions near the lake. In the 1990s, farmers from Central Ethiopia heard about the usage of *L. camara* as an effective field boundary (common practice in parts of the tropics, (Howes, 1946)) and obtained the shrub and planted it to protect their crops from their grazing animals (Binggeli and Desalegn Desissa, 2002). An *Opuntia* taxon (probably *O. monacantha*) was introduced into southwest Madagascar to defend a French fort from attack by the natives. The local tribes rapidly realized the potential usage of this plant as a live-fence, and a source of food and water, subsequently spreading the cactus to much of southern Madagascar, and by 1900 this plant dominated much of the southern Malagasy countryside (Binggeli, 2003a).

The majority of plant species that are now spreading in tropical Africa and the surrounding islands have been introduced intentionally and purposefully from one biogeographical region to another. A few notable exceptions among the woody plants include *Clidemia hirta* in Madagascar (Cabanis et al, 1969) and many agricultural weeds such as *Parthenium hysterophorus* (CABI, 2005). The history of introductions of woody plants is better documented than that of other life-forms. The purpose of introducing those species that have become invasive is broadly similar to the rest of the world, in decreasing order of importance they are: amenity planting; forestry; agriculture; landscape management; and botanic collections (see Binggeli, 1996 and Binggeli et al, 1998 for details). In many instances secondary unintentional introductions by humans have spread woody species within new biotic regions. This section illustrates the diversity of purposes involved in plant introductions.

Agriculture

Especially during the exploration and conquest of the world by western European powers, essential vegetables were planted wherever they would grow and fruit trees were widely dispersed. Scores of introduced species were planted to provide impenetrable hedges to keep livestock in or out of fields or demarcate field boundaries (Howes, 1946). Many species were introduced for various agricultural purposes including as foodstuff for livestock, preferably with both foliage and fruit being edible, for example *Prosopis* spp to East Africa (Pasiecznik et al, 2001) and *Azadirachta indica* in West Africa (Judd, 2004). Some weeds 'followed' crops as they were dispersed across the planet (e.g. weeds with rice from Asia to Africa, (Alpern, 2008)). More recently, germplasm exchanges between agricultural institutions, at both the national and international level, have favoured the spread of these species (Dantsey-Barry, 2003).

Although we do have a satisfactory chronology of when crop species were introduced (Alpern, 2008), the pathways of introductions are rather sketchy.

Some unusual examples include the introduction of *Chromolaena odorata* to West Africa to control weedy grasses, but *Imperata* spp in particular (Chevalier, 1952). It is alleged that a European planter introduced the weedy *Rubus mollucanus*, called 'vigne marrone' (wild vine), from Réunion to Madagascar because he mistook it for a grape vine (Koechlin et al, 1974).

Traditional uses

The introduction of plants for traditional uses, such as medicinal and religious practices, has received scant attention in the literature. Morat (1972) reported that the seeds of *Albizia lebbeck*, introduced from Asia via Mauritius in 1814, were widely used in divination (Sikidy) in western Madagascar. The use of plants for spiritual and medicinal purposes is central to the practice of Candomblé, a religion of the Yoruba that originated in West Africa and brought to Brazil with the slave trade; leading to the translocation of plant species across the Atlantic. The return of first-generation slaves to Africa is thought to be responsible for the introduction of now widespread neotropical species, such as *Sida rhombifolia*, *Petiveria tetrandra* and *Solanum paniculatum* (Voeks, 1990).

Wood production and environmental management

Exotic species have always fascinated people and the term exotic has often conveyed a degree of superiority. Research funding was easier to obtain to work on exotics than on indigenous species and among foresters the word exotic can become a deciding factor in choosing species. Practical considerations have also been important, such as lack of knowledge of native species and ease of propagation and establishment of fast-growing trees in plantations (Zobel et al, 1987; Persson, 1995).

Numerous woody species were introduced throughout the 20th century in the large-scale planting of trees for timber production (Richardson, 1998) and experimental plots involving scores of species were established at many forestry stations (e.g. for Madagascar see Binggeli, 2003b). Species that tolerate low rainfall, heat and poor or saline soils have been widely used in the dry tropics to halt desertification. In more recent decades the emphasis has shifted from forest timber trees to agroforestry species. Woody legumes have been widely introduced to tropical Africa and the reasons for the introduction of these taxa have been to supply fuelwood, prevent desertification, restore degraded lands, and provide fodder to livestock. More recently, the provision of food for humans to complement existing agricultural production has been a further aim (Harwood et al, 1999). Forestry introductions have occasionally resulted in the translocation of non-forestry species. Seeds of *Chromolaena odorata* were inadvertently introduced into southeast Nigeria from Sri Lanka in 1936–1937 during the importation of *Gmelina arborea* seeds to establish timber plantations (Moder, 1996).

In the East Usambara Mountains (Tanzania), an IUCN conservation project included the demarcation of forest reserves to prevent forest encroachment.

In order to make the demarcation clear, use was made of noticeable exotics including species known to be invading the natural forest, and species such as *Cedrela odorata* were planted (Binggeli et al, 1998). In some regions seeds of various tree species, including the weedy *Acacia dealbata* in Madagascar, were broadcast over large expanses of the landscape from the air (Le Bourdiec, 1972).

'Botanic gardens' and amenity planting

Large numbers of ornamentals have been widely introduced to much of sub-Saharan Africa. These introductions were carried out by botanic gardens, missions and a large number of private individuals (see Binggeli et al, 1998 for further details). Botanic gardens, which were originally chiefly concerned with the introduction of economic plants, were established in a number of European colonies mainly in the 1890s. These included Limbe, formerly Victoria, (Cameroon), Fouta-Djalon (Guinea), Eala (Congo), Lagos and Calabar (Nigeria), Entebbe (Uganda), Amani (Tanzania) and Tsimbazaza (Madagascar). Pineapple on the oceanic island of Mauritius was established much earlier in 1729. In every instance some of the planted species began to spread into the surrounding vegetation and the case of Amani Botanic Gardens has been closely examined (Dawson et al, 2008). Many of these undertakings have been largely forgotten and have sometimes vanished and this West African example illustrates this point. In 1908, the Dalaba Garden (in former French Guinea) was established with the planting of around 950 species and six years later a tree nursery was planted with forest species originating from Indochina. At the onset of World War I both sites were abandoned and by 1947 the gardens had disappeared, whereas the tree nursery had grown into a small forest. The structure and species composition of this stand was similar to a southeast Asian forest and many of the species were reported to be regenerating in the undergrowth (Adam, 1957). A botanic garden near Toliara in southwest Madagascar is protected from cattle intrusions by a dense and slowly spreading live fence of *Opuntia ficus-indica*.

Countries dominated and populated by the British, such as Kenya, rapidly built up a large catalogue of extensively planted exotics (see Jex-Blake, 1934). However, urban ornamentals have been planted throughout Africa's urban areas. In rural areas, trees such as *Azadirachta indica* have been planted in villages for shade (Judd, 2004).

Scientific

Although scientists rarely write about their motives and the historical perspectives of their work, some facts can be gleaned from the literature. The French botanist Auguste Chevalier wrote a paper on human roles in the dispersal and spread of tropical plants (Chevalier, 1931), and regarded *Chromolaena odorata* as a weed (Chevalier, 1949), yet he recommended its introduction to West Africa in order to control weedy grasses (Chevalier, 1952). The exchange of contaminated crop germplasm between research institutions appears to have led to the spread of weedy species (Dantsey-Barry, 2003).

Impacts on humans

Rural communities

In rural Africa, some species have become dominant over large areas and have a number of significant impacts, often both deleterious and beneficial to local communities. Severe negative impacts on humans are usually widely reported but other effects tend not to be so widely known.

A major feature of the majority of the most invasive species is that they form impenetrable monotypic stands and often thrive thanks to human-induced disturbances. Furthermore a number of taxa may be spiny, as in many woody plants, or have deleterious health effects. Thus, a number of species do have some highly negative and widespread effects on rural populations. Dense mats of *Eichhornia crassipes* greatly hinder or even prevent boating by fishermen and may halt fishing altogether. Transport of goods on Lake Victoria was seriously hindered during a massive bloom in 1994–1995 (CABI, 2005). The dense stands of thorny and prickly *Opuntia* spp and many other thorny taxa prevent access to large areas of the countryside in semi-arid zones. Paths situated in stands of prickly shrubs, such as *Lantana camara* and *Rubus* spp, require regular maintenance to keep access open (Binggeli et al, 1998; CABI, 2005). Thorns of various species may readily cause punctures to bicycle tyres.

A number of species can be poisonous to livestock. Ingestion of *Lantana camara* fruits may result in the death of cattle and sheep. In Kenya, varying impacts have been reported following the ingestion of *P. juliflora* pods by livestock, including diarrhoea, toxic reactions and even death (Anttila et al, 1993; Anon., 1997). Both thorns and hairs *O. monacantha* induced intestinal inflammation in Malagasy cattle, often resulting in death (Binggeli, 2003a). Local people from Kenya's Baringo region claimed that *P. juliflora* seeds stick to goats' gums, eventually causing their teeth to fall out. They marched a toothless goat into court to demonstrate their case (Mawathe, 2006).

Some invasive species favour the spread of taxa that can be highly detrimental to humans. The tree locust (*Anacridium melanorhodon arabafrum*), a major pest in Africa, and hitherto not a problem in the Lake Turkana region of Kenya, was found to be feeding on *Prosopis juliflora* introduced in the early 1980s. The subsequent spread of the shrub has given this potentially devastating pest the opportunity to become established in the area (Anon., 1997). In Madagascar, *Opuntia* infestations became refuges for a number of mammal species, introduced rat species in particular (Binggeli, 2003a).

A number of widespread species that become dominant in the landscape produce masses of fruits which are highly sought after by humans. The shrubby tree *Psidium guajava* is often considered a pest, but its fruit is highly valued. In Mauritius *Psidium cattleianum* smothers the vegetation of the National Park, yet it is the focus of large-scale fruit-collecting by thousands of people and fruit vendors. The monetary value of fruit sold by about 1000 traders over a six-month period was estimated to be around US$3 million per season (Pappiah, 2001). In the drier region of Madagascar *Opuntia* fruits are often the only ones found in local markets and are viewed as a key food by local human populations. Prior to the eradication of *Opuntia monacantha* in the 1920s, local tribes were also dependent on this cactus for water for much of the year (Binggeli, 2003a).

Some invasive species provide resources that are exploited by livestock and sometimes harvested for future use. Pods of *Prosopis juliflora* are collected and stored for feeding livestock by some Kenyan tribes where the shrub is prevalent (Anttila et al, 1993). Many invasive species are viewed as important to bee-keeping as they attract large numbers of bees (e.g. *Prosopis juliflora* in Kenya, (Gichora, 2003)). *Azadirachta indica* appears to have clear insecticidal properties beneficial to farmers (CABI, 2005).

In West Africa the grass *Imperata cylindrica* is a field weed and Chikoye et al (1999) found that over 80 per cent of farmers had to manually control the grass three to four times per season to sustain crops. The appearance of new species may result in a shift in farmers' behaviour, in most instances increasing the amount of weeding. With the appearance of *Chromolaena odorata* the traditional slash-and-burn practices could not be sustained and novel agricultural practices had to be initiated (Hauser and Mekoa, 2009). In other cases local people's reluctance to clear weeds may lead to land-use changes. As soon as *Lantana camara* became a noxious weed on the eastern coast of Madagascar, the local Betsimisaraka people preferred to abandon infested areas rather than clear land smothered by the shrub's spiny and intertwined stems (Binggeli, 2003c).

Forestry and erosion

A number of invasive tree species can provide timber products. However, the timber and wood value of invasive trees varies widely. In Malawi the timber of *Pinus patula* is worth only 5 per cent of the native *Widdringtonia cupressoides*, which it is displacing. Small tree species such as *Psidium cattleianum* in Mauritius, *P. guajava*, a neotropical species invading much of the tropics, and *Prosopis juliflora* supply valuable firewood, especially in regions that have suffered from acute deforestation (CABI, 2005). Some shrubby species have an impact on forestry operations, access to plantations and natural forests may be seriously hindered by shrub species such as *Lantana camara* and *Rubus* spp. Natural regeneration of native trees and young plantations can be suppressed by *Lantana camara* and *Chromolaena odorata. In* some mountainous regions of Tanzania and Mauritius the presence of *L. camara* and *Rubus* spp was once viewed as a good erosion-preventing ground cover (e.g. Strahm, 1993).

Public health

Some introduced species are a cause for concern and threaten public health. As many invasive species are thorny and form extensive impenetrable thickets, their sharp spines often puncture people's skin and increase the likelihood of infections. The wind-blown fruit hairs of *Opuntia monacantha* caused lung problems and conjunctivitis (Binggeli, 2003a). The fruits of a number of species are edible, but in some cases they can create health issues and/or even be fatal. The unripe fruits of the ubiquitous *Lantana camara* eaten by children have resulted in some fatalities (CABI, 2005). Pods of *Prosopis juliflora* have been found by Kenya's Turkana people to be a tasty food that has, at times, caused stomach problems (Anon., 1997). The seeds appear to vary greatly in palatability or even toxicity and this resource can only be described as famine food. Pollen and debris (trichomes) of *Parthenium hysterophorus*, a species rapidly spreading in East Africa, has been reported as causing severe allergenic reactions. In India *Parthenium*-contaminated animal feed led to tainted milk that causes Indian childhood cirrhosis (CABI, 2005).

Indirect health effects are also well reported. In Tanzania *L. camara* thickets provide breeding grounds for tsetse flies infected with trypanosomes of domestic animals, so the species is considered a serious health threat (CABI, 2005). By promoting stagnant water, *Eichhornia crassipes* favours mosquitoes and other insects as well as snails, and these can propagate serious and widespread diseases such bilharzia, filariasis and malaria (CABI, 2005).

Since the 19th century introduced plants have been used by specialist and non-specialist African healers and include invasive species such as *Argemone mexicana*, *Datura stramonium*, *Lantana camara*, and *Solanum mauritianum* (Dold and Cocks, 2000). Introduced and invasive species can represent a substantial proportion of the plants used in traditional medicine (e.g. for Madagascar see Pernet, 1957, 1959; for West Africa, Burkill, 1985–97; Alpern, 2008). However, many of these species have been introduced only a few decades ago and it is generally unclear how these plants should be identified for the cure of particular ailments. In South Africa, Dold and Cocks (2000) reported recent new usage of introduced plants and concluded that African traditional healing is dynamic. In most cases information relating to the use of new species came from family members and for one specialist four species were identified by means of promoting dreams! Much remains to be understood about how a recently introduced plant species gains an important role in traditional health care and whether they actually provide any health benefit.

Perceptions of introduced and invasive plants

Although academic journals give the impression that invasive plants are chiefly, and sometimes only, a concern of scientists and conservationists, the grey literature and the media reveal that society has, at times, great interest in these issues. Cultural and political aspects of non-native species and their effect on people's perception, including that of scientists, are sometimes important and the following reported examples will illustrate this point.

Species names

Vernacular names assigned locally to new species give a good indication of people's perception of their status (Table 13.1). *Chromolaena odorata* has been given many names in tropical Africa and this reflects the species speed of spread and negative impacts on countryside, often using political leaders for inspiration.

Table 13.1 *Examples of common names of invasive plants referring to their weediness*

Scientific name	*Common name*	*Meaning*	*Region*	*Source*
Chromolaena odorata	Acheampong	Military head of state	Ghana	Timbilla and Braimah (1996)
Chromolaena odorata	Adiawuo	Killer	Ghana	Timbilla and Braimah (1996)
Chromolaena odorata	Woafa me fuo	You have taken my farm	Ghana	Timbilla and Braimah (1996)
Chromolaena odorata	Wo amma me gye	I am taking over if you are not coming	Ghana	Timbilla and Braimah (1996)
Chromolaena odorata	Bokassa	Ruling president	Central African Republic	Loumeto (1998)
Chromolaena odorata	L'envahisseur	The invader	Cameroon	Baxter (1995)
Chromolaena odorata	Oiabantou	Toxic	Congo	Banil and Le Gall (1996)
Chromolaena odorata	Rawlings	Species' forceful growth recalls Rawlings' repeated seizures of power	West Africa	Loumeto (1998)
Lantana camara	Curse of India	Indicates a pest introduced from India	East Africa	Pratt and Gwynne (1977)
Solanum verbascifolium	Fiente de sauterelle or Kondogbo	Grasshopper's excrement	Sierra Leone	Portères (1959)

Rural populations

There appears to be a clear tendency for rural populations to favour the planting of introduced trees. They are seen as faster growing and requiring less maintenance. Many species are also viewed as unpalatable by livestock and thus more likely to survive. When Judd (2004) was trying to grow native species in Gambia some incredulous farmers remarked 'Why do you grow these? We already have them!' In much of rural Ethiopia there is no tradition of growing trees from seeds and farmers will then readily buy seedlings of easily grown exotics (Binggeli, pers. obs.). Farmers do not differentiate between native and introduced weeds, but rapidly identify and assess new weeds. In West Africa, *Imperata cylindrica* and *Chromolaena odorata* are widespread invasive plants of agricultural land and are ranked, respectively by 50 per cent and 39 per cent of farmers, as the two most important weeds (Chikoye et al, 1999). Around Lake Victoria surveys identified a number of important effects of *E. crassipes* on lake-shore communities. The main social effects were difficulty in accessing water points, increase in vector-borne diseases, and migration of communities. Reduced fish catches, increase in transportation costs, and difficulties in water extraction were perceived as the main negative economic consequences of the weed (Mailu, 2001).

Campaigns to educate the public have been initiated for some time. After World War II a poster with the bold caption: 'Wanted for stealing the land! Kill Lantana on sight' was circulated in southern Africa. In 2009 posters entitled 'Warning! Dangerous weed invasion – Uproot and destroy it on spot!' could be seen here and there in Uganda. They also warned the population against the deleterious properties of *Parthenium hysterophorus*. These belated campaigns are initiated when the introduced species is beyond the point of eradication and probably have no impact on the course of the invasion.

City-dwellers

Little can be gleaned about city-dwellers' perception of invasives from the literature. One exception is the importance of *Psidium cattleianum* to people on Mauritius. The species is ubiquitous in the green parts of the island and, when it bears fruit, it provides Mauritians an opportunity to visit the countryside and gather ample delicious fruit. Not only does it provide a sought-after food, it also turns into a major family occasion that often results in massive traffic jams.

Media

A decade before biological invasions became a global issue the *Maesopsis eminii* invasion of the Est Usambaras made the front page of the Tanzanian *Daily News*. Two articles provided a good popular account of the issues relating to the tree invasion of an important biodiversity site (Mwalubandu, 1989a, b). When *Prosopis juliflora* became a widespread weed at the turn of the millennium it made headlines in newspapers and radio programmes in much of Africa and Europe. Some of these headlines are more sensational than factual (Table 13.2), though no different from an article entitled '*Chromolaena odorata*: The benevolent dictator?' written by and for biologists (Norgrove et al, 2008).

The scientific community

Interest in invasive species stems back to the mid 19th century (de Candolle, 1855), yet most scientists neglected this topic until the 1980s. In the 1950s, 'ecologists worked mainly in natural systems, often avoiding human-modified systems and alien organisms as if these were "noise"' (Richardson, 2000) and many phytosociologists have maintained this tradition much longer. Until quite recently invasive plants have received greater attention in Anglo-Saxon countries than elsewhere. This emphasis reflected differing scientific traditions relating to the perception of vegetation as well as scientific method. Scientists' perceptions of introduced species and of the impacts of invasives have varied

Table 13.2 *Selected media headlines depicting the spread of invasive plants in Africa*

Prosopis juliflora	
Kenya's imported dream tree becomes a nightmare.	*Mail and Guardian*, 16 April 2006
Devil of a problem: the tree that's eating Africa.	*The Independent*, London, 2006
Saviour shrub turns killer.	*Kenya Times Magazine*, 2006
Au Kenya, l'arbre miracle est devenu fléau.	*La Libre Belgique*, 19 April 2005
Killer weed hits Kenyan herders.	*BBC News*, 7 August 2006
Goats: heroes of drought-stricken Africa.	*The Times*, 18 February, 2006
Residents file fresh suit over 'toxic weed'.	*The East African Standard* (Nairobi), 24 August 2006
Namibia: alien trees pose invader problem.	*The Namibian*, 27 July 2006
Chromolaena odorata	
Acheampong weed is killing Ghana.	*Accra Mail*, 14 October 2002

widely and appear to be strongly affected by their area of biological specialization and background.

In conservation circles invasive species are widely viewed as deleterious additions to ecosystems. However, in countries such as Ethiopia, where biodiversity is highly threatened by wood harvesting and deforestation, the spread of invasive harvestable woody species could be viewed as an potential answer to prevent species extinction (Ensermu, pers. comm., 2001). In Madagascar invasive plants were only viewed as a problem if they were deemed to be harmful to human activities and especially if they were economically detrimental (Perrier de la Bâthie, 1928). Carrière et al (2008) found that 37 per cent of people involved in nature conservation in Madagascar had no opinion on the subject of bio-invasions or thought they were not a problem. On the other hand managers of conservation areas considered invasive species a major issue but were unable to take action as the official policy of National Parks is to allow nature to take its course. NGOs were considering what to do about the problem, but thought it not to be significant because they consider invasions to occur only in disturbed areas, hence outside protected areas.

The eradication of *Opuntia monacantha* from southern Madagascar in the early 20th century caused a major controversy and has been the focus of renewed interest in recent years. Three reviews of the subject have resulted in differing conclusions probably reflecting the researchers' respective backgrounds and training (see Middleton, 1999; Kaufmann, 2001; Binggeli, 2003a).

Much of nature conservation focuses on taxonomic groups that are attractive to the public, such as most bird and many mammal species. In order to enhance the prospects of rare species, habitats are managed in order to build up their populations. In some cases invasive species are recognized as supplying resources deemed to be essential to threatened species. In Kenya lantana thickets provide shelter, now readily available in a human-dominated countryside, to a threatened bird species Hinde's Babbler (*Turdoides hinduei*) (Njoroge and Bennun, 2000). Presumably Kenyan bird conservationists would now view lantana most favourably. A Malagasy lemur species can spend up to 95 per cent of its time looking for fruits of the invasive *P. cattleianum* and the control of the tree could be detrimental to lemurs (Carrière et al, 2008). In southern Madagascar Long and Racey (2007) suggested that plantations of exotic species may viewed as a keystone resource for endemic fruit bats. It appears that researchers involved in single taxa conservation may overlook the issues of ecosystem conservation by focusing on the value of an invasive to their subject of interest.

Agroforesters have generally viewed the natural spread of woody plants into semi-natural vegetation and highly degraded habitats as a bonus (Richardson et al, 2004). Baumer (1990) stated in reference to Africa that, 'one would be only too happy in certain very degraded, not to say denuded, regions to find an invasive plant with as many qualities as *Prosopis*' and this view is largely supported by others (e.g. Coppen, 1995). These authors were also of the opinion that the spread of *P. juliflora* had to be checked in some areas to prevent any negative spread through good management (i.e. 'necessary precautions'). Similar statements have often been made in the agroforestry literature (e.g. BOSTID, 1980); however, workable guidelines have never been provided to indicate how 'good management' could be achieved and more importantly how 'necessary precautions' could be defined and successfully implemented. As highlighted by Mack (2008) agroforestry researchers knew that the species they were promoting were a hazard, but still favoured their introduction, and the current interest in biofuel crops appears to suffer from the same issues. Furthermore, most authors have largely ignored the environmental and human impacts of spreading agroforestry species. In the Sudan the spreading *P. juliflora* was viewed as having a great potential for rural people and became a key component of agro-ecosystems, and it was actively promoted. 'The natural spreading of this exotic tree species can be viewed as of great value in areas empty of natural vegetation and common in Sudan today' (El Fadl, 1997); and the author hoped 'that the results of this work will be of benefit to mankind'. Within a decade the emphasis had shifted to containing the weed's spread. The invasive potential of *Prosopis* species has long been known and this issue was ignored when the taxa was considered for large-scale planting in Africa. In Kenya *Prosopis* taxa were widely informally used and planted along the coast and inland throughout the 20th century. It became a major focus of an agroforestry project in the Bura region and it was rapidly realized that the shrub had strong weedy tendencies. Apart from a short note by concerned entomologists (Anon., 1997) the scientific community, and agroforesters in particular, did not confront the issue until this decade when forced by complaints from the general public that were widely reported in the media. In South Africa, a country faced with similar problems, Zimmermann et al (2006) concluded that, 'It is unfortunate that the agroforestry fraternity is not becoming more involved in finding solutions to the problems that they have created and in preventing future invasions from new species.'

Traditionally, attributes such as ease of propagation and fast growth rates have been favoured in agroforestry species selection. However, any detailed species specific descriptions of desirable traits are unusual. A recent, well described example was *Acacia colei* identified as the most promising species among a number of Australian taxa introduced to Niger. Harwood et al (1999) identified the most significant traits of *A. colei* as:

- prolific seeding;
- early fruiting (from age of two years onwards);
- foliage that is unattractive to livestock.

They also noted that:

- The species grew well on disturbed sites.
- It was self-fertile.
- Genetic changes could lead to high-yielding individuals.
- To reduce costs direct seeding would be an advantage.

They were aware of the invasive issue and swiftly dealt with the risk of invasiveness by stating that:

- Many of the useful plants in the region 'were also exotic at one time.
- Australian acacias do not sucker, seldom regenerate naturally from seed under Sahelian conditions, and have shown no ability to spread as weeds in the 20 years following their introduction to the Niger.

The view that *A. colei* will not become troublesome because it has shown no sign of spread after 20 years since its introduction is misguided. Most introduced woody plant species that have become invasive exhibit a time lag between their introduction and subsequent spread, typically in the tropics of between 40 and 70 years (Binggeli, 2001b). The duration of these time lags is determined by:

- biotic factors (e.g. change in grazing regime or dominant herbivore species, introduction of pollinator);
- abiotic factors (e.g. unusual and large disturbance events such as hurricane, flood, logging).

Another important factor is the issue of secondary introductions. When a species is first introduced to a region in a habitat inappropriate for natural regeneration (e.g. botanic gardens, trial plots) and is subsequently moved intentionally or accidentally to a habitat favourable to the species spread, it may become invasive (Binggeli, 2001b). In fact, the positive traits identified by Harwood et al (1999) are indicative of a high invasive potential and are widely reported in the literature. This suggests a lack of awareness of these basic ecological facts among agroforesters.

At the onset of the 21st century the extent and impacts of the *P. juliflora* invasion on the drier regions of Kenya were revealed to the world via the media (Table 13.2). Yet the spread of the species had been well-known by agroforesters for over a decade and had been widely commented upon in 1993 in a special issue of the *East African Agriculture and Forestry Journal*. In fact, the whole scientific community failed to take notice and consider what may happen in the future. In the African literature there appears to be only one clear report of an individual identifying a species that could become troublesome and that required immediate action. Perrier de la Bâthie (1928), a leading Malagasy naturalist denigrated by Middleton (1999) as the 'peripheral amateur', identified *Eichhornia crassipes* as a potential threat to the freshwater bodies of Madagascar two decades after its introduction as an ornamental. His advice to eradicate the species was not heeded and by the late 20th century *E. crassipes* became an obnoxious weed (Binggeli, 2003d). This call for action, in view of the threat of potential extensive spread by a species, is in stark contrast to the widespread current advice for species monitoring.

Conflicts of interest

Clearly many invasive plants have both beneficial and deleterious effects on humans and of course often have major negative impacts on biodiversity. In many instances the conservation of biodiversity can only be successful if the invasive species is controlled and preferably eradicated. This is readily conceivable as long as the invader does not have any clear benefits. From the above review it is clear that a substantial number of invasive plants in Africa have benefits for some human groups. Various stakeholder groups will value a particular species differently. Conflicts of interest arise between those groups that want to either control a nuisance or maintain a precious resource. Species that produce vast quantities of edible or valuable fruits (*Opuntia* spp, *Psidium* spp, many legumes) will have supporters (e.g. farmers, fruit traders, single-species conservationists) who will object to the control of the weed, whereas for others, such as *Parthenium hysterophorus*, few or no objections will be made. For widespread invasive plant eradication is impossible and sustained control can potentially be achieved only through the introduction of biological control agents. These agents often target fruit and/or seed production and such a course of action will prove impossible where plants are highly valued for their fruits. For woody plants primarily used as a source of fuelwood this will not be an issue but may not be effective (e.g. for a South African programme; see Zimmermann et al (2006)). Other control methods include the use of the species in novel ways. In Mauritius, a project was envisaged to use *Psidium cattleianum* wood from the National Park to generate electricity (Pappiah, 2001).

Species with both positive and negative impacts, such as *Chromolaena odorata*, which are perceived differently in different locations and by different stakeholder groups, are more difficult to deal with. Many compromises have to be made to try to bridge the gap between conflicting opinions. The likely outcome will be more a change in people's habits and traditions than a serious and effective control programme. In the case of C. *odorata*, it was suggested that if the species' positive effects on soil fertility in fallows proved to be correct the promotion of this plant with small-scale farmers might even be considered (Baxter, 1995).

In many instances conflicts of interest are most likely to arise between conservation priorities and local priorities. *Senna spectabilis* is considered a threat to the Gombe National Park and attempts at controlling it have been made. A recent survey of villagers adjoining the park revealed that *Senna* spp were the fifth most frequently listed tree taxa; it is a valued firewood (Chepstow-Lusty et al, 2006), and it is most unlikely that villagers would condone the disappearance of the resource unless an alternative were provided. Kull et al (2007) and Carrière et al (2008) have extensively discussed the conflicts of interest associated with *Acacia* species and *P. cattleianum* in Madagascar.

Conclusions

Introduced and invasive plants in tropical Africa have many facets. This review shows that the history and causes of introductions and people's perceptions of invasive plants as well their impacts are highly variable. For most of the poor rural Africa, where focusing on day-to-day survival is often the sole concern for much of the population, the status and origin of species is of little importance. In fact rural people's only concerns are the beneficial and deleterious effects of plants, whether they are native or not. People rapidly accept new species if beneficial and when confronted with undesirable plants they quickly try to make use of them. In parts of rural India, when confronted with serious environmental problems, villagers rely on outside help and development assistance has, as is traditionally the case, focused on the introduction of alien plants, even when the indigenous population is aware of the limited merit provided by the new species (Binggeli, 2001a). All these aspects have been studied too little in Africa and warrant further attention.

Conflicts of interest between various sectors of society are inevitable. Sharp differences of opinion and perception exist as to the value of a particular plant within the agricultural sector. A species may be viewed as a weed in large commercial monocultures whereas it may be perceived to be beneficial in traditional small-scale agriculture (and vice versa). Among conservationists similar conflicting points of view also exist. More difficult to solve are going to be those conflicts of interest between conservationists, applied scientists, horticulturalists and more particularly rural populations. As African countries are likely to face drastic changes in their social fabric and economic structures these conflicts of interests are likely to change in line with changes in perception and usage of introduced species. These changes have already been observed in the developed world but how extensive this might be is unclear as only anecdotal evidence has so far emerged (Binggeli, 2001a). In Africa woodfuel harvesting is extensive but will this resource be used as extensively in future? If harvesting of wood for fuel decreases, perception of invasive woody species used for that purpose may well change radically.

The introduction of many plant species has been essential to the development of modern societies and the survival of many rural communities, yet too many species with major deleterious effects have been intentionally introduced to tropical Africa. These introductions have been carried out with little concern for potential problems that they may cause. Wherever a species started spreading, no action was taken until the plant had become a major pest; by that stage any hope of eradicating the problem would have vanished. The introduction of ornamentals via the horticultural trade has hitherto been limited compared to developed countries. How to prevent the introduction and spread of ornamentals introduced as well as the unintentional spread of seeds of weedy species is a major challenge facing Africa. Aid projects have to demonstrate the need for and long-lasting beneficial effects on human societies of introducing and promoting new species in the African tropics. It is also obvious that risk assessments should be mandatory for new introductions to a country and the translocation of rare or uncommon introduced species known to have invasive tendencies within a country. Above all more emphasis should be placed on enhancing the potential value of native species and plant communities, and providing environmental management systems appropriate to local conditions. This would help combat future negative impacts of invasive exotics.

Acknowledgements

Thanks to Jim Paterson and Dick Mack who swiftly provided comments on the MS. A number of long-suffering friends provided essential technical and material support during the write-up.

References

Adam, J. G. (1957) 'Le jardin Chevalier ... Dalaba (Fouta-Djalon, Guinée française)', *Bulletin I.F.A.N., Sér. A*, 19, 1030–1046

Alpern, S. B. (2008) 'Exotic plants of western Africa: Where they came from and when', *History in Africa*, 35, 63–102

Anon. (1997) 'Prosopis ... a desert resource or a menace?', *National Museums of Kenya Horizons*, 1, 14

Anttila, L. S., Alakoski-Johansson, G. M. and Johansson, S. G. (1993) 'Browse preference of Orma livestock and chemical composition of *Prosopis juliflora* and nine indigenous woody species in Bura, eastern Kenya', *East African Agriculture and Forestry Journal*, 58, 83–90

BaniI, G. and Le Gall, P. (1996) '*Chromolaena odorata* (L.) R. M. King and H. Robinson in the Congo', in Prasad, U. K., Muniappan, R., Ferrar, P., Aeschliman, J. P.and de Foresta, H. (eds) *Distribution, Ecology and Management of* Chromolaena odorata. *Workshop Report,* ORSTOM, ICRAF and University of Guam, pp25–28

Baumer, M. (1990) '*The Potential Role of Agroforestry in Combating Desertification and Environmental Degradation with Special Reference to Africa*', CTA, Wageningen, The Netherlands

Baxter, J. (1995) '*Chromolaena odorata*: Weed for the killing or shrub for the tilling?', *Agroforestry Today*, 7(2), 6–8

Binggeli, P. (1996) 'A taxonomic, biogeographical and ecological overview of invasive woody plants', *Journal of Vegetation Science*, 7, 121–124

Binggeli, P. (2001a) 'The human dimensions of invasive woody plants', in McNeely, J. A. (ed) *The Great Reshuffling – Human Dimensions of Invasive Alien Species*, IUCN, Gland, Switzerland, 145–159

Binggeli, P. (2001b) 'Time-lags between introduction, establishment and rapid spread of introduced environmental weeds', in Proceedings of the III International Weed Science Congress, International Weed Science Society, Corvallis, OR, USA

Binggeli, P. (2003a) '*Opuntia* spp., prickly pear, raiketa, rakaita, raketa', in Goodman S. M. and Benstead, J. P. (eds) *The Natural History of Madagascar*, University of Chicago Press, Chicago, USA, pp335–339

Binggeli, P. (2003b) 'Introduced and invasive plants', in Goodman S. M. and Benstead, J. P. (eds) *The Natural History of Madagascar*, University of Chicago Press, Chicago, USA, pp257–268

Binggeli, P. (2003c) '*Lantana camara*, fankatavinakoho, fotatra, lantana, mandadrieko, rajejeka, radredreka, ramity', in Goodman S. M. and Benstead, J. P. (eds) *The Natural History of Madagascar*, University of Chicago Press, Chicago, USA, pp415–417

Binggeli, P. (2003d) '*Eichhornia crassipes*, water hyacinth, Tetezanalika, Tsikafokafona', in Goodman S. M. and Benstead, J. P. (eds) *The Natural History of Madagascar*, University of Chicago Press, Chicago, USA, pp476–478

Binggeli, P. and Desalegn, D. (2002) '*Lantana camara* – the invasive shrub that threatens to drive people out of their land', *Newsletter of the Ethiopian Wildlife and Natural History Society,* April–June 2002, 4–6. http://members.multimania.co.uk/ethiopianplants/invasives/lantana.html

Binggeli, P., Hall, J. B. and Healey, J. R. (1998) '*A Review of Invasive Woody Plants in the Tropics*', School of Agricultural and Forest Sciences Publication, Number 13. University of Wales, Bangor, UK, www.safs.bangor.ac.uk/iwpt

BOSTID (1980) '*Firewood Crops'*, National Academy of Sciences, Washington, DC, USA

Bourdiec, P. Le (1972) 'Accelerated erosion and soil degradation', in Battistini, R. and Richard-Vindard, G. (eds) *Biogeography and Ecology in Madagascar*, Junk, The Hague, The Netherlands, pp227–259

Burkill, H. M. (1985–1997) '*The Useful Plants of West Tropical Africa'*, Vols 1–4, 2nd edn, Royal Botanic Gardens, Kew, London, UK

Cabanis, Y., Chabouis, L. and Chabouis, F. (1969) '*Végétaux et groupements végétaux de Madagascar et des Mascareignes, Volume 2*', Bureau pour le Développement de la Production Agricole, Tananarive, Madagascar

CABI (2005) '*Crop Protection Compendium'*, CAB International, Wallingford, UK. www.cabicompendium.org/cpc

Candolle, A. de (1855) '*Geographie Botanique, Tome 1 and 2'*, Masson, Paris, France

Carrière, S. M., Randrianasolo, E. and Hennenfent, J. (2008) 'Aires protégées et lutte contre les bioinvasions: des objectifs antagonistes? Le cas de *Psidium cattleianum* Sabine (Myrtaceae) autour du parc national de Ranomafana à Madagascar', *Vertigo*, 8, 1–14

Chepstow-Lusty, A., Winfield, M., Wallis, J. and Collins, A. (2006) 'The importance of local tree resources around Gombe National Park, Western Tanzania: Implications for humans and chimpanzees', *Ambio*, 35, 124–129

Chevalier, A. (1931) 'Le role de l'homme dans la dispersion des plantes tropicales', *Revue de Botanique appliquée et d'Agriculture tropicale*, 11, 633–650

Chevalier, A. (1949) 'Sur une mauvaise herbe qui vient d'envahir la S. E. de l'Asie', *Revue de Botanique appliquée et d'Agriculture tropicale*, 29, 539–537

Chevalier, A. (1952) 'Deux Composées permettant de lutter contre l'*Imperata* et l'empèchant la dégradation des sols tropiquaux qu'il faudrait introduire rapidement en Afrique noire', *Revue de Botanique appliquée et d'Agriculture tropicale*, 32, 494–497

Chikoye, D., Ekeleme, F. and Ambe, J. T. (1999) 'Survey of distribution and farmers' perceptions of speargrass [*Imperata cylindrica* (L.) Raeuschel] in cassava-based systems in West Africa', *International Journal of Pest Management*, 45, 305–311

Coppen, J. J. W. (1995) '*Gums, Resins and Latexes of Plant Origin*', FAO, Rome, Italy

Dantsey-Barry, H. (2003) 'Problems of introduced weeds in West Africa', in Labrada, R. (ed) *FAO Expert Consultation on Weed Risk Assessment*, FAO, Rome, Italy, 45–53

Dawson, W., Mndolwa, A. S., Burslem, D. F. R. P. and Hulme, P. E. (2008) 'Assessing the risks of plant invasions arising from collections in tropical botanical gardens', *Biodiversity and Conservation*, 17, 1979–1995

Dold, A. P.and Cocks, M. L. (2000) 'The medicinal use of some weeds, problem and alien plants in Grahamstown and Peddie districts of the Eastern Cape, South Africa', *South African Journal of Science*, 96, 467–473

El Fadl, M. A. (1997) 'Management of *Prosopis juliflora* for use in agroforestry systems in the Sudan', *University of Helsinki Tropical Forestry Reports*, 16, 1–107

Gichora, M. (2003) 'Towards realization of Kenya's full beekeeping potential: A case study of Baringo District', *Ecology and Development Series*, 6

Grove, R. H. (1995) '*Green Imperialism: Colonial Expansion, Tropical Island Edens and the Origins of Environmentalism, 1600–1860*', Cambridge University Press, Cambridge, UK

Harwood, C., Rinaudo, T. and Adewusi, S. (1999) 'Developing Australian acacia seeds as a human food for the Sahel', *Unasylva*, 50 (196), 57–64

Hauser, S. and Mekoa, C. (2009) 'Biomass production and nutrient uptake of *Chromolaena odorata* as compared with other weeds in a burned and a mulched secondary forest clearing planted to plantain (*Musa* spp.)', *Weed Research*, 49, 193–200

Howes, F. N. (1946) 'Fence and barrier plants in warm climates', *Kew Bulletin*, 1, 51–87

Jex-Blake, A. J. (1934) '*Gardening in East Africa*', Longmans Green, London, UK

Judd, M. P. (2004) 'Introduction and Management of Neem (Azadirachta indica) in Smallholder's Farm Fields in the Baddibu Districts of The Gambia, West Africa', Unpublished MSc thesis, Michigan Technological University, USA

Kaufmann, J. C. (2001) 'La question des raketa: Colonial struggles with prickly pear cactus in southern Madagascar, 1900–1923', *Ethnohistory*, 48, 87–121

Koechlin, J., Guillaumet, J.-L. and Morat, P. (1974) '*Flore et Végétation de Madagascar*', Cramer, Vaduz, Lichtenstein

Kull, C. A., Tassin, J. and Haripriya, R. (2007) 'Multifunctional, scrubby, and invasive forests? Wattles in the highlands of Madagascar', *Mountain Research and Development*, 27, 224–231

Long, E. and Racey, P. A. (2007) 'An exotic plantation crop as a keystone resource for an endemic megachiropteran, *Pteropus rufus*, in Madagascar', *Journal of Tropical Ecology*, 23, 397–407

Loumeto, J. J. (1998) 'Des présidents ... en herbe/What's in a weed's name?', *Spore*, 75, 10

Mack, R. N. (2004) 'Global plant dispersal, naturalization and invasion: Pathways, modes and circumstances', in Ruiz, G. and Carlton, J. (eds) *Global Pathways of Biotic Invasions*, Island Press, Washington, DC, USA, pp3–30

Mack, R. N. (2008) 'Evaluating the credits and debits of a proposed biofuel species: Giant Reed (*Arundo donax*)', *Weed Science*, 56, 883–888

Mailu, A. M. (2001) 'Preliminary assessment of the social, economic and environmental impacts of water hyacinth in the Lake Victoria Basin and the status of control', in Julien, M. H., Hill, M. P., Center, T. D.and Ding Jianqing (eds) *Biological and Integrated Control of Water Hyacinth*, Eichhornia crassipes – *ACIAR Proceedings*, 102, Australian Centre for International Agricultural Research, Canberra, Australia, pp130–139

Mawathe, A. (2006) 'Killer weed hits Kenyan herders', *BBC News* 7 August 2006. http://news.bbc.co.uk/2/hi/africa/5252256.stm

McNeely, J. A. (ed) (2001) '*The Great Reshuffling – Human Dimensions of Invasive Alien Species*', IUCN, Gland, Switzerland

Middleton, K. (1999) 'Who killed "Malagasy cactus"? Science, environment and colonialism in southern Madagascar (1924–1930)', *Journal of Southern African Studies*, 25, 215–248

Moder, W. W. (1996) 'Does the African Grasshopper, *Zonocerus variegatus*, need *Chromolaena odorata?*', in Prasad, U. K., Muniappan, R., Ferrar, P., Aeschliman, J. P. and de Foresta, H. (eds) *Distribution, Ecology and Management of* Chromolaena odorata. *Workshop Report*, ORSTOM, ICRAF and University of Guam, pp156–164

Morat, P. (1972) 'Les savannes de l'Ouest de Madagascar', *Mémoire ORSTOM*, 68, 1–235

Mwalubandu, C. (1989a) 'Alien species "threaten forests"', *Daily News*, 14 March, p1

Mwalubandu, C. (1989b) 'Tree species generates controversy', *Daily News*, 29 March, p4

Nightingale, G. M. (1981) 'Experience with growing multipurpose trees', in Buck, L. (ed) *Proceedings of the Kenya National Seminar on Agroforestry*, ICRAF, Nairobi, Kenya, 325–327

Njoroge, P. and Bennun, L. A. (2000) 'Status and conservation of Hinde's Babbler *Turdoides hinduei*, a threatened species in an agricultural landscape', *Ostrich*, 71, 69–72

Norgrove, L., Tueche, R., Dux, J. and Yonghachea, P. (2008) '*Chromolaena odorata*: The benevolent dictator?', *Chromolaena odorata Newsletter*, 17, 1–3

Pappiah, H. (2001) '*Forestry Outlook Studies in Africa (FOSA), Mauritius*', Ministry of Natural Resources and Tourism, Mauritius

Pasiecznik, N. M., Felker, P., Harris, P. J. C., Harsh, L. N., Cruz, G., Tewari, J. C., Cadoret, K. and Maldonado, L. J. (2001) '*The* Prosopis juliflora – Prosopis pallida *Complex – A Monograph*', HDRA, Coventry, UK

Pernet, R. (1957, 1959) 'Les plantes médicinales malgaches – Catalogue des connaissances chimiques et pharmacologiques', *Mémoire de l'Institut des Sciences de Madagascar*, 8B, 1–143 + 9B, 217–303

Perrier de la Bâthie, H. (1928) 'Les pestes végétales à Madagascar', *Revue de Botanique Appliquée et d'Agriculture Tropicale*, 8, 36–42

Persson, A. (1995) 'Exotics – prospects and risks from European and African viewpoints', *Buvisindi*, 9, 47–62

Portères, R. (1959) 'Une plante pionnière américaine dans l'ouest-africain (*Solanum verbascifolium* L.)', *Journal de Botanique appliquée et d'Agriculture tropicale*, 6, 598–600

Pratt, D. J. and Gwynne, M. D. (eds) (1977) '*Rangeland and Management and Ecology in East Africa*', Hodder and Stoughton, London, UK

Richardson, D. M. (1998) 'Forestry trees as invasive aliens', *Conservation Biology*, 12, 18–26

Richardson, D. M. (2000) 'On global ecology', *Global Ecology and Biogeography Letters,* 9, 182–184

Richardson, D. M., Binggeli, P. and Schroth, G. (2004) 'Invasive agroforestry trees: Problems and solutions', in Schroth, G., de Fonseca, G. A. B., Harvey, C. A., Gascon, C., Vasconcelos, H. and Izac, A.-M. N. (eds) *Agroforestry and Biodiversity Conservation in Tropical Landscapes*, Island Press, Washington, DC, pp371–396

Strahm, W. A. (1993) 'The Conservation of the Flora of Mauritius and Rodrigues', Unpublished PhD thesis, University of Reading, UK

Timbilla, J. A.and Braimah, H. (1996) 'A survey of the introduction, distribution and spread of *Chromolaena odorata* in Ghana', in Prasad, U. K., Muniappan, R., Ferrar, P., Aeschliman, J. P. and de Foresta, H. (eds) *Distribution, Ecology and Management of* Chromolaena odorata. *Workshop Report*, ORSTOM, ICRAF and University of Guam, pp6–18

Voeks, R. (1990) 'Sacred leaves of Brazilian Candomblé', *Geographical Review*, 80, 118–131

Zimmermann, H. G., Hoffmann, J. H. and Witt, A. B. R. (2006) 'A South African perspective on *Prosopis*', *Biocontrol News Information*, 27, 6N–10N

Zobel, B. J., van Wyk, G. and Stahl, P. (1987) '*Growing Exotic Forests*', Wiley, New York, USA

14
The Rise and Fall of Japanese Knotweed?

John Bailey

Introduction

Japanese knotweed is the plant that every one loves to hate; it is regularly the subject of newspaper articles, where it attracts epithets such as monster, triffid barbarian and so on. The general public are terrified that it is going to turn up in their gardens and developers worry about potential costs of infestations on their sites. Environmentalists despair as continuous stands cover mile after mile of picturesque and ecologically sensitive rivers, while politicians look aghast at the vast sums spent on herbicides and removal of this plant. Japanese knotweed is indeed a very successful plant in Europe and is particularly at home in the more western regions. In its native Japan it is just one component of a giant herb community, where it must compete with other giant herbs such as Miscanthus. It is rarely considered to be a problem in Japan, where it is preyed upon by a large range of invertebrates, and struggles to find a niche along riverbanks and forest edges. Each year it grows to about 3m in height and by autumn is covered with small white nectiferous flowers; it is then cut back to the ground by the first frost, and overwinters with its extensive system of woody rhizomes. What might appear to be profligate vigour in the plant's growth in Europe is the bare minimum needed to compete successfully in its native regions. Much has been written on theories of invasion and invasability; but the consequences of transferring a heavily predated plant from an intensively competitive giant herb community to an environment where it towers over the native herbs and is ignored by the native invertebrates are scarcely surprising! Due to its rhizome (small fragments of which can produce a new plant), the plant is spread vegetatively along watercourses and by people's earth-moving activities. On account of the rhizome, well-established stands are extremely difficult to eradicate by herbicide alone.

By the late 1970s the true extent of the spread of Japanese knotweed in the UK was becoming apparent, and there was a growing feeling that something ought to be done about it. The 1981 Wildlife and Countryside Act made it a criminal offence to 'plant or otherwise cause the plant to grow in the wild'. In reality this had little effect on its continued spread, and it was only when the Environmental Protection Act of 1990 placed a duty of care obligation on the disposal of Japanese knotweed rhizome that any effective measures were introduced. In some ways the pendulum then swung too far in the other direction, for in the UK a whole industry has sprung up dedicated to the total eradication of Japanese knotweed from development sites. Various high-profile projects, such as the new Wembley Stadium and the 2012 London Olympics, have been hit by the high removal costs of Japanese knotweed infestations. The costs of physical removal and the financial and ecological costs of herbicide treatment have led to the development of a biological control programme for the plant. While the problems caused by Japanese knotweed are well-documented, in this chapter I want to look more at how Japanese knotweed came to be here in the first place and what exactly it is doing, in what must now constitute one of the classic cases of exotic plant invasion.

Japanese knotweed was introduced to gardens from Japan in the mid 19th century and its powers of vegetative reproduction enabled it to spread widely throughout its adventive range, where it is a serious pest. Molecular work indicates that only a single male-sterile clone was involved in this massive invasion. This has resulted in a huge inadvertent breeding experiment, the resultant hybrids and backcrosses compensating for its own lack of genetic diversity. Some of these hybrids may be worse than the parents. The origins of the plant in Japan have been located using molecular markers and this information has informed collection policies for a biological control programme that has successfully identified a control agent.

The 'Inwards Books' of the Royal Botanical Gardens at Kew are a series of large handwritten ledgers recording the species of plants received by this important centre. On pages 120 and 121 of the 1848–1858 volume a neat copperplate hand lists some 40 plants received from a certain Mr Siebold of Leyden on 9 August 1850. Accompanying this list is the following note: 'They are intended to be in exchange for new China and Japan plants.' But on account of the bad selection he is written to, saying that only six of them are probably new to us (Bailey and Conolly, 2000). Bad selection is an apt description for item 34, *Polygonum Sieboldii* 'Reinw.', for this particular unwanted gift was none other than Japanese knotweed. Using a combination of historical sources, herbarium specimens, cytological, morphological and molecular data, this chapter attempts to reconstruct the history and unravel the consequences and impacts of this particular introduction to the West.

While we are all familiar these days with the threats posed when alien plants and animals are transported from one part of the world to another according to the various whims of mankind, we are not so aware of the potential end point of such a process. Charles Elton, an acute and early observer of this phenomenon, in 1958 chillingly talks of 'one world with the remaining wild species dispersed up to the limits set by their genetic characteristics, not to the narrower limits set by mechanical barriers as well'. While in a biological sense we have to recognize this would be the ultimate expression of 'survival of the fittest', it is not a world that we would wish to contemplate! I well recall being told by Herbert Sukopp that in a remnant of the concrete debris of post-World War II Berlin, long preserved by its proximity to the then Berlin wall, the climax vegetation was *Robinia pseudacacia* with a *Fallopia japonica* understorey. While it is self-evident that the long-isolated faunas of Australia and New Zealand are no match for exotic advanced placentals, the threat posed by exotic plants is perhaps less clear-cut.

Japanese knotweed was described as *Polygonum cuspidatum* by Siebold and Zuccarini in 1846; for a fuller account of its complex early taxonomic history see Bailey and Conolly (2000). Two years later it appeared in the 1848 sales list of Von Siebold and Co. of Leiden as *Polygonum sieboldii* 'Reinw.' (Siebold, 1848). It was under the heading '*Plantes nouvellement importées du Japon*' [plants recently imported from Japan]. This carries a long footnote extolling the virtues of the plant and informing the reader that in 1847 it won a gold medal for the most interesting new ornamental plant at a horticultural show in Utrecht. The plant is brought to greater prominence by an illustrated piece by de Vriese in 1849. While this is little more than a glowing advertisement for the plant, it confirms that *P. sieboldii* and *P. cuspidatum* are the same thing, that Siebold has sole rights to the plant, and most importantly the illustration of the plant clearly shows it to be male-sterile with truncate leaf bases.

It is important to take into account the context of the arrival of Japanese knotweed in Britain. At that time there was nothing like the range of exotic garden plants that we take for granted today, and the polluted urban air would have taken its toll on more delicate plants. In Japanese knotweed you had a plant that 'did what it said on the box' – plant the rhizome and soon strong green shoots would emerge, grow rapidly to a height of 2–3m and be covered by tiny white flowers in autumn. Its evolution in the sulphurous volcanoes of Japan certainly held it in good stead in soot-covered Salford or similar cities, but it was another consequence of its evolutionary history that was to give rise to most of the problems. As an early colonizer of volcanic lava fields it was necessary to cope with sudden heavy deposits of hot volcanic ash, and this was achieved by sequestering precious reserves deep underground in a woody rhizome (Bailey, 2003). Possession of such a rhizome gives it prodigious powers of vegetative reproduction.

F. japonica var *japonica* is currently found in 2761 of the 3859 10km recording squares in the British Isles (Preston et al, 2002), and it has achieved this distribution (Figure 14.1) by vegetative propagation alone. An often-quoted figure of 0.7gm of rhizome being enough to produce a new plant (Brock and Wade, 1992) goes part of the way to explaining this. However, the primary spread was from garden to garden as an initially highly valued ornamental plant. Once its true nature became apparent, rhizome material was abandoned in the time-honoured manner of gardeners – such abandoned roadside or riverside plants perhaps attracting the attention of less affluent

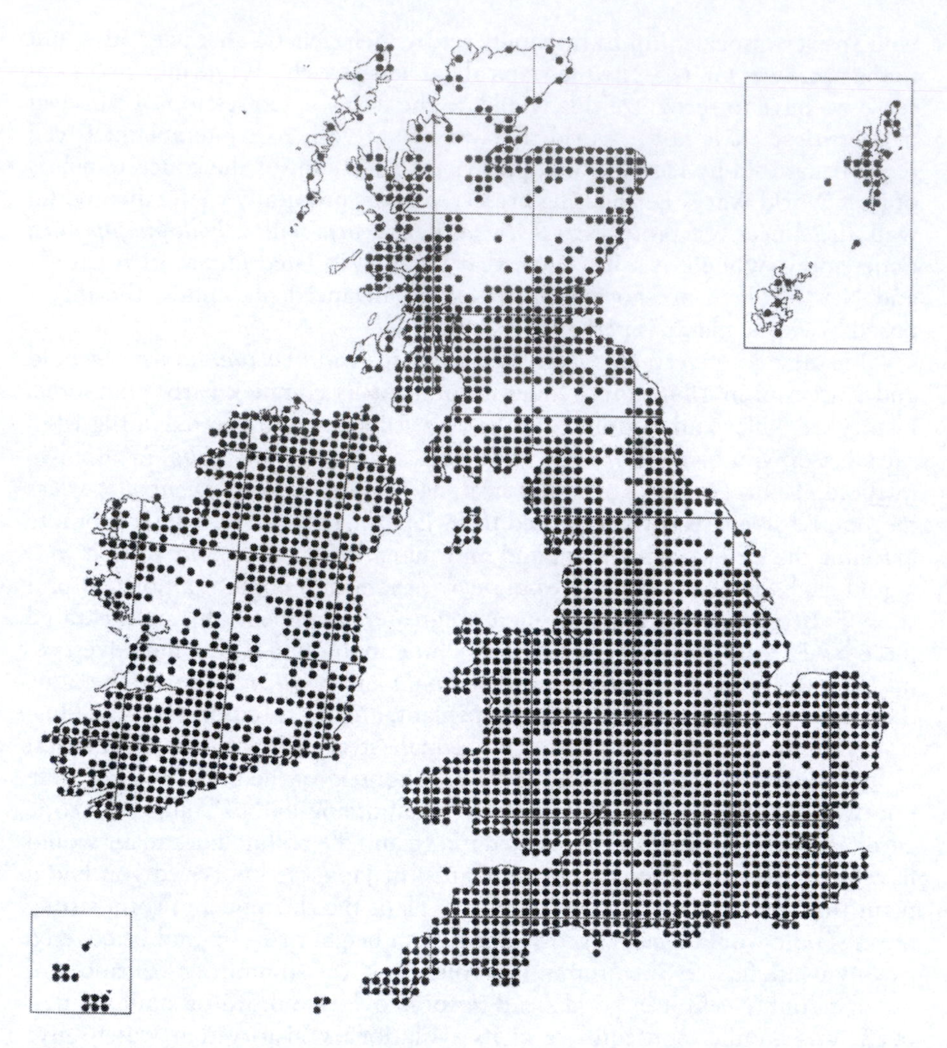

Figure 14.1 *Distribution of* Fallopia japonica *in the British Isles (Preston et al, 2002)*

gardeners also anxious for something green and reliable. The mid 20th century saw major advances in the technology for dispersing Japanese knotweed. Local authorities and developers would carefully remove and store topsoil containing Japanese knotweed rhizome from a site during development, only to grade it and spread it back over the whole area! Both in their native and adventive ranges, riverside plants are effectively dispersed downstream.

While the speed and breadth of spread of a male-sterile plant is remarkable enough (Conolly, 1977), it is by no means unique; take for instance the case of another exotic invasive water plant in Britain, *Elodea canadensis* (Simpson, 1984). Accidentally introduced to Britain in the mid 19th century, this was also all female and with prodigious powers of vegetative reproduction. Removed from all its natural predators it grew unchecked in a series of booms and busts. Nowadays it is very much reduced and is being overtaken by *Elodea nutalli*, a fellow invasive exotic. By comparison Japanese knotweed shows no signs of disappearing, and in spite of its clonal nature, it is an extremely effective colonizer. A major difference is that Japanese knotweed, unlike *Elodea canadensis,* was able to add to its genetic diversity by hybridizing with a series of related alien species, with varying degrees of success. The introduction of a single male-sterile genotype of a dioecious species has had the unintended consequence of setting up a massive inadvertent breeding experiment on a world scale!

When Conolly (1977) published her ground-breaking paper on the history and distribution of Japanese knotweed in the UK, the assumption was that, while both sexes of Japanese knotweed occurred, the male-fertile was very much rarer than the ubiquitous male-sterile plants. However, once chromosome counts were performed on these plants in the early 1980s, a very different picture emerged (Bailey and Conolly, 1985). All male-sterile plants with small leaves and truncate leaf bases proved to be 2n=88, while all male-fertile plants and some of the larger leaved female plants were found to be either 2n=44 or 2n=66. It gradually became apparent that we had a widespread male-sterile octoploid (8x) *F. japonica* var *japonica* clone and various hybrids. The 2n=66 hybrids could be explained as straightforward F1 crosses between female *F. japonica* var *japonica* (2n=88) and male-fertile *F. sachalinensis* (2n=44). The tetraploids were assumed to be crosses between 4x *F. japonica* var *compacta* and the 4x *F. sachalinensis*, which can be produced reciprocally due to the availability of both sexes of both taxa. Hybrids between *F. japonica* (either variety) and *F. sachalinensis* are known as *Fallopia x bohemica*, on account of their discovery in the Czech Republic (Chrtek and Chrtkova, 1983). But this still left the question of the 6x male-fertile plant from Buryas Bridge (Cornwall), which had smaller leaves, truncate bases and crimped edges. Was this some additional import from the Far East or had it arisen in the UK as a cross between the two varieties of Japanese knotweed present? To test these various hypotheses, a range of artificial hybrids was made at Leicester in the autumn of 1982. All cross combinations were successful, and while chromosome number confirmation occurred shortly afterwards, the plants needed to be grown to maturity for morphological comparison. This confirmed the putative origins of the 4x and 6x *F. x bohemica* hybrids (Bailey, 1989). Crosses between *F. japonica* var *japonica* and *F. japonica* var *compacta* produced strongly growing hexaploid plants of both sexes, morphologically identical to the Buryas Bridge individual.

A single 8x *F. x bohemica* was found in Wales in 1982, but was considered an isolated oddity until substantial populations of this ploidy were discovered in mainland Europe in the 1990s. It is not possible to recreate this ploidy level by artificial hybridization in a single step and there are several possible origins (Bailey and Wisskirchen, 2006), but the easiest route is by pollination of 8x *F. japonica* var *japonica* with an unreduced *F. sachalinensis* gamete. While *F. x bohemica* can spontaneously occur from seed (Pashley et al, 2003), most are spread vegetatively like their parents. Their distribution is independent of the parents and indeed, the further east one goes into Europe, the higher the proportion of the Japanese knotweed s.1. population they constitute. There is also increasing evidence that *F. x bohemica* is a greater threat than its parents (Bímová et al, 2001).

An investigation into the seed being produced by *F. japonica* var *japonica* plants, many miles from the nearest male-fertile *F. sachalinensis* plant, led to the discovery of extensive hybridization with another exotic garden plant, *Fallopia baldschuanica* or Russian Vine (Bailey, 1988, 1992). This was an unlikely hybrid for a number of reasons, a cross between a diploid (2n=20) and an octoploid (2n=88) involving plants with conflicting overwintering strategies, *F. japonica* a rhizomatous herb and *F. baldschuanica* a woody climber lacking rhizomes. The latter was actually something of a blessing in disguise since the undoubted vigour of each parent was cancelled out rather than enhanced, and the *F. x conollyana* hybrids were neither fully woody nor fully rhizomatous. It also went against the prevailing taxonomy, since the two species were in different genera. For Ronse DeCraene and Akeroyd (1988) working on the comparative floral morphology of this group, this hybridization was to reinforce the conclusion that they had come to from their anatomical research; that the genera *Reynoutria* and *Fallopia* should be merged under the older name *Fallopia*. When grown from seed in cultivation *F. x conollyana* is often a rather weak plant, though there is clearly some genetic variation given the presence of the established stand in Railway Fields, North London (see Figure 14.2a, b). In spite of the huge amount of viable seed of this constitution on *F. japonica* plants around Europe, spontaneous germination and establishment is still a rare event, though several new plants have recently been found in Southern England (Bailey, 2001; Bailey and Spencer, 2003).

Investigation of open pollinated seed from 8x *F. japonica* var *japonica* from New Zealand has recently revealed evidence of an even more extraordinary hybridization involving the Southern Hemisphere endemic genus of *Muelhenbeckia*. The genus *Muelhenbeckia* is essentially confined to Australasia and parts of South America, and so has clearly been long isolated from the Asiatic genus Fallopia. Seed collected from introduced male-sterile *F. japonica* var *japonica* plants (of the standard European 8x male-sterile form) growing in New Zealand and sent to Leicester by Tim Senior readily germinated to produce a range of semi-climbing plants with 2n=54, indicating hybridization with a male parent with 20 chromosomes. The most likely parent was *Muehlenbeckia australis*, which was growing in the same area. While these hybrids look superficially like *Fallopia x conollyana*, this is not altogether surprising since *M. australis* itself can be mistaken for *F. baldschuanica*. Careful examination of the NZ hybrids revealed various intermediate characters between *F. japonica* and *M. australis*, not found in *F. baldschuanica* or *F. x conollyana*. A major difference was the monoecious arrangement of flowers on the one plant that has so far flowered (male and female flowers in separate clusters), and the fact that the ends of the shoots can develop roots in the autumn. Neither of these features is known in the *Fallopia* taxa under discussion. In *F.*

Figure 14.2a Fallopia x conollyana *with the eponymous Ann Conolly (second from left) in its type locality at Railway Fields, Haringey, North London (photograph courtesy David Bevan)*

Figure 14.2b *Close-up of the Railway Fields with* F. x conollyana, *2 October 1987 (photograph courtesy David Bevan)*

x conollyana the conflicting growth forms of the two vigorous parents cancel each other out. But in the NZ hybrids the stolons and rhizomes of the two parents are much more complementary – possibly even to the point of synergy. Examples of exotic plants hybridizing with native species to produce even more serious pests are fortunately very rare (the amphiploid *Spartina anglica* perhaps being the best example (Marchant, 1968)), but it is clear that the more that is known about Japanese knotweed and its origins the better it will be in terms of developing effective responses!

Fingerprinting

The application of DNA fingerprinting technologies to plants in the early 1990s made it possible to probe more deeply into population structure than cytological and morphological approaches allowed. That said it is still the case in this group of plants that molecular data is of greatest value when used in conjunction with the morphological and cytological data. The most interesting notion was: 'could all the male-sterile 8x *F. japonica* var *japonica* plants throughout Europe really be clonal descendants of the plant sent to Kew by Siebold in 1850?' Another important question was the degree to which the distribution of *F. x bohemica* around the UK was clonal versus sexual. Application of RAPDs (Randomly Amplified Polymorphic DNAs) was one of the early methods for producing genotype specific banding patterns, and we were able to confirm that all the UK material tested plus smaller samples from mainland Europe and North America were all a single genotype (Hollingsworth and Bailey, 2000b).

Initial work using RAPDs on *F. x bohemica* (Hollingsworth and Bailey, 2000a) indicated considerably more diversity in these plants, but the use of RAPDs was superseded by the successful application of inter simple sequence repeats (ISSRs). Working on the so-called knotweed 'hotspot' at Dolgellau (Wales), Pashley et al (2003) were able to demonstrate that 6x *F. x bohemica* plants had originated in the overgrown gardens at Brithdir Hall and then spread down the River Clewdog to the River Wnion and that some genotypes had actually been moved to a site well away from the river. Incidentally this is a most atypical phenomenon in Britain, as only a limited number of sites where male-fertile *F. sachalinensis* and *F. japonica* var *japonica* co-occur are known.

It is indeed remarkable that a single clone of an Asiatic plant should be able to make such inroads into the West, invading the entire British Isles (Figure 14.1), and extending to continental Europe, North America and New Zealand. Much learned debate has occurred on the characteristics of both invader and the environment invaded. Suffice to say, Japanese knotweed is an octoploid that has evolved as one component of a giant herb community in Japan, it has escaped from the numerous invertebrates that predate it at home and is more than a match ecologically for most of our herbaceous communities. The fact that it was a male-sterile plant has had the unexpected consequence that wherever it has found itself it has hybridized with fellow introduced species or native species. While many of these hybrids are little more than botanical curiosities (Figure 14.3), others such as *F. x bohemica* and its backcrosses, not only increase

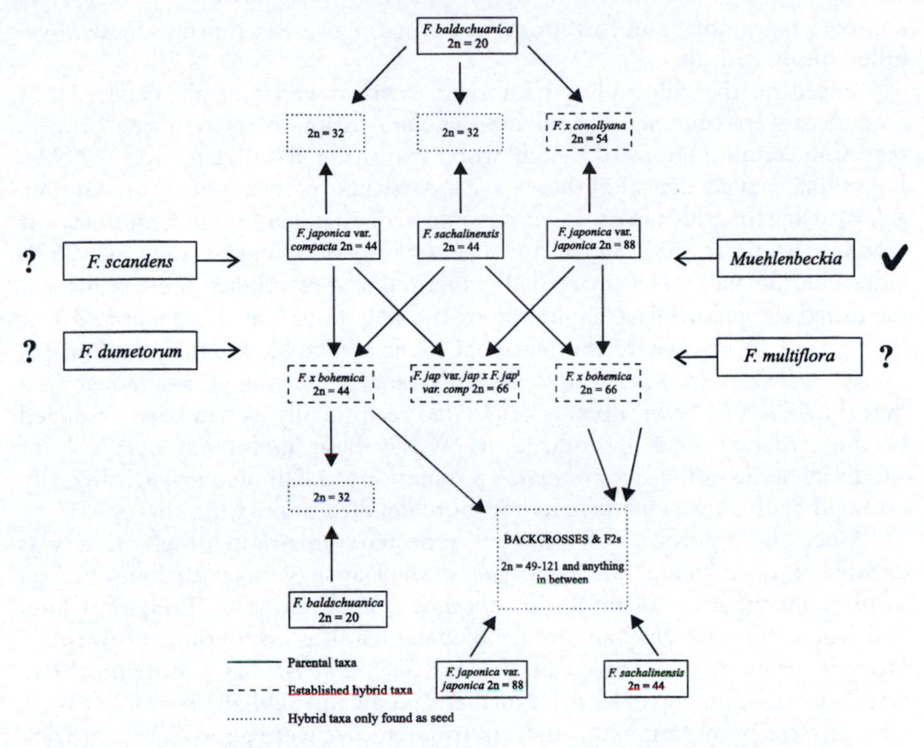

Figure 14.3 *The extent of hybridization within Japanese knotweed s.1. in Europe (centre), four additional taxa potentially able to hybridize along the edges; ticks and question marks indicate current knowledge*

genetic diversity and produce a plant potentially more difficult to control than its parents, but also hold out the possibility of genome rearrangements producing individuals better tailored to their various adventive environments. The newly discovered *Muehlenbeckia* hybrids are a further potential threat to an environment already decimated by alien introductions, and yet what more 'hopeful monsters' are waiting in the wings? In a sense these plants are immune from the genetic constraints of producing seed in order to spread and survive; for such vigorous perennials reproducing asexually, a single individual no matter what bizarre mixture of genomes it contains is potentially immortal.

I do not intend to go into detail here on the considerations, costs and difficulties encountered with eradication of exotic invasive Japanese knotweed. However, I would like to point out that due to the Wildlife and Countryside Act (1981) and the duty of care regulations consequent on it, a whole business sector has grown up in the UK dedicated to the eradication of this species from development sites. While this has had the beneficial effect of improving the botanical knowledge of would-be developers, it all comes at a considerable cost. The ecological and monetary costs of chemical and physical approaches to knotweed elimination have renewed interest in the possibility of a biological control programme, and fortuitously our ongoing research programme dovetailed nicely with this.

Based on the chloroplast haplotype work of Ferris et al (1993), DNA sequences were obtained by polymerase chain reaction (PCR) using primers targeting certain regions of the chloroplast transfer RNA sequences (tRNA). Following enzyme digestion these gave a particular pattern when run out on a gel. By using the trnK intron it was possible to distinguish British *F. japonica* var *japonica* from var *compacta* and *F. sachalinensis* (Hollingsworth et al, 1999). Since chloroplast DNA is maternally inherited in these plants, these sequences identified the chloroplast haplotype of the female parent. So hexaploid *F. x bohemica* hybrids, where the maternal plant was always female in Europe, shared the 8x *F. japonica* var *japonica* haplotype. It was also demonstrated that the 4x *F. x bohemica* plants originated reciprocally as had been predicted by the earlier artificial hybridizations. While these haplotypes were able to subdivide some of the *F. x bohemica* populations, the limited genetic diversity revealed by them was inadequate for more detailed population surveys.

Since the chloroplast haplotypes reflected the origin of plant, it was decided to try and match the haplotypes of the British plants with plants in Asia to pinpoint the areas of origin. *F. japonica* occurs naturally in Japan, China and Korea, but historical and morphological features both pointed towards a Japanese origin for the European plants. *F. sachalinensis* has a more northerly distribution, occurring in Korea, Northern Japan and Sakhalin Island (Russia). Late into the preparations for the trip to Japan, we were approached by Dick Shaw of CABI Bioscience, who was interested in instigating a biological control programme for Japanese knotweed. Since the basic tenets of biological control require a detailed knowledge of the plant and the collection of biocontrol agents from the area of origin of the introduction, what had begun as a purely academic study acquired an important application. That UK *F. japonica* var *japonica* was a single clone made it an ideal candidate for what would be the first application of biological control against an alien plant in Britain.

After extensive sampling throughout Japan, multiprimer haplotypes (mph) were compiled using trnC-trnD and trnF-trnV, from both Japanese and UK accessions. Most of the *F. sachalinensis* (of either sex) was of mph 6 which matched that of plants collected from Hokkaido (Pashley et al, 2007). The second British haplotype (mph 1) matched that of *F. sachalinensis* plants from Niigata, Northern Honshu. Interestingly this haplotype was mainly restricted to plants found growing in an abandoned nursery garden in Colchester, underlining the role of such businesses in the procurement of new genotypes. Though we did not have access to authentic Sakhalin material, the absence of an unmatchable mph suggests that only Japanese material entered the UK.

The single clone of *F. japonica* var *japonica* was also matched to a specific region of Japan (Pashley, 2003), and the CABI team were able to centre their collecting activities in this area. In 2009, after extensive testing, two suitable biocontrol agents were identified, a leaf spot fungus: *Mycosphaerella polygoni-cuspidati* and a psyllid sap-sucking insect: *Aphalara itadori*. The specific names of both of these taxa relate to Japanese knotweed, *Polygonum cuspidatum* being a synonym of *F. japonica*, and *itadori* is the Japanese name for Japanese knotweed. It has been decided to seek permission to release the insect first, and although the proposed release has satisfied the scientific case, the government is moving extremely carefully as this would be first release of a biocontrol organism in Britain. A period of public consultation on the release commenced on 22 July 2009 and ended on 19 October 2009. Satisfactory passage through this process allowed the Aphalara to be released under licence in carefully monitored sites from April 2010 onwards. Classical biological control does not seek the eradication of the targeted organism. However, Japanese knotweed has so far enjoyed freedom from the many organisms that prey on it in its native range, and exposure to one or two carefully chosen members of this group would do much to reduce its vigour and spread. Hopefully this will increase the effectiveness of existing methods of control. Even with the approval of the release programme, somehow I don't think that we have seen the end of Japanese knotweed as an invader in Great Britain.

Acknowledgements

I would like to thank Michelle Hollingsworth and Catherine Pashley for all their hard work, the late Ann Conolly for her unstinting support over the years, and Tim Senior for sending me the seed from New Zealand.

References

Anon. (1981) Wildlife and Countryside Act 1981, HMSO, London

Bailey, J. P. (1988) 'Putative *Reynoutria japonica Houtt. x Fallopia baldschuanica* (Regel) Holub hybrids discovered in Britain', *Watsonia*, 17, 163–164

Bailey, J. P. (1989) 'Cytology and Breeding Behaviour of Giant Alien Polygonum Species in Britain', Unpublished PhD thesis, University of Leicester, Leicester

Bailey, J. P. (1992) 'The Haringey knotweed', *Urban Nature Magazine*, 1, 50–51

Bailey, J. P. (2001) '*Fallopia x conollyana* the railway-yard knotweed', *Watsonia*, 23, 539–541

Bailey, J. P. (2003) 'Japanese knotweed s.1. at home and abroad', in Child, L., Brock, J. H., Prach, K., Pysek, P., Wade, P. M. and Williamson, M. (eds) *Plant Invasions: Ecological Threats and Management Solutions*, M. Backhuys Publishers, Leiden, pp183–196

Bailey, J. P. and Conolly, A. P. (1985) 'Chromosome numbers of some alien Reynoutria species in the British Isles', *Watsonia*, 15, 270–271

Bailey, J. P. and Conolly, A. P. (2000) 'Prize-winners to pariahs: A history of Japanese knotweed s.1. (Polygonaceae) in the British Isles', *Watsonia*, 23, 93–110

Bailey, J. P. and Spencer, M. (2003) 'New records for Fallopia x conollyana: Is it truly such a rarity?' *Watsonia*, 24, 452–453

Bailey, J. P. and Wisskirchen, R. (2006) 'The distribution and origins of *Fallopia x bohemica* (Polygonaceae) in Europe', *Nordic Journal of Botany*, 24, 173–200

Bímová, K., Mandák, B. and Pyšek, P. (2001) 'Experimental control of Reynoutria congeners: A comparative study of a hybrid and its parents', in Brundu, G., Brock., J., Camarda, H. I., Child, L. E., Wade, P. M. and Williamson, M. (eds) *Plant Invasions: Species Ecology and Ecosystem Management*, Backhuys Publishers, Leiden, pp283–290

Brock, J. H. and Wade, M. (1992) 'Regeneration of *Fallopia japonica*, Japanese Knotweed, from rhizome and stems: Observations from greenhouse trials', *IXe Colloque International sur la Biologie des Mauvaises Herbes*, Dijon, France, pp85–94

Chrtek, J. and Chrtoková, A. (1983) '*Reynoutria x bohemica*, novy krinzinec z celedi rdesnovitych', *Casopis narodniho muzea v Praze r. prir.*, 152, 120

Conolly, A. P. (1977) 'The distribution and history in the British Isles of some alien species of Polygonum and Reynoutria', *Watsonia*, 11, 291–311

Elton, C. S. (1958) *The Ecology of Invasions by Animals and Plants*, Methuen, London

Ferris, C., Oliver R. P., Davy, A. J. and Hewitt, G. M. (1993) 'Native oak chloroplasts reveal an ancient divide across Europe', *Molecular Ecology Notes*, 2, 337–343

Hollingsworth, M. L., Bailey, J. P., Hollingsworth, P. M. and Ferris, C. (1999) 'Chloroplast DNA variation and hybridisation between invasive populations of Japanese Knotweed and Giant Knotweed (Fallopia, Polygonaceae)', *Botanical Journal of the Linnean Society,* 129, 139–154

Hollingsworth, M. L. and Bailey, J. P. (2000a) 'Hybridisation and clonal diversity in some introduced Fallopia species', *Watsonia*, 23, 111–121

Hollingsworth, M. L. and Bailey, J. P. (2000b) 'Evidence for massive clonal growth in the invasive weed *Fallopia japonica* (Japanese Knotweed)', *Botanical Journal of the Linnean Society*, 133, 463–472

Marchant, C. J. (1968) 'Evolution in Spartina (Gramineae): II Chromosomes, basic relationships and the problem of *S. x townsendii* agg.', *Botanical Journal of the Linnean Society*, 60, 381–409

Pashley, C. H. (2003) 'The Origin and Evolution of Japanese Knotweed s.1.', Unpublished PhD thesis, University of Leicester, Leicester

Pashley, C. H., Bailey, J. P. and Ferris, C. (2003) 'Further evidence of the role of Dolgellau in the production and dispersal of Japanese Knotweed s.1.', in Child, L., Brock, J. H., Prach, K., Pysek, P., Wade, P. M. and Williamson, M. (eds) *Plant Invasions: Ecological Threats and Management Solutions*, M. Backhuys Publishers, Leiden, 197–211

Pashley, C. H., Bailey, J. P. and Ferris, C. (2007) 'Clonal diversity in British populations of the alien invasive Giant Knotweed, *Fallopia sachalinensis* (F. Schmidt) Ronse Decraene, in the context of European and Japanese plants', *Watsonia*, 26, 359–371

Preston, C. D., Pearman, D. A. and Dines, T. (2002) *New Atlas of the British and Irish Flora*, Oxford University Press, Oxford

Ronse Decraene, L. P. and Akeroyd, J. R. (1988) 'Generic limits in Polygonum and related genera (Polygonaceae) on the basis of floral characters', *Botanical Journal of the Linnean Society*, 98, 321–371

Shaw, R. S., Bryner, S. and Tanner, R. (2009) 'The life history and host range of the Japanese Knotweed pysllid, Aphara itadori Shinji: Potentially the first classical weed control agent for the European Union', *Biological Control*, 49, 105–113

Siebold, P. F. (1848) 'Extrait du catalogue et du prix-courant des plantes du Japon et des Indes-Orientales et Occidentales Neerlandaises', *Jaarboek van de Koninklijke Nederlandsche Maatschappij tot Aanmoediging van den Tuinbouw*, 38–49

Siebold, P. F. and Zuccarini, J. G. (1846) 'Polygonum cuspidatum', Thesis in the Mathematical-Physics Class of the Royal Bavarian Academy of Science, 4(2), 208 (*Observations on the Flora of Japan*, 2:84)

Simpson, D. A. (1984) 'A short history of the introduction and spread of *Elodea Michx.* in the British Isles', *Watsonia*, 15, 1–9

Vriese, W. H. De (1849) 'Polygonum cuspidatum', *Jaarboek van de Koninklijke Nederlandsche Maatschappij tot Aanmoediging van den Tuinbouw*, 30–32

15
History and Perception in Animal and Plant Invasions – the Case of Acclimatization and Wild Gardeners

Ian D. Rotherham

Summary

The roles of both people and nature have been identified as interactions at the core of biological invasions (Rotherham, 2005a, b; Rotherham et al, 2005). This chapter focuses on two specific aspects of the invasion paradigm, first that of the deliberate introduction of plants and animals around the world by the Victorian acclimatization societies, and second the Wild Garden Movement. These two 19th-century phenomena led directly to many of the issues and challenges that face conservation today. Important within this consideration are the changing perceptions and attitudes of people towards nature, and especially to the exotic, over the period from the early 19th century to the end of the 20th century. While Davis et al (2001) addressed the changing attitudes towards exotic plants and animals in Britain consequent on the writings and broadcasts of Charles Elton, this wider influence of fashion and taste in shaping responses to aliens has generally been overlooked. Furthermore, the critical role of the practical manifestations of fashion such as the accidental or even deliberate introduction of now invasive species to the countryside has not been recognized. This chapter provides an insight into the importance of the human cultural role in facilitating invasions, and how perceptions and attitudes have affected this. Crossing boundaries of ecological science and history it considers two specific British examples that together created many of the invasions of the 20th and 21st centuries.

The problem of aggressive and invasive plants and animals is not new but the scale of impact combined with rapid climate change and other environmental fluxes is dramatic. Starting points for discussion must be deciding what is alien and what is a problem (Gilbert, 1989, 1991, 1992). An alien species is a plant, animal, or microorganism not 'native' to an area, but which has been accidentally or deliberately introduced by humans. It may or may not be invasive and in fact only about 0.1 per cent of aliens are damaging. Furthermore, the spread of species across the planet is not new but recent horror stories have stirred up a debate among ecologists, politicians, industry and the public. Indeed there are good reasons why this is so. Some 15 per cent of Europe's 11,000 aliens have environmental or economic impacts and damage to the UK economy is estimated at £2bn per annum. However, underlying the headlines are deep-seated questions of what is native and where, what is alien and when. From Spanish bluebell, to eagle owls and Canada geese, and from big cats, beavers and signal crayfish to wild boar, which ones should get a free pass?

These are issues frequently overlooked or ignored, but history is informative in such debates. In particular, it is worth considering how human actions have triggered invasions and indeed, how many of these were deliberate acts. It then quickly becomes apparent that many invasive alien species were initially welcomed and furthermore, that human attitudes to these plants and animals are not fixed in time. Our responses to what we now see as problem alien species are often subjective, not objective, and those regarded as problems are chosen very selectively. The past roles of two particular English or British groups are especially informative when we try to understand both today's problems (some of which cannot be doubted) and the causes of many invasive colonizations. These two examples are the 'Victorian Wild Garden Movement' and the 'Acclimatization Societies'.

The problems and challenges of the exotic

Few doubt that adverse impacts of invasive alien plants and animals pose some of the most serious threats to nature conservation (see for example Simberloff and McNeeley, Chapters 8 and 2). There is a large and expanding literature based on research into the issues, impacts and effects of exotic species. Overall the adverse effects on wild animals and plants across the world have been compared in severity to the probable impacts of human-induced climate change in terms of their severity and significance in the 21st century. In particular, once-isolated faunas and floras of islands have suffered most and in today's globalizing world ecology they are threatened and in many cases have been destroyed. Often these ecosystems have simply been erased by European colonization over the last few hundred years. With the chronic environmental disruption of industrialization, intensive farming, urbanization and now climate change, those species unable to adapt are declining dramatically, and those able to exploit changed conditions are spreading around the globe (Rotherham, 2009a, b). For nature conservationists this is potentially the stuff of nightmares. Indeed a superficial consideration can result in a view of the processes that blurs and blends essential aspects of both the science and the cultural aspects of this hugely important interaction of people and nature.

The immediate knee-jerk response to exotic species which dare to naturalize is first that we know which they are, and second that they should be eradicated. However, in recent years there have been some significant publications which broaden the issues and debates and raise serious questions about how and why we label certain species as 'good' or 'bad', as 'alien' or 'native'. The *Landscape Research* volume (Volume 28, 2003) edited by Coates and Hall is a seminal point in the debate (Coates, 2003). Along with this are other contributions such as Davis et al (2001) and Grime (2003). Smout (2003) and Warren (2002, 2007) are others who have weighed in and in doing so have presented major challenges to many precepts of conservation management. Some of the underpinning ecological issues and questions are raised by Macdonald et al (2006) and they present a very useful series of insights and case studies to demonstrate the serious adverse effects of many introductions, but also the difficulties in embedding actions within objective scientific frameworks.

Therefore, while the seemingly reasonable response to aliens seems to be backed by rigorous scientific evidence, and the issues seem to all intents and purposes clear-cut (Anon., 2007), the reality is more complex. Engaging the media and others involved in a wider public debate and dialogue is also challenging. The media loves 'sound bites' and 'sexy issues' so an alien invasion with its clear and simple message is easy to cover, whereas the more subtle and complex debate is less newsworthy. In essence it seems that alien species which establish and then naturalize may have the potential to wreak havoc among native ecosystems. Simple observation also suggests that with a very few exceptions humans have been unwilling or unable to do much about this, and most bad invasions with harmful effects have gone unchecked and their effects have run their natural course. Many of the world's ecologies have already become 'Disneyfied'; local or regional characters and distinction lost or diluted. However, this process goes on against the uncomfortable backdrop of an evolving and changing landscape and environment, with some of the flux natural and some anthropogenic. As with climate change, not all the effects we see today are human-induced, but a subtle mix of natural and inevitable change now dramatically catalysed by human intervention. In this context, it is worth recognizing that species and ecologies are not static but dynamic. They ebb and flow, flux and change, over periods from decades to centuries or millennia. Changes during the period from around AD 1400 to the 1800s provide a good example of this as Northern Europe was gripped by the harsh chill of the so-called 'Little Ice Age'.

Into the mixing pot with empires, acclimatization and wild gardening

The consequences of all this are especially pertinent to Britain and to our attitudes. Indeed, observations in Europe and discussions with ecologists and landscape historians from continental Europe are informative. In many cases, it seems that Europeans worry more about 'problem species' than necessarily 'alien' ones. In Britain, we are an island race (or at least a collection of races now on a collection of islands); the boundaries of land and sea give clear definition of what should be 'in' and what should be 'out'. Yet our perceptions of this and the attitudes and responses that stem from them are surprisingly recent in origin. Until the 1940s, we generally went out from our islands and travelled the globe collecting and selecting plants and animals to bring back to Britain. Not merely content with collecting this mixed bag of species, we then deliberately set about their release into the landscape to 'improve it' for people and for economic benefit too. This movement spread around the world: European acclimatization societies were established first and sought particularly to introduce and test new crops for economic purposes and especially their potential for food. However, these organizations developed in other ways and in Britain and the colonies, they looked to the introduction of animals and birds to new places in order to improve economies, gastronomies and landscapes. Veitch and Clout (2001) described the impacts of acclimatization societies and the huge effect they had on New Zealand ecology and today on New Zealand conservation. Similar issues are well known from Australia too.

The acclimatization societies and the Victorian Wild Gardeners (Rotherham, 2005c, d) were manifestations of processes that had occurred to greater or lesser degrees for centuries. We know that through history the waves of settlers or conquerors of Britain had done just the same. The Romans and then the Normans imported huge numbers of animals and probably plants too, many of which are keystone species in the modern ecology. The most obvious example is the humble rabbit, but we can add the fallow deer and the brown hare to the list to consider later when we examine issues of perception and attitude. We also accidentally, unwittingly and uncaringly released many other animals and plants to make their own ways in the world. Many of these species went on to become an intimate part of what we now see as 'British' ecology. The Romans and Normans imported herbs and food plants from southern Europe and the Mediterranean, as did the returning Crusaders and the various monastic dynasties that controlled much of the productive landscape for several centuries. Many of these species have been absorbed into the mix of native ecology. Most are now tolerated and many are celebrated and conserved.

By the 1500s, seafarers from Britain and Holland for example, were beginning to chart their ways around the globe. From all the corners of the world they brought back exotic plants and sometimes animals; many of these introductions perished but others did not. Accidental imports had already included black rat and brown rat, plus a dash of bubonic plague, and in return later explorers spread these around the planet along with dogs, cats and much more. The cultural homogenization of ecology was speeding up. The collection and dissemination of alien species increased more rapidly as travellers went in search of exotic plants and animals for gardens and menageries. As landscaping, forestry and gardening emerged in Britain through the 1700s and 1800s, the impacts on the environment increased. This process continues today with catastrophic consequences. However, it should not be imagined that these changes to 'native' ecology were isolated from other impacts. At about the same time, in the 18th and 19th centuries, the wider landscape was traumatized by the parliamentary enclosures with common land wrested from the commoners, the peasants and the poor and converted into intensive food production units (Rotherham, 2009b). Much of the more natural landscape and its ecology were swept away by this change. Traditional coppice woods were converted to high forest plantations and industrial cities began to sprawl across the countryside. Lands which remained relatively untouched by this were often blended into leisurely landscapes for the pleasure of landowners and industrialists, and were often populated by the exotic plants and animals being introduced from around the world. Associated with these changes was also a seminal undercurrent of transformation from ecology dominated by native 'stress tolerators' to often exotic species of 'ruderals' and 'competitive' plants. This is noted in the 20th century by Grime et al (2007), but in reality it began much earlier as the landscape flexed and changed, and disturbance plus nutrient enrichment came to the fore.

Acclimatization and naturalization through wild gardening

The two movements which lie at the epicentre of biological invasions in Britain were the acclimatization societies as described by Lever in *The Naturalized Animals of Britain and Ireland* (1977, 2009) and *They Dined on Eland: The Story of the Acclimatization Societies* (1992). These seminal publications provided a fresh view of the cultural processes underlying biological invasions in Britain. In particular, Lever provides the historical insight that places the British Acclimatization Societies within their wider global context and also shows how, though only short-lived as a functioning entity, they both reflected attitudes of the time and also acted as a platform from which to launch introductions and invasions over the following century. 'Acclimatization' had a more formal basis first in Europe and Britain, and then around the world, than did the 'Wild Garden Movement'. The European basis for the acclimatization of plants and animals was led largely by a desire to introduce production crops and livestock in a very pragmatic approach (see Borowy, Chapter 10).

Lever (1977) describes how in 1860 'The Society for the Acclimatisation of Animals, Birds, Fishes, Insects and Vegetables within the United Kingdom' was formed under the enthusiastic guidance of Frank Buckland. While the Society was short-lived, merging into the 'Acclimatisation and Ornithological Society' by 1866 and closing by the 1870s, it spawned others across the Empire and the movement for promoting introductions carried on well into the 20th century. The reasons and personalities behind the establishment and the demise of this organization are entertainingly described by Lever (1977, 1992). They included its narrow base of wealthy landowners and competition with the wider-based and more knowledgeable Zoological Society of London.

The introduction of plants to Britain had long been stimulated by global exploration and the twin desires to bring back species of economic importance and ornamental potential. Kew Gardens was established during the 1800s and become the greatest centre for economic botany in the world; at this time the idea and passion for domestic gardening grew among the emerging middle classes in urban areas. In the wider countryside and on great estates interest in creating landscapes and in introducing or naturalizing species to benefit game management became of huge importance. Again this continued through until the 1940s.

Plants such as rhododendron, Himalayan balsam, Japanese knotweed and giant hogweed all spread rapidly across Britain, aided by Victorian 'Wild Gardeners' and often by foresters and estate managers. While the use of such exotics was undoubtedly widespread already, William Robinson formalized and popularized the idea of 'naturalizing' exotic plants into the landscape rather than merely planting them for effect. This was especially during the 1800s and early 1900s. These species recognized as invaders at a relatively early date have now been joined by a whole raft of others such as: montbretia, buddleia and cherry laurel. As a general pattern, rates of spread build slowly over perhaps 50–100 years through suitable environments. They then move into an exponential and explosive expansion. A more recently recognized invasive plant is the variegated yellow archangel, which I now believe to be a Victorian cultivar. Spreading rapidly wherever it is introduced to suitable habitat, woodland or hedgerow, it can cover several hundred square metres of woodland ground flora in only 10–15 years. As it doesn't produce viable seed, this spread is totally by vegetative spread, and its introduction is always originally by direct human agency. The role of the 'Wild Gardeners' has been less recognized for its role in triggering the plant invasions of the 20th century but it was at the heart of most terrestrial plant naturalizations with the exception of the massive establishment of exotic conifers for forestry. The first significant recognition of this importance was by Rotherham (2005c, d), but the reader is directed to the hugely important book by William Robinson: *The Wild Garden*, published in 1870. As probably the most influential popular writer on English gardening in the late 19th century this landmark publication triggered a massive response for 50 years or more. Robinson's life and work was described in detail by Allan (1982) – see also Figure 15.1.

When William Robinson published his book in 1870, the practices he described and advocated were already well-established. These included in particular the embellishment of woodlands, utilizing the naturalization of exotics in the landscape, along with the admixture of plants from the subtropical garden, with an emphasis on foliage groupings. This was the basis for wild gardening throughout the 1870s and 1880s (Elliott, 1986). To be practised on a large scale, wild gardening required species capable of spreading themselves in large masses. Plants such as *Rhododendron ponticum*, *Heracleum mantegazzianum* (giant hogweed), *Polygonum* spp (Japanese and giant knotweeds), and of course Himalayan balsam, were all ideal. Such wild garden favourites are widely recognized as later becoming the scourge of conservationists. Locations such as Chatsworth Park in Derbyshire were used to juxtapose the garden flowers, forest trees and wild undergrowth. At Chatsworth, the rhododendrons, balsam and indeed the giant hogweed remain and have gone on to bigger and better things! Robinson stated that the principle of wild gardening was 'naturalizing or making wild innumerable beautiful natives of many regions of the earth in our woods, wild and semi-wild places, rougher parts of pleasure grounds, etc.'

Figure 15.1 *William Robinson aged 84 in 1922*

An assessment of British flora confirms that many of todays most problematic alien 'weeds' come from deliberate introductions. These releases were into domestic and landscape gardens, and to forest and woodland estates, from the 1700s to the early 1900s. Inspired by William Robinson, The Victorian Wild Garden movement was responsible for the introduction and subsequent escape of many plants. Some of these are our most spectacular invaders such as giant hogweed, and giant knotweed, stunning garden plants for the 'Wild Garden' of the late 1800s (Figures 15.2 and 15.3). Over the same period, however, there was a change from traditional coppice management of native woods to high forestry. This was often with exotic tree species, and supplemented by introduction of exotic trees and shrubs such as rhododendron, snowberry, mahonia and Gaultheria, for woodland management for game preservation. In many ways, the seal was now set on the alien plant issues for our contemporary landscape – and there is no going back. Since this 'golden age of plant introductions' there has been a continual release of new exotic species into the Britain's landscape. This has been both from gardens and from modern plantings of

Figure 15.2 *Giant hogweed or giant cow parsley, from William Robinson* (The Wild Garden)

Figure 15.3 *Japanese knotweed, from William Robinson* (The Wild Garden)

exotic trees and shrubs. From the latter a wide range of berry-bearing trees and shrubs is rapidly establishing throughout our woods and forests. Ratcliffe (1984) notes the beginnings of spread by shrubs like berberis and cotoneaster. Exotic cultivars of holly and of sorbus are also spreading into woodlands old and new.

Ecology transformed

The uncomfortable result of these observations and the compounding effects of the abandonment of traditional countryside practices throughout the 19th century, continued until the late 20th century, was a radically transformed ecology. 'Cultural severance' (Rotherham, 2009b), which is this ending of traditional uses, values and management of the landscape and its ecological resources, has been a final compounding factor. Many areas, realized from the subsistence exploitation of centuries, have rapidly gained biomass and nutrients and the stress-tolerant species that are often of high conservation value slip quietly away. Micro-disturbance associated with traditional management is replaced by either abandonment or pulses of macro-disturbance. These stresses in the ecosystem are most obvious in urbanized zones where, combined with the exotic species described earlier, they are forming new ecological associations, the so-called 'recombinant ecology' (Barker, 2000); different and distinctive from what went before. There are glimmers of the former landscapes in what we generally call 'seminatural' habitats but even here the cultural drivers for these areas over centuries of human exploitation have changed and often ceased. One result is subtle, long-term blurring of the ecology; in other cases the consequences are rapid and dramatic. One remarkable fact is that so much of the former landscape and its ecology are indeed visible through the modern veneer. Some aspects of ancient ecosystems are surprisingly resilient unless totally swept aside by modern mechanization. But there are major issues for conservation such as the latter-day recognition for example, of the importance of remnants of medieval parks and their links to the 'Frans Vera primeval landscape' (Vera, 2000); for decades these areas received little recognition or protection. We are now searching for so-called 'shadow woods' etched in the landscape possibly before the Norman conquest, yet still surviving though often unrecognized. Similarly, many heaths and commons hark back to this antique ecology and yet are sadly abandoned and neglected. A point to emerge from these observations is that what we value is not necessarily an ecology which is truly native but one that is perceived to be such. Some of our most ancient landscapes still have little protection and often very unsympathetic management. On the other hand some of the landscape features and their ecology that are passionately protected, such as 18th- and 19th-century enclosure hedgerows, are actually imposed exotic features. Many of the 'native' oak woods from which school children carefully collect 'local' acorns to grow and then plant into 'local provenance' woods were actually not native at all. These are frequently imports from Dutch nurseries in the 18th century, as estate accounts confirm. Indeed, a forester today can often spot the distinctive manifestations of genetic traits that distinguish native trees from Dutch. There are wonderful ancient hedges from pre-Domesday and these cross ancient landscapes to link patches of wood, common and heath, but they are different and distinct from the imposed barriers that separated commoner from common.

The genie out of the box: Politics and ecology in late Victorian and Edwardian Britain

During the late 19th century, there were major upheavals in British politics and a long-term change in the balance of power between landowner and factory owner, between rural and urban, and between great houses and estates and the urban gentry. Much of this was played out over the period 1880–1940, with the obvious complications of World War I and the subsequent Depression. One of the main effects on British ecology was facilitated by the collapse of many rural estates, both large and small, as grand houses and smaller mansions were abandoned. In the urban fringes, gardens and landscaped areas were subsumed into expanding cities, and in the wider countryside, they were taken into productive farming. Huge numbers of gardeners and other estate workers never returned from the war effort, and many who did return found their jobs gone so they left for the town and city. For two centuries, these estates dotted across the British landscape had been storehouses of exotic plants and animals. Some remained fixed in situ, unable to move or to naturalize; they are indicators of a time and a society now long gone. For species able to adapt and to naturalize into the British landscape their time had come. With the controlling hands of thousands of gardeners now removed and a landscape modified in their favour (as discussed earlier), a host of exotic and often aggressive plants leapt the garden wall to freedom. (A rather smaller and often less successful group of animals followed the plants to freedom.) Rhododendron, giant hogweed, Japanese knotweed, giant knotweed, Himalayan balsam, Portuguese laurel and many others began their move across the British landscape and into conservation folklore. The genie was out of the bottle but surprisingly it was not until 50 to 60 years later that ecologists and then conservationists realized the effects and scale of the invasion. The genie had escaped and the die was cast.

A dawn of realization

Some voices of dissent over the importation and release of exotic species came from the game conservation interests, with campaigns in the 1930s to eradicate the imported little owl, 'Frenchie', or 'Frenchman'. A detailed ecological investigation (Hibbert-Ware, 1938) exonerated this bird from the accusations of harming game birds. However, the tide was beginning to turn with a realization of the impacts and effects, actual or perceived, on native fauna and flora and on other countryside interests such as game.

Charles Elton's post-World War II writing on invasive species had a critical effect on this issue. Having worked as part of the war science effort in the team charged with preventing the importation of foreign pests and diseases into the wartime agricultural economy of Britain, Elton was charged with the fervour of patriotic isolation that permeates his ecological output. His three seminal books on animal and plant ecology (Elton, 1927, 1958, 1966) had a huge impact on probably every British ecologist of note in the later 20th century. Davis et al (2001) very eloquently assess how his style and language changed from pre-war to post-war as we became an island, a race and ecology under siege. The result was a change from the attitude in the period before 1940 when exotic species were generally tolerated and often even welcomed, to one more xenophobic. In this case simply being alien, or perceived to be alien, is enough to warrant outcry.

It was during the 1950s and then the 1960s that concerns began to emerge about the adverse impacts of invasive alien plants such as *Rhododendron ponticum*, though again this was mostly in terms of their effects on commercial forestry rather than on wildlife. It was not until the 1960s and 1970s that real worries over damage to native wildlife became a serious consideration. From this time onwards, armed with Elton's observations of impacts on ecology around the world, and especially on the unique faunas and floras of remote islands, ecologists began to show more interest in both invasions and in their threats. It was clear, and the science is unequivocal, that the worst impacts were ecosystem devastation and local extinctions.

However, it is also abundantly obvious that not all introduced species are the cause of major problems, or that early problems persist over a longer timescale.

Invasive aquatic weeds such as Canadian pondweed (*Elodea canadensis*) seem to 'settle down' after a period of aggressive invasion and impact. Furthermore, many invasive and problem species are not exotic but are natives liberated by a change in landscape management. One of the most pernicious and problematic invaders of heaths, moors, woods and commons is bracken (*Pteridium aquilinum*), once managed by cattle herdsmen and harvested by farmers and others as bedding, fuel and a source of potash for glass-making. A cultural change from cattle farming to sheep grazing in the late 1800s and early 1900s precipitated much of the contemporary problem.

Science, politics and environmental democracy

It is important to set the problems of alien invasive and exotic species into the broader contexts of environmental change, conservation and politics. Low (2001) provided a broad overview of the interrelationships between ecology and politics in terms of reactions to biological invasions. He also makes the important point about the long history of human interference with species distributions.

In bringing together science and politics it is necessary to acknowledge that many conservation decisions are not based on 'truths' and often not even on science, and are not objective. They are subjective decisions based on the best scientific understanding we have blended with an emotional response to a situation based on and twisted by many social, cultural and historical influences. Unfortunately, this applies to both the professional conservation manager and to the wider public alike. Even the language used to define and to describe the issues is loaded with bias, as discussed by Keulartz and van der Weele (2009), and the decisions are political and social. Why for example do we seek to eradicate Himalayan balsam as a riverside and roadside invader but not sweet cicely, an alien from the mountains of central Europe first recorded wild in Britain in 1777? *The New Atlas of the British and Irish Flora* (2002) suggests that it has not changed its distribution significantly since 1962, but in the Peak District and South Pennines it is spreading rapidly and its impact is dramatic. So if control is based on science and objectivity then why one and not the other? *Buddleia davidii* causes millions of pounds of damage to services and buildings, and is now expanding into woods, hedgerows and other habitats such as cliff tops, but we welcome it as 'the butterfly bush'. In contrast, conservationists dislike rhododendron, which they seek to bash and eradicate.

Devolution in the (dis)-United Kingdom raises further issues, as discussed by Warren (2002), where conservation managers are trying to decide whether a species should be native to England, Wales, Scotland or some lesser region. In the face of climate change and the inevitable fluxing of species distributions, this is a nonsense and misunderstanding of the serious matters at stake. It is also totally missing the point about the palimpsest nature of historic landscapes and the value of cultural and historical aspects of the environment. Is it relevant that a plant found in Carlisle was not 'native' in Gretna or beyond? Should it therefore be eradicated if it does spread north? This is a formalizing of the old idea of beech being only native to southern England and so treated as an alien in the northern regions. It has now been found in the early pollen records for North Yorkshire, and a further point is that in the centuries since the closing of the English Channel surely beech would have made its way northwards anyway. In that case, it would now be native in the north as well. There is certainly a case for celebrating and conserving where possible local and regional distinctiveness and character, but regional ecological xenophobia is a dangerous route down which to travel.

Another major problem in dealing with the apparently simple matter of alien and exotic invaders is the difficult relationship between conservation and (i) the cultivation of exotic trees for forestry and amenity, and (ii) farming, horticulture and gardening. In all these situations there is a blend of nature and culture that makes the assertion of native or exotic status somewhat fraught with problems. Both (i) and (ii) are major causes of the undoubted problems being caused by alien invasions. Nevertheless, this does not mean that all the effects are negative, or even that the bad effects are significant or important in all cases and in all situations.

Conclusions

A lesson of the British experience is also that perceptions of what is a problem, what is alien or native, and even who is responsible for any management or control varies dramatically over decades and even over centuries. Furthermore, issues of exotic and invasive cannot be separated from the wider fluxes of society, economy and environment; so where and when we consider a particular species has a huge effect on our perception of it as a positive or negative influence on ecology, economy and society. It can also be argued that as western imperialism spread around the globe during the 19th and early 20th centuries, many of these attitudes and a lot of 'native' plants and animals from 'home' were also exported. The twin desires to improve and to adorn created many of the invasion and extinction problems witnessed today.

In Britain, the effects of the two major movements, acclimatization and wild gardening, should not be judged in isolation. They emerged from social, economic and political trends in the preceding centuries and they echo throughout the 20th and now into the 21st century. The two clearly interacted with each other and with other parallel ideas and forces. Their combined effects have been massive and generally, and rather surprisingly, overlooked. An understanding of these interactions between people and nature is hugely informative in increasing our awareness of the subjectivity of perceptions and values and the fluxes of these human facets over time.

References

Allan, M. (1982) *William Robinson 1838–1935: Father of the English Flower*, Faber and Faber, London

Anon. (2007) *Protecting Our Native Wildlife: Guidance for the Control of Non-Native Invasive Weeds in or Near Fresh Water*, Environment Agency, Bristol

Barker, G. (ed) (2000) *Ecological Recombination in Urban Areas: Implications for Nature Conservation,* English Nature, Peterborough

Coates, P. (2003) 'Editorial postscript: The naming of strangers in the landscape', *Landscape Research*, 28(1), 131–137

Davis, M. A., Thompson, K. and Grime, J. P. (2001) 'Charles S. Elton and the dissociation of invasion ecology from the rest of ecology', *Diversity and Distribution*, 7, 97–102

Elliott, B. (1986) *Victorian Gardens*, B. T. Batsford, London

Elton, C. S. (1927) *Animal Ecology*, Sidgwick and Jackson, London

Elton, C. S. (1958) *The Ecology of Invasions by Animals and Plants,* Methuen, London

Elton, C. S. (1966) *The Pattern of Animal Communities*, Methuen, London

Gilbert, O. L. (1989) *The Ecology of Urban Habitats,* Chapman and Hall, London

Gilbert, O. L. (1991) 'The ecology of an urban river', *British Wildlife Journal*, 3, 129–36

Gilbert, O. L. (1992) *The Flowering of the Cities: The Natural Flora of 'Urban Commons',* English Nature, Peterborough

Grime, J. P. (2003) 'Plants hold the key: Ecosystems in a changing world', *Biologist*, 50(2), 87–91

Grime, J. P., Hodgson, J. G. and Hunt, R. (2007) *Comparative Plant Ecology. A Functional Approach to Common British Species,* 2nd edn, Castlepoint Press, Dalbeattie, Scotland

Hall, M. (2003) 'Editorial: The native, naturalized and exotic – plants and animals in human history', *Landscape Research*, 28(1), 5–9

Hibbert-Ware. A. (1938) *Report of the Little Owl Food Inquiry 1936–37*, H. F. and G. Witherby, London

Keulartz, J. and van der Weele, C. (2009) 'Framing and reframing in invasion biology', *Configurations*, 16, 93–115

Lever, C. (1977) *The Naturalized Animals of Britain and Ireland*, Hutchinson, London

Lever, C. (1992) *They Dined on Eland: The Story of the Acclimatisation Societies*, Quiller Press, London

Lever, C. (2009) *The Naturalized Animals of Britain and Ireland*, New Holland (UK) Ltd, London

Low, T. (2001) 'From ecology to politics: The human side of alien invasions', in McNeeley, J. A. (ed) *The Great Reshuffling: Human Dimensions of Invasive Species,* IUCN, Gland, Switzerland

Macdonald, D. W., King, C. M. and Strachan, R. (2006) 'Introduced species and the line between biodiversity conservation and naturalistic eugenics', in Macdonald, D. and Service, K. (eds) *Key Topics in Conservation Biology*, Wiley-Blackwell, London

Preston, C. D., Pearman, D. A. and Dines, T. D. (2002) *New Atlas of the British and Irish Flora*, Oxford University Press, Oxford

Ratcliffe, D. A. (1984) 'Post-medieval and recent changes in British vegetation; the culmination of human influence', *New Phytologist*, 98, 73–100

Robinson, W. (1870) *The Wild Garden*, The Scolar Press, London

Rotherham, I. D. (1986) *The introduction, spread and current distribution of* Rhododendron ponticum *in the Peak District and Sheffield area. Naturalist*, 111, 61–67

Rotherham, I. D. (2000) 'People, perception and fashion', in Barker, G. (ed) *Ecological Recombination in Urban Areas: Implications for Nature Conservation,* English Nature, Peterborough, pp21–24

Rotherham, I. D. (2001) 'Himalayan Balsam – the human touch', in Bradley, P. (ed) *Exotic Invasive Species-Should we Be Concerned?* Proceedings of the 11th Conference of the Institute of Ecology and Environmental Management, Birmingham, April 2000, IEEM, Winchester, pp41–50

Rotherham, I. D. (2001) 'Rhododendron gone wild', *Biologist*, 48(1), 7–11

Rotherham, I. D. (2002) 'Aliens and woodlands', *Quarterly Journal of Forestry*, 96(2), 128–129

Rotherham, I. D. (2003) 'Alien, invasive plants in woods and forests ecology, history and perception', *Quarterly Journal of Forestry*, 97(3), 205–212

Rotherham, I. D. (2004) 'The wild Rhododendron: The history and ecology of an invasive alien and neglected native', in Rotherham, I. D. (ed) *The Ecology and Management of* Rhododendron ponticum – *Invasive Alien or Neglected Native?* Conference Pre-proceedings and notes, *Journal of Practical Ecology and Conservation* Special Series, No 2

Rotherham, I. D. (ed) (2005a) 'Loving the Aliens??!!? Ecology, history, culture and management of exotic plants and animals – issues for nature conservation', *Journal of Practical Ecology and Conservation* Special Series, No 4

Rotherham, I. D. (2005b) 'Loving the aliens?' ECOS, 26(3–4), 1–2

Rotherham, I. D. (2005c) 'Invasive plants – ecology, history and perception', *Journal of Practical Ecology and Conservation* Special Series, No 4, 52–62

Rotherham, I. D. (2005d) 'Alien plants and the human touch', *Journal of Practical Ecology and Conservation* Special Series, No 4, 63–76

Rotherham, I. D. (2009a) 'Exotic and alien species in a changing world', ECOS, 30(2) 42–49

Rotherham, I. D. (2009b) 'The importance of cultural severance in landscape ecology research', in Dupont, A. and Jacobs, H. (eds) *Landscape Ecology Research Trends*, Nova Science Publishers, New York, USA. Ch 4, pp1–18

Rotherham, I. D., Minter, R. and Spray, M. (eds) (2005) 'Alien and Native – what belongs?', *ECOS*, 26(3–4), 1–119

Smout, T. C. (2003) 'The Alien species in 20th-century Britain: Constructing a new vermin', *Landscape Research*, 28(1), 11–20

Veitch, C. R. and Clout, M. N. (2001) 'Human dimensions in the management of invasive species in New Zealand', in McNeeley, J. A. (ed) *The Great Reshuffling: Human Dimensions of Invasive Species,* IUCN, Gland, Switzerland

Vera, F. W. M. (2000) *Grazing Ecology and Forest History*, CABI Publishing, Wallingford

Warren, C. (2002) *Managing Scotland's Environment*, Edinburgh University Press, Edinburgh

Warren, C. (2007) 'Perspectives on the "alien" versus "native" species debate: A critique of concepts, language and practice', *Progress in Human Geography*, 31(4), 427–446

16

Factors Affecting People's Responses to Invasive Species Management

Paul H. Gobster

Introduction

Natural areas managers contend with an increasingly diverse array of invasive species in their mission to conserve the health and integrity of ecosystems under their charge. As users, nearby neighbours and de facto 'owners' of the lands where many significant natural areas reside, the public is often highly supportive of broad programme goals for management and restoration, but becomes less enamoured when specific actions such as prescribed burning or biocide applications are called for (Barro and Bright, 1998). Frequently these actions are aimed at invasive plants and animals, mostly non-native exotics but at times invasive natives that due to their abundance or location may also interfere with restoration objectives.

On the surface, negative public sentiment surrounding invasive species management might be explained as a reaction to having been sold a false bill of goods: restoration connotes life and putting back in order to make ecosystems whole again, yet invasive species management is by and large an activity of death and taking away (Gobster, 2005). But most people have no compunction about swatting flies in their house or plucking dandelions from their yard in order to maintain a sense of order and quality of life, and in managing the natural world they may also accept the need to value some species over others. How such distinctions are made is critical to many natural areas programmes that take place on public lands since the success of managers in gaining social acceptance for their programmes will often determine effectiveness in attaining ecological goals (McNeely, 2001).

A more nuanced view is needed on how people perceive invasive species and the actions to control them if managers are to better gear their natural areas programmes to meet public acceptance and reduce conflicts and controversy. In this chapter I outline a heuristic model for thinking about how people respond to a given invasive species based on a number of factors and their relationships. The model draws from a wide variety of examples where there has been public reaction to invasive species management and is presented with the intent to promote discussion, systematic study and more accurate specification. After discussing the model factors and their relationships I apply the model to case studies of Asian longhorned beetle eradication in Chicago and feral cat control in rural Wisconsin. I conclude by considering the model in the context of natural areas restoration and suggest its potential use in future research and decision making.

Surveying the evidence

The impetus for this work stems from my study of stakeholder conflicts in the restoration of natural areas in Chicago and San Francisco (Gobster and Hull, 2000; Gobster, 2007). The control of invasive species has played a central role in both conflicts, pitting nearby residents, recreational users and other interest groups against natural areas managers and volunteer restorationists who on the whole all share a common love for nature and desire to see it protected. In Chicago, prairie and savannah ecosystems are primary targets of restoration efforts, and the removal of trees to restore the structure and function of these open communities has been contentious. This has been especially true in the removal of large diameter non-native trees, which are seen as having high aesthetic value, but in some cases concerns have also been voiced about removing saplings and immature trees such as black cherry that are native to the region but not a part of the ecosystem being restored. Restoration critics in Chicago have been less concerned with the removal of invasive ground-cover plants such as garlic mustard and shrub species such as European buckthorn, though the use of herbicides and prescribed fire for controlling these plants has often been controversial because of perceived dangers to people and wildlife. Moreover, while some shrubs like buckthorn are seen as problematic, they may provide value as bird cover and for the visual screening of development, both of which can be compromised if the buckthorn is not replaced with a comparable native species. Finally, while white-tailed deer are native to the Chicago region, actions by restorationists to control their numbers to protect vegetation has been highly controversial, particularly the use of rocket nets and sharpshooting which have ethical implications for an increasing proportion of urbanites.

I began studying the Chicago restoration conflict in 1996, and when another conflict broke out in San Francisco six years later it provided an opportune comparison. In examining the two conflicts I noticed that, while the botanical aspects of the native ecosystems and their unwanted invaders were quite different between the two cities, there were some strong similarities in the overall landscape structure, in patterns of cultural landscape change, and in how critics responded to invasive species management. Like Chicago's prairies, San Francisco's dominant native coastal scrub and dune ecosystems are open, largely treeless landscapes, but fire suppression and a long history of dedicated afforestation have made both the non-native Eucalyptus and nearby native Monterey pine and cypress trees familiar elements in the landscape. Removal of mature tree cover has been a major point of conflict among those who value its aesthetic and historic qualities, and while many residents find dense mid-storey growths of Himalayan blackberry unattractive and difficult to traverse, the plant does have wildlife value as a food source. In fact, it has been argued that several non-native shrubs and ground plants have considerable functional value for wildlife, to stem hillside erosion, and other purposes. Like Chicago, the use of herbicides has been contentious in some areas, and concerns over air quality and a collective memory of the fires that devastated the city after the Great Earthquake of 1906 have generated acute opposition to the use of prescribed fire as a management tool. While deer overpopulation is not an issue in the city, domestic animal control has been highly disputed. The reduction of feral cat populations has raised many ethical issues of protecting the life of a species to which people hold strong bonds, and while the right for people to run their dogs off leash in public parks has been mostly cast as an access issue, restorationists also see off-leash dogs as an unwanted invader that tramples native plants, adds excess nitrogen to the soil, and disturbs birds and other sensitive wildlife.

From my study of these two conflicts I began to notice some common patterns in how people responded to an invasive species issue based on characteristics of the species itself, the management situation and other factors. If data from additional case studies were compiled, I thought, perhaps one could develop a systematic way to anticipate what people's reaction might be and in so doing, gear management strategies in ways that would accomplish restoration objectives and reduce social conflicts. While few social science studies have examined the human dimensions of invasive species (e.g. Fischer and van der Wal, 2007; Norgaard, 2007), I have found considerable indirect information in ecological studies that mentioned social repercussions of managing for invasive species in popular books about invasives, newspapers and magazine articles. Table 16.1 shows a sampling of individual invasive species that have been the subject of these accounts, from diseases and insects to small and large plant species to small and large mammals. These accounts do not provide sufficient detail for a case-by-case analysis, but in looking across them they begin to suggest a set of factors and relationships. What follows is a preliminary attempt to bring these emergent patterns into model form.

Specifying the model

Proposing a model or theory without rigorous study is a risky proposition, yet a first approximation can help reveal relationships and suggest hypotheses for further study and testing. The model (Figure 16.1) considers people's response (R) to an invasive species management scenario as a function of the perceived value of the invasive (V), the perceived threat (T) and impact (I) it has on a native species or ecosystem, and the perceived benefits and costs of management

$$R = \sum C_i S_i \left(\frac{V}{T + I}\right) - M$$

(WARNING: a first approximation)

- R = response to the exotic/invasive species
- V = value of the exotic/invasive
- T = threat
- I = impact or value of the exotic/invasive on the species/ecosystem/object of concern
- M = impact of management control mechanisms
- C = context
- S = stakeholder group factors

Notes:
$R = W_1(C,S)V + TW_2(C,S)I + W_3(C,S)M$
R = Response to an invasive species management scenario
R > 0 ⇒ acceptable, stakeholder–manager agreement
R = 0 ⇒ neutral, unconcerned
R < 0 ⇒ unacceptable, stakeholder–manager conflict
V = Value or likeability of the invasive
V > 0 ⇒ liked, valued, hold positive feelings for
V = 0 ⇒ neutral, no value one way or the other
V < 0 ⇒ disliked, disgust sensitivity, hold negative feelings for
I = Impact of invasive on a native species/ecosystem/other object of concern
I > 0 ⇒ beneficial impact (e.g. assists in endangered species recovery)
I = 0 ⇒ benign (no effect)
I < 0 ⇒ negative impact (loss, degradation, cost of replacement)
T = Threat – imminence of or adaptation to invasive impact
$0 \leq T \leq 1$
T = 0 ⇒ no threat or foreseeable impact
T = low ⇒ low threat or adaptation to impact
T = 1 ⇒ imminent threat, certain impact
M = Management actions – benefits/costs
M > 0 ⇒ positive/beneficial outcomes (effective control, positive externalities)
M = 0 ⇒ benign or without impact
M < 0 ⇒ negative/deleterious outcomes (ineffective, negative externalities)
C = Context factors (e.g. physical, social setting)
S = Stakeholder factors (e.g. centrality, education)
W_1, W_2, W_3 = Weights that depend on Context (C) and Stakeholder (S) factors

Figure 16.1 *Model for predicting human response to management of an invasive species*

actions (M) instituted to deal with the invasive. This additive model recognizes that each value, impact and management action for a given scenario can be perceived as positive, negative or neutral in nature, and that people's responses are influenced by these factors as well as by various contextual (C) and stakeholder group (S) factors that recognize the heterogeneity of responses to a given situation. People's responses can thus vary depending on the nature of the individual factors within this relationship, and for each factor I outline of some of the major issues and questions that should be considered in determining people's responses to a given management scenario (see references in Table 16.1 for further information).

Value (of the species of concern)

The value of an invasive species can be thought of as its likeability, irrespective of the damage it might do to a natural area or protected species. While many people dislike or are even disgusted by bacteria, nematodes, insects and snakes (e.g. Bixler and Floyd, 1997), the value of an invasive seems to increase as species get larger in size and their physical qualities are more readily perceptible. As I mentioned in the Chicago and San Francisco case studies, big trees hold important aesthetic value for many people regardless of whether they may be invasive, and they may also provide important recreational (e.g. shady sites), functional (e.g. cooling), and economic (e.g. enhancing property value) values. Colourful plants and animals such as purple loosestrife and monk parakeets may also endear people because of their aesthetic qualities, and animals with furry coats and big eyes such as deer, grey squirrels, feral cats and wild horses can have considerable charismatic appeal. Despite their invasive and destructive tendencies, some purposefully introduced exotics herbs such as Italian fennel and animals such as feral pigs hold cultural value for particular ethnic groups, and maintenance of sufficient populations may be important to keep alive cultural traditions. Finally, the value of protecting native biodiversity and ecosystem integrity is sometimes weighed against the value of alternative ecologies that non-native species provide. Here again trees can be effective in moderating urban heat island effects, filtering air pollutants, sequestering carbon and performing other ecosystem services, and it is hard to argue that these values are less important than those provided by native ecosystems.

Impact (of an invasive on a native species or ecosystem)

Native species and ecosystems can also be valued for their aesthetic, recreational and functional qualities, but because their use-oriented values are sometimes less readily apparent, restoration groups and natural areas managers often argue for the protection of native species and ecosystems on the basis of more intrinsic values such as biodiversity or ecosystem health, or for educational, option and existence values (Millennium Ecosystem Assessment, 2005). Beyond these values, the sheer rarity of some native species and ecosystems can heighten their value, particularly if they are endemic or endangered. The impact of an invasive is thus the loss or degradation in the value(s) of a native species or ecosystem or the cost of its replacement (e.g. Pimentel et al, 2005). While the impact of an invasive species is usually thought to be negative, some species may serve beneficial functions for native species and ecosystems – such as providing wildlife habitat or food supply (e.g. Shapiro, 2002) – which should be considered in a management assessment.

Table 16.1 *Selected non-agricultural invasive species, characteristics and case studies*

Species	*Latin name*	*Origin*	*Place of introduction*	*Year of intro*	*Ecosystem type*	*Documentation*
Chestnut blight	*Cryphonectria parasitica*	Europe	Northeastern USA	1906	Cities and forests	Friederici, 2006; Freinkel, 2007
Dutch elm disease	*Ophiostoma ulmi*	Europe	USA	1931	Cities and forests	Campanela, 2003
Earthworms	e.g. *Lumbricus terrestris*	Europe, Asia	USA	1800s	Northern forests	Bohlen et al, 2004; Friederici, 2004
Asian longhorned beetle	*Anoplophora glabripennis*	Asia	USA	1996	Cities (NY and Chicago)	Antipin and Dilley, 2004
Emerald ash borer	*Agrilus planipennis*	Asia	Midwestern USA	2002	Cites and forests	BenDor et al, 2006
Gypsy moth	*Lymantria dispar*	Europe, Asia	Northeast USA	1860s	Cities and forests	Liebhold and Muzika, 1996; Pearson, 2002
Garlic mustard	*Alliaria petiolata*	Europe	Midwestern USA	1800s	Forest	Neumann, 1999
Purple loosestrife	*Lythrum salicaria*	Europe	USA	1800s	Wetlands	Hager and McCoy, 1998; Slobodkin, 2001
Italian fennel	*Foeniculum vulgare*	Europe	California	1800s	Hillsides	Shapiro, 2002
Cattail	*Typa* spp	native	USA	native	Wetlands	Apfelbaum, n.d.
Zebra mussels	*Dreissena polymorpha*	Europe Asia	Midwestern USA	1988	Lakes and streams	Johnson and Padilla, 1996
European buckthorn	*Rhamnus cathartica*	Europe	Midwestern USA	1800s	Forest, savannah, prairie	Whelan and Dilger, 1992; Gobster and Hull, 2000
Eucalyptus	*Eucalyptus spp.*	Australia	California	1700s	Scrub, dune	Williams, 2002; Slack, 2004
Black cherry	*Prunus serotina*	USA	Central Europe	1600s	Forest	Starfinger et al, 2003

Table 16.1 *(Continued)*

Species	*Latin name*	*Origin*	*Place of introduction*	*Year of intro*	*Ecosystem type*	*Documentation*
Brown tree snake	*Boiga irregularis*	Australia	Guam	1949	Forests	Burdick, 2005
Northern snakehead	*Channa argus*	China	Maryland	2002	Lakes	Dolin, 2003
Brown trout	*Salmo trutta*	Germany	USA	1880s	Rivers	Todd, 2001
Pigeon (rock dove)	*Columba livia*	Europe	USA	1600s	Urban and rural	Todd, 2001; Humphries, 2008
Parakeets	e.g. *Myiopsitta monachus*	S. America	USA	1960s	Cities	Todd, 2001; Bittner, 2004
English house sparrow	*Passer domesticus*	Europe	USA	1851	Parks and woodlands	Burdick, 2005; Mann, 1948
American grey squirrel	*Sciurus carolinensus*	USA	England, Italy	1800s	Woodlands and cities	Max, 2007; Bertolino and Genovesi, 2003
Feral cats	*Felis catus*	Egypt?	Worldwide	1600s in USA	everywhere	Beversdorf, 2008; Robertson, 2008
Domestic off-leash dogs	*Canis lupis familiaris*	Eurasia, N. America	USA	?	Urban natural areas	Forrest and St Clair, 2006; Banks and Bryant, 2007
Mountain goat	*Oreamnos americanus*	W. USA and Canada	Olympic Peninsula, Washington	1920s	Alpine	Todd, 2001
Feral pigs, European wild boar	*Sus scrofa*	Polynesia, Europe	Hawaii, many places in USA	1500, 1912	Forests	Van Driesche and Van Driesche, 2004; Burdick, 2005
White-tailed deer	*Odocoileus virginianus*	native	Eastern USA	native	Woodland and savannah	Kilpatrick et al, 2007; Lauber and Brown, 2006
Wild horses	*Equus ferus caballus*	Europe, Asia	Western and Eastern USA	1500s	Rangelands, seashore	Flores, 2008; Welch, 2009

Threat

The degree to which an invasive is seen as a threat to a native species or ecosystem must also be considered along with impact in gauging people's response to a management scenario. Threat might best be construed as the perceived probability or likelihood that an impact will occur, and the gypsy moth exemplifies how the element of threat can operate. This invader came to the USA in the 1860s and has been widespread throughout the north-east for many years. People who have lived with it have to some degree come to accept the damage from its tree defoliation as well as developed some scepticism over the effectiveness and consequences of management (see below). However, where the gypsy moth is just arriving in some areas of the Midwest, unfamiliarity with it has raised the perceived threat and the urgency with which some people feel control efforts should be instituted. Perceived threat thus may have temporal and spatial dimensions that relate to the imminence and rate of spread of invasions. Knowledge of how threat perception operates can be helpful in spurring public awareness about invasives, but overplaying the 'fear factor' can also have negative repercussions such as loss of trust (Mackenzie and Larson, 2010) and maladaptive responses (Gobster, 2005).

Management actions (benefits and costs of intervention)

As mentioned previously, factors related to the management of an invasive species can work to heighten conflicts between managers and the public even if the latter group holds little or no value for the invasive or shows little concern about its impact on the native species or ecosystem. These factors have to do with the perceived impact of the management action, its duration of implementation, and its probability of success. In some cases, people may be willing to accept a management treatment even if it is perceived as harsh or has significant negative externalities as long as it is applied only once or a few times before an invasive is eradicated. People may grow disenchanted, however, if treatments occur again and again without clear success. This may be the case with gypsy moth control programmes in the northeastern USA, where annual aerial applications of the bacterial pesticide *Bacillus thuriengensis* or 'Bt' lessens tree defoliation but also kills other leaf-eating caterpillars, including rare and valued butterfly species.

Context factors

The context of a natural area can modify how people respond to an invasive management scenario. Context factors that influence people's responses include the physical and social setting (e.g. land use, proximity to residences, and location along an urban-to-wilderness remoteness gradient) and previous site disturbances. The effect of context is a complicated one, for even if an invasive is seen as intruding upon a person's home environment, such as an insect that defoliates a backyard tree or nearby natural area, that person may be less willing to put up with a management activity such as insecticidal spraying than if that action were to occur in a more remote location. Because of this heightened sensitivity, urban natural areas management programmes are often the locus of intense social conflicts over such issues as animal control, herbicide application and prescribed burning. But this does not mean that remote natural areas are exempt from invasive species management conflicts, for if those sites are viewed as pristine or free from human interference people may be less willing to accept management interference and instead favour 'allowing nature to take its course'.

Stakeholder factors

A range of individual and stakeholder group factors can also modify how people respond to an invasive species management scenario. If an invasive is valued in some way, by definition it has some measure of centrality to particular stakeholders, and those who may depend on it or similar species for income (e.g. pet or horticulture industry) or as part of a cultural tradition may be more resistant to its control than those whose value is more casually aesthetic or recreational in nature. The heterogeneity and strength of stakeholder interests can also complicate public responses, and in some cases a dominant but weakly supportive majority can be upset by a small group of highly vocal critics who makes it difficult for managers to reach public consensus on how an invasive should be managed (Bertolino and Genovesi, 2003). Other factors that can influence individual and stakeholder views on invasive species are many and include education, urban versus rural residency, regional and cultural factors and differing meanings of nature. Knowledge and expertise often play important roles in challenging stakeholder positions on issues, and while they may be helpful in negotiating consensus for managing invasive species, they are also open to challenge and alternative interpretations that further complicate resolution of contested issues.

Response

The factors described above work in combination to describe people's response to a management scenario. A positive response is one where people find the management scenario acceptable; for example, the invasive has low value relative to its impact on a native species or ecosystem, and the impact of managing the invasive is low or justified in order to achieve reasonable success. In contrast, a negative response points to an unacceptable management scenario or one where people's values conflict with natural areas managers'. In either case, the value of the response is mediated by context and stakeholder factors; thus the weights (W) placed on V, I and M can change the acceptability of a management scenario depending upon where it is implemented and who is included in the assessment.

Testing the model

As can be seen by this brief discussion, the factors that influence people's response to an invasive species are numerous and may relate to each other in complex ways. Yet a thoughtful analysis of a given situation may illuminate potential pathways and roadblocks to successfully garnering public favour in addressing management issues. To illustrate this potential I apply the model to two recent invasive species control efforts, one that was largely successful and the other that in hindsight was doomed to failure from the very start.

Asian longhorned beetle in Chicago

Antipin and Dilley (2004) describe a successful effort in the city of Chicago to eradicate the Asian longhorned beetle, *Anoplophora glabripennis*. The fairly large (2–3cm), slow-moving adult insect lays its eggs underneath the bark of a variety of tree species and the hatched larvae worm their way up and down the tree, feeding on the phloem and eventually killing the tree in the process. The beetle is thought to have arrived accidentally in Chicago on packing crates delivered from China to a business on the city's north side, and was discovered in the nearby Ravenswood residential neighbourhood in 1998. The pest had been discovered in a separate outbreak in New York City two years earlier and thus the city was aware of the potential magnitude of the problem. It mobilized a team of local, state and federal officials who developed a plan to quarantine affected areas and prohibit moving wood materials out of the zone. Then by an intensive system of detection the team identified and destroyed more than 450 infested and suspected trees within the first year and another 1000 in the following five years until the pest was effectively eliminated. Minor outbreaks in four other neighbourhoods within the metro area were similarly contained, and ten years after the 1998 discovery the beetle was officially declared eradicated.

Looking at the case study within the model framework reveals how the various factors contributed to a positive response (R) to support the effort's success. First, the beetle was an accidental introduction and although one could consider it somewhat attractive (in China it is known as the starry night beetle for the whitish speckles that dot its glossy black shell), there was no expressed value (V) in maintaining its presence in the city. In contrast, the threat (T) of the beetle was imminent and its impact (I) would be highly negative; if it were permitted to spread, its potential losses to industry and tourism were estimated at more than US$41 billion. While the management treatment (M) needed to eradicate the beetle also had negative externalities – in the most affected part of Ravenswood nearly all trees were cut and destroyed within an eight square block area – the effectiveness of the identification and control effort was successfully demonstrated within the first few years. The initial strategy to also remove healthy trees within a 0.2km radius of infected trees was very unpopular with residents, but when a new chemical prevention treatment for injecting healthy trees proved effective in tests it was quickly adopted. The severity of this treatment strategy was also tempered by an aggressive reforestation programme which planted large-calliper replacement trees, and a very proactive public relations programme that included numerous public meetings, a supportive media and enlistment of numerous civic groups that offered financial, labour and even emotional support to grieving residents. The success of this effort was also bolstered by the isolated residential context (C) and a homogeneous stakeholder constituency (S), which led to a high degree of engagement and united cooperation in working toward a solution.

Feral cats in Wisconsin

Beversdorf's (2008) video documentary vividly captures the social dynamics of a failed 2005 proposal to control feral cats in Wisconsin and stands in stark contrast to the beetle's success story on nearly every factor. Feral domestic cats (*Felis catus domesticus*) are common worldwide due to abandonment by pet owners, and if not sterilized their local populations can increase rapidly. In both urban and rural areas, cats prey upon birds and small mammals, and while estimates of damage vary considerably, a study in rural Wisconsin (Coleman and Temple, 1996) calculated that a population of 1.4–2 million feral cats and pet cats whose guardians leave them outside kill between 7.8 and 219 million birds annually. Such information spurred a 2004 proposal by a Wisconsin hunter and trapper that the state delist feral cats as a protected species, making an uncollared cat in a rural area not under an owner's direct control fair game to shoot or trap and kill. When the citizen's advisory Conservation Congress decided to put this proposal on the ballot for public vote in hearings held in each county of the state the following spring, it set off a campaign among cat activists that quickly grabbed national and international attention. The Congress, traditionally attended by a small percentage of the state population and dominated by hunting and fishing enthusiasts, voted 57 per cent to 43 per cent in favour of delisting feral cats, but the broader sentiment of state was so far in the other direction that the state's governor urged legislators not to move forward on the issue: 'I don't think Wisconsin should become known as a place where we shoot cats ... everybody is laughing at us.' For the first time in its 71-year history, the Executive Board of Conservation Congress voted against the wishes of its constituents and the proposal died a quick death.

In this case it is clear that feral cats hold considerable positive value (V) to many people. While the negative impacts (I) of cat predation on bird populations are recognized, the wide variance in estimates of the magnitude of the problem may have served to question the reliability of the research. Also, the long-standing presence and acceptance of free-roaming cats around farms and even within the urban landscape probably served to lower their perceived threat (T). The proposed management control mechanism (M) of having hunters shoot feral cats on sight was highly negative to the point of being outrageous to many people, and even scientists who advocated cat control maintained such an approach would not significantly curb cat populations. Cat advocates on the other hand argued that trap-neuter-release programmes provide an effective and humane alternative to curbing populations, though the research backing this approach has also been questioned. Finally, with a resident population that is 74 per cent urban, 88 per cent non-hunting, and 30 per cent cat owners, the context (C) and stakeholder (S) factors likely skewed public response strongly against the proposal.

Conclusion

In this chapter I have argued that, if natural areas managers are to effectively address invasive species in their restoration efforts, sensitivity to the social dimensions of the issue may be as important as understanding the ecological and technical aspects of management. By surveying a range of examples where the social aspects of invasive species management have been dealt with, one begins to gain an appreciation for what factors might be important in anticipating public response to an invasive and its control. While my model of these relationships is certainly a work in progress, it can serve as a useful heuristic device to begin thinking systematically about how one might deal with a particular species.

The case studies of Chicago Asian longhorned beetles and Wisconsin feral cats provide ideal illustrations for how the model might be applied, for both involve a single species and have a clear beginning and end. This unfortunately may not always be the case in addressing invasive species in the context of natural areas restoration and management. In my work in Chicago and San Francisco, conflicts involving invasive species were embedded in larger questions of what restoration means and how it should be approached in urban settings, power and interest group relationships, access and use of public space, and other issues. So while the approach outlined here may be a necessary step forward in addressing the social aspects of an invasive, by itself it may not be sufficient in ensuring the success of a restoration programme.

Although they are not as fully documented as the case studies of Antipin and Dilley (2004) and Beversdorf (2008), many real-world examples of people's response to invasive species management scenarios exist. Systematic documentation of a varied selection of these examples could provide a valuable data base for further study, and with the employment of techniques such as conjoint analysis (e.g. Champ et al, 2005) the model proposed here could be empirically tested. Further work examining how people negotiate or trade off various values and impacts in arriving at a socially acceptable or appropriate alternative (e.g. Brunson et al, 1996) could shed further light on the resolution of management conflicts. The Reasonable Person Model proposed by Kaplan and Kaplan (2003) suggests ways to meaningfully engage stakeholders in such complex decision making, and along with a thoughtful analysis of the factors outlined here, may hold promise in achieving social success in the implementation of natural areas restoration (Phalen, 2009).

Acknowledgements

Many thanks to Herb Schroeder for his assistance on model formulation and Bob Haight and Brendon Larson for their helpful critiques and insights.

References

Antipin, J. and Dilley. T. (2004) 'Chicago vs. the Asian Longhorned Beetle: A Portrait of Success', Report MP-1593, USDA Forest Service, Washington, DC: USDA Forest Service

Apfelbaum, S. (n.d.) 'Cattail (*Typha spp.*) management', Unpublished paper, Applied Ecological Services, Broadhead, WI

Banks, P. B. and Bryant, J. V. (2007) 'Four-legged friend or foe? Dog walking displaces native birds from natural areas', *Biology Letters,* 3 (6) 611–613

Barro, S. C. and Bright, A. D. (1998) 'Public views on ecological restoration: A snapshot from the Chicago area', *Restoration & Management Notes*, 16 (1), 59–65

BenDor, T. K., Metcalf, S. S., Fontenot, L. E., Sangunett, B. and Hannon, B. (2006) 'Modelling the spread of the Emerald Ash Borer', *Ecological Modelling*, 197, (1–2), 221–236

Bertolino, S. and Genovesi, P. (2003) 'Spread and attempted eradication of the grey squirrel (*Sciurus carolinensis*) in Italy, and consequences for the red squirrel (*Sciurus vulgaris*) in Eurasia', *Biological Conservation*, 109(3), 351–358

Beversdorf, A. (2008) 'Here, kitty kitty: They shoot cats in Wisconsin, don't they? (video documentary)', Madison, WI, Prolefeed Studios, www.prolefeedstudios.com.

Bittner, M. (2004) *The Wild Parrots of Telegraph Hill: A Love Story with Wings*, Harmony Books, New York

Bixler, R. D. and Floyd, M. F. (1997) 'Nature is scary, disgusting, and uncomfortable', *Environment and Behavior*, 29(4), 443–467

Bohlen, P. J., Pelletier, D. M., Groffman, P. M., Fahey, T. J. and Fisk M. C. (2004) 'Influence of earthworm invasion on redistribution and retention of soil carbon and nitrogen in Northern Temperate forests', *Ecosystems*, 7(1), 13–27

Brunson, M. W., Kruger, L. E., Tyler, C. B. and Schroeder, S. A. (eds) (1996) 'Defining social acceptability in ecosystem management: A workshop proceedings', General Technical Report PNW-GTR-369, USDA Forest Service Pacific Northwest Research Station, Portland, OR

Burdick, A. (2005) *Out of Eden: An Odyssey of Ecological Invasion*, Farrar, Straus, and Giroux, New York

Campanela, T. J. (2003) *Republic of Shade: New England and the American Elm,* Yale University Press, New Haven, CT

Champ, P. A., Alberini, A. and Correas, I. (2005) 'Using contingent valuation to value a noxious weeds control program: The effects of including an unsure response category', *Ecological Economics*, 55(1), 47–60

Coleman, J. S. and Temple, S. A. (1996) 'On the prowl: In suburban backyards and rural fields, free-roaming cats are pouncing on songbird populations', *Wisconsin Natural Resources Magazine,* www.wnrmag.com/stories/1996/dec96/cats.htm, accessed 13 November 2009

Dolin, E. J. (2003) *Snakehead: A Fish out of Water*, Smithsonian Books, Washington, DC

Fischer, A. and van der Wal, R. (2007) 'Invasive plant suppresses charismatic seabird – The construction of attitudes towards biodiversity management options', *Biological Conservation*, 135(2), 256–267

Flores, D. (2008) 'Bringing home all the pretty horses', *Montana: The Magazine of Western History*, 58(2), 3–21

Forrest, A. and St. Clair, C. C. (2006) 'Effects of dog leash laws and habitat type on avian and small mammal communities in urban parks', *Urban Ecosystems*, 9(1), 51–66

Freinkel, S. (2007) *American Chestnut: The Life, Death and Rebirth of a Perfect Tree*, University of California Press, Berkeley, P. (2004) 'Earthwormed over', *Audubon*, 106(1), 28–31

Friederici, P. (2006) *Nature's Restoration: People and Places on the Front Lines of Conservation*, Island Press, Washington, DC

Gobster, P. H. (2005) 'Invasive species as ecological threat: Restoration as an alternative for fear-based resource management', *Ecological Restoration*, 23(4), 260–269

Gobster, P. H. (2007) 'Restoring urban natural areas: Negotiating nature in San Francisco', Unpublished final report, agreement no. 04-CO-11231300–004 with the University of California-Berkeley, Department of Landscape Architecture and Environmental Planning, USDA Forest Service Northern Research Station, Evanston, IL

Gobster, P. H. and Hull, R. B. (eds) (2000) *Restoring Nature: Perspectives from the Social Sciences and Humanities*, Island Press, Washington, DC

Hager, H. A. and McCoy, K. D. (1998) 'The implications of accepting untested hypotheses: A review of the effects of purple loosestrife (*Lythrum salicaria*) in North America', *Biodiversity and Conservation*, 7(8), 1069–1079

Humphries, C. (2008) *Superdove: How the Pigeon took Manhattan and the World*, Smithsonian Books, New York

Johnson, L. E. and Padilla, D. K. (1996) 'Geographic spread of exotic species: Ecological lessons and opportunities from the invasion of the Zebra Mussel *Dreissena polymorpha*', *Biological Conservation*, 78(1), 23–33

Kaplan, S. and Kaplan, R. (2003) 'Health, supportive environments, and the Reasonable Person Model', *American Journal of Public Health*, 93(9), 1484–1489

Kilpatrick, H. J., Labonte, A. M. and Barclay, J. S. (2007) 'Acceptance of deer management strategies by suburban homeowners and bowhunters', *Journal of Wildlife Management*, 71(6), 2095–2101

Lauber, T. B. and Brown, T. L. (2006) 'Learning by doing: Policy learning in community-based deer management', *Society and Natural Resources*, 19(5), 411–428

Liebhold, A. and Muzika, A. M. (1996) 'Here come the Gypsy Moth brigades', *Inner Voice*, March/April 1996

Mackenzie, B. F. and Larson, B. M. H. (2010) 'Participation under time constraints: Landowner perceptions of rapid response to the emerald ash borer', *Society and Natural Resources*, 23(10), 1013–1022

Mann, R. (1948) 'The English Sparrow', *Nature Bulletin*, no 139, Forest Preserve District of Cook County (Illinois) www.newton.dep.anl.gov/natbltn/100–199/nb139.htm, accessed 13 November 2009

Max, D. T. (2007) 'The squirrel wars', *New York Times*, 7 October 2007 www.nytimes.com/2007/10/07/magazine/07squirrels-t.html?pagewanted=all, accessed 13 November 2009

McNeeley, J. A. (2001) '*The Great Reshuffling: Human Dimensions of Invasive Alien Species*', International Union for the Conservation of Nature and Natural Resources (IUCN), Gland, Switzerland

Millennium Ecosystem Assessment (2005) 'Ecosystems and human well-being: Biodiversity synthesis', World Resources Institute, Washington, DC

Neumann, S. (1999) 'An evening with Garlic Mustard', *Chicago Wilderness Magazine*, http://chicagowildernessmag.org/issues/spring1999/garlicmustard.html, accessed 13 November 2009

Norgaard, K. M. (2007) 'The politics of invasive weed management: Gender, race, and risk perception in rural California', *Rural Sociology*, 72(3), 450–477

Pearson, A. (2002) 'Gypsy moths and Bt: A double scourge', *Chicago Wilderness Magazine* http://chicagowildernessmag.org/issues/summer2002/gypsymoths.html, accessed 13 November 2009

Phalen, K. B. (2009) 'An invitation for public participation in ecological restoration: The Reasonable Person Model', *Ecological Restoration*, 27(2), 178–186

Pimentel, D., Zuniga, R. and Morrison, D. (2005) 'Update on the environmental and economic costs associated with alien-invasive species in the United States', *Ecological Economics*, 52(3) (special issue), 273–288

Robertson, S. A. (2008) 'A review of feral cat control', *Journal of Feline Medicine and Surgery*, 10(4), 366–375

Shapiro, A. M. (2002) 'The Californian urban butterfly fauna is dependent on alien plants', *Diversity and Distributions*, 8(1), 31–40

Slack, G. (2004) 'Monarchs hail the Eucalyptus', *California Wild: The Magazine of the California Academy of Sciences,* 58(1), http://research.calacademy.org/calwild/2004summer/stories/habitats.html. accessed 13 November 2009

Slobodkin, L. B. (2001) 'The good, the bad and the reified', *Evolutionary Ecology Research*, 3(1), 1–13

Starfinger, U., Kowarik, I., Rode, M. and Schepker, H. (2003) 'From desirable ornamental plant to pest to accepted addition to the flora? The perception of an alien tree species through the centuries', *Biological Invasions*, 5(4), 323–335

Todd, K. (2001) *Tinkering with Eden: A Natural History of Exotic Species in America*, W. W. Norton, New York

Van Driesche, J. and Van Driesche, R. (2004) *Nature Out of Place: Biological Invasions in the Global Age*, Island Press, Washington, DC

Welch, W. M. (2009) 'Wild horse debate gallops on', *USA Today*, 27 October, www.usatoday.com/tech/news/2009–10–26-wild-horses_N.htm, accessed 13 November

Whelan, C. J. and Dilger, M. L. (1992) 'Invasive, exotic shrubs: A paradox for natural area managers?', *Natural Areas Journal*, 12(2), 109–110

Williams, T. (2002) 'America's largest weed', *Audubon*, 104(1), http://magazine.audubon.org/incite/incite0201.html accessed 13 November 2009

17
The Paradox of Invasive Species in Ecological Restoration: Do Restorationists Worry about Them Too Much or Too Little?

Stuart K. Allison

Introduction

Recently I was cycling with some friends along a bike trail built on an old railroad bed. As is common in the American Midwest, the railroad right-of-way land along the bike trail was home to remnant prairie and savannah. Very early in the trip, I spotted a huge patch of garlic mustard (*Alliaria petiolata*) that was just starting to flower. Because garlic mustard is considered an aggressive invasive species in the Midwest, I stopped to pull up the plants. Once we got biking again, I saw another large patch of garlic mustard that I also stopped to pull. Upon restarting I soon saw more large patches of garlic mustard all along the trail and it was obvious that I had a choice – I could either continue biking with my friends or spend the day pulling up garlic mustard. One of my friends observed my dilemma about what to do and asked, 'Do you ever feel like you're just spitting into the wind?'

Many times, I and other restoration ecologists with whom I have spoken do feel like we are just spitting into a wind of invasive species blowing across our restoration sites. Most of my experience with invasives has arisen in my role as the director of Knox College's Green Oaks Field Research Station. Garlic mustard is an increasingly large problem in our forest habitats. We also have 40 acres of restored prairie (Allison, 2002), where I have continual problems with both woody invasives (mainly black locust *Robinia pseudoacacia* and autumn olive *Elaeagnus umbellata* – see Figure 17.1) and herbaceous invasives (mostly yellow and white sweet clover, *Melilotus officinalis*

Figure 17.1 *A large pile of autumn olive stems (an invasive species) that were removed from the prairie forest edge at Green Oaks*

and *M. albus* respectively). At times I have been so frustrated by my lack of success in limiting their spread in the prairies that I have become extremely angry with these species, although when I'm being more reflective my anger does not make sense to me. After all those plants that are frustrating me are just doing the job they evolved to do and doing it rather well, although too well from my perspective.

In August 2008, I attended a workshop hosted by the Grassland Restoration Network in Madison, Wisconsin. The participants at the workshop spent a lot of time discussing problem plants in their restorations. Many of the problem plants were members of invasive species that I have also struggled with. Most were originally native to other regions of the world such as Eurasia (autumn olive, white and yellow sweet clover, and common buckthorn *Rhamnus cathartica*), the southern USA (black locust), but some were native to the Midwestern region and simply had become too abundant in some restorations and thus were deemed problematic. Most of the native group were woody plants considered to be undesirable in prairies, such as smooth sumac (*Rhus glabra*), blackberry (several species of *Rubus* but usually *R. allegheniensis*), and grey dogwood (*Cornus racemosa*), but some plants on the problem list surprised me, such as Canada goldenrod (*Solidago canadensis*). What was obvious to me was that while problem plants in prairie restorations generated a lot of passionate discussion, often accompanied by frustration and anger, different restorationists had very different responses to particular species. For example, for some restorationists Canada goldenrod was a huge problem, occupying large areas of prairie and crowding out other species, but for other restorationists it was just a typical part of the prairie ecosystem. Others worried that all the time and effort directed at eliminating or at least limiting the spread of problem plants was time and effort that could be better directed at more important work such as increasing the size and number of restoration projects. After listening to the discussion at the GRN workshop and reflecting on my own experiences and feelings, I was left wondering whether restorationists spend too much or too little time worrying about invasive plants. And how do we decide the best course of action with respect to invasive plants in restorations?

The scope of the problem

In order to fully appreciate why invasive species are a problem in ecological restoration, we first need to consider (briefly) what restorationists are trying to do through ecological restoration and also what makes a species invasive. Ecological restoration has been defined in various ways over the years but the most widely accepted definition is as follows:

> *Ecological restoration is the process of assisting the recovery of an ecosystem that has been degraded, damaged, or destroyed.* (Society for Ecological Restoration Science and Policy Working Group, 2002)

Humans can damage ecosystems by many different activities such as agriculture, urbanization, logging, mining, road construction, damming rivers and so on, but in almost all instances human damage to an ecosystem changes the environment in ways that are detrimental to local, native biota. Damaged ecosystems are frequently colonized by species tolerant of human disturbance regimes and often these colonizing species include many that are not native to the local area. Sometimes the mere growth of non-native species such as *Myrica faya* in Hawaii (Vitousek and Walker, 1989) is the major source of damage to the local ecosystem. Thus one of the main goals in any restoration project is to promote the growth of native species and eliminate or limit the growth of non-native species.

As many chapters in this volume make clear, there are complex issues involved with the naming and defining of non-native or introduced species. It is not my intention to enter into the debates about definitional issues, but I prefer to use the term 'non-native species' when discussing species that have been introduced by human activity to an ecosystem; even though it is a circular definition, I think of non-native species as any species not native to the ecosystem under consideration (following the definition in President Clinton's Executive Order 13112 of 3 February 1999).

However, not all non-native species attract the same amount of attention from restorationists and site managers. Non-native species can be categorized in various ways but the most important distinction is between current problems and not problems (Table 17.1). It is also important to note that some native species may become problems in restorations. The key issue is recognizing problem populations of non-native species. Problem populations are

Table 17.1 *Characteristics of problem and non-problem species in the context of ecological restoration*

Problem invasive species	Not native to ecosystem. Increase and spread in ecosystem, offspring easily dispersed. Lead to decline in native species via negative interactions such as competition, predation or parasitism. Cause changes in ecosystem properties such as hydrology, nutrient cycles, energy flow, fire regime. Tolerant of human disturbance regimes.
Problem native species	Native to ecosystem. Effects are ecologically similar to problem invasive species. Frequently they are edge species tolerant of human disturbance.
Non-problem native species	Native to ecosystem. Populations exhibit minor fluctuations in size and distribution. Presence enhances other native species. Maintain ecosystem properties. Frequently sensitive to human disturbance regimes.
Non-problem non-native species	Not native to ecosystem. Reproduce and survive in ecosystem. Populations exhibit minor fluctuations in size and distribution. Presence at least is not leading to decline in native species abundance or distribution. Do not lead to changes in ecosystem properties. Have potential to become problematic as climate changes.

usually considered invasive and the properties that make them invasive are best defined as follows:

> *Biotic invaders are species that establish a new range in which they proliferate, spread and persist to the detriment of the environment.* (Mack et al, 2000)

Problem populations of invasive species typically change ecosystem properties such as local hydrology, fire regime, nutrient cycles and energy flow and as a result often lead to a reduction or loss of populations of local native species (Mack et al, 2000). Because invasives alter ecosystem properties, their removal is critical at the beginning of any restoration project and is often the first step in restoration. Restoration projects (certainly those I am most familiar with in American Midwestern prairies and savannas) almost always require continued vigilance and removal of invasives in order to maintain the integrity of the restoration for the entire life of the restoration (Norton, 2009).

The goal of ecological restoration is usually to return a damaged ecosystem to its previous condition, although how to choose that previous condition is a challenge (in North America we frequently use the condition of the ecosystem prior to the arrival of Euro-Americans as the desired end point for restoration) (Allison, 2007). Recently restorationists have realized that along with trying to match an historical ecosystem, it is critical to plan for restorations that accommodate natural changes due to typical ecosystem processes, dynamics and evolution (Higgs, 2003). If we truly want to match the restoration to historical ecosystems that existed prior to modern human disturbance, then we would almost certainly have to eliminate all non-native species from the restoration and there are purists who argue that non-native species have no place in a restoration. But given that there are so many non-native species in any particular ecosystem today, it may be impossible to eliminate all such populations. New Zealand has the best documented flora of native and non-native species: it is home to 2065 native plant species and is also home to 24,774 non-native plant species, of which at least 2200 have become naturalized (able to survive outside of human cultivation) (Duncan and Williams, 2002; Norton, 2009). Even allowing for the persistence of a few dozen beneficial naturalized species such as some of our food and forage plants, would it be possible to eliminate about 2200 naturalized species today? If as many as 52 per cent of the species in an ecosystem are non-native species the task of creating completely native restored ecosystems may be extremely difficult to achieve.

Due to limited time and budgets, restorationists have to focus on populations of truly invasive species. Which invasives to focus on will vary from site to site, but will almost certainly be those that cause the most change to environmental properties such as fire regime, hydrology, nutrient cycling, and have the greatest negative affect on native species. Essentially restorationists must practise a form of triage in which they identify invasive populations of greatest concern and work to control them (Hiebert, 2001). Early in the history of prairie restoration, Kentucky bluegrass (*Poa pratensis*) was frequently an invasive problem but once restorationists began to use prescribed fire on a regular basis they found that Kentucky bluegrass was easily controlled and no longer a problem (Blewett and Cottam, 1984). Today biennial species like yellow and white sweet clovers can be problems at some prairie restorations, at least temporarily, but usually changes to the prescribed fire regime will greatly reduce their abundance (Illinois Natural History Survey, www.inhs.uiuc.edu/chf/outreach/VMG/wysclover.html). The greatest problems are woody species that change the fire regime and shade out native herbaceous grass and forb species. Black locust and common buckthorn spread aggressively by vegetative growth. Autumn olive is spread by animals consuming the fruit and passing the seeds, but can also expand via vegetative growth. All three of those species sprout from burned and cut stumps and have very persistent root systems making them difficult to eliminate once they are well established in a prairie. So a site manager would probably attempt to control less problematic populations of Kentucky bluegrass and the sweet clovers simply by prescriptive fire, but would have to use more direct and individualized approaches to handle populations of woody invasives.

The way forward

Ecological restoration is likely to be one of the most important, if not the most important, land management practice in the 21st century as we attempt to repair human-caused damage to the Earth's ecosystems (Hobbs and Harris, 2001). Indeed, some groups have identified ecosystem restoration as the primary strategy for addressing global climate change (e.g. Ecosystem Restoration Associates). Yet just as ecological restoration is becoming a more widespread and accepted practice, the continued spread of invasive species in almost all global ecosystems (Chornesky and Randall, 2003) raises questions about how successful ecological restoration will be in meeting the goals of both restoration to a previous condition and repairing human-caused damage to ecosystems. Habitat destruction is the greatest threat to locally rare native species, but populations of invasive species and their impact on local biota are the second leading cause of reductions in populations of endangered species, leading to declines in 57 per cent of endangered plant species in the USA (Wilcove et al, 1998). Moreover human habitat modifications and spread of non-native species are leading to the world becoming increasingly domesticated (Kareiva et al, 2007) and homogenized (McKinney and Lockwood, 1999). I once heard Daniel Janzen give a lecture in which he said he feared that the entire planet would come to resemble Iowa – a comment meant to generate mental images of extensive rural areas dominated by domesticated corn (*Zea mays*), soybeans (*Glycine max*), cattle and pigs, and urban areas consisting of expanses of Kentucky bluegrass, Norway maples (*Acer platanoides*), house sparrows and starlings. This was not meant to disparage the good people of Iowa. Given the reality of global climate change and spread of non-native species, ecological restoration is likely to be an increasingly important tool, more so than habitat preservation, as we attempt to maintain high levels of species and ecosystem diversity (Hobbs and Harris, 2001). Predictions that ecosystems will become dominated by just a few species that tolerate human disturbance regimes and climate change, while most species decline due to sensitivities to both human disturbance and climate change, are particularly disheartening (McKinney and Lockwood, 1999). Ecological restoration may be our most effective tool for preventing the development of an ecologically boring world dominated by a few hardy cosmopolitan species.

But some authors question the recent focus on non-native and invasive species and don't think those species cause many problems. In fact they claim that with plants in particular the arrival of non-native species has resulted in an increase of local species diversity with no or extremely few instances of non-native plants driving native plant species extinct (Brown and Sax, 2004, 2005; Sax and Gaines, 2008). However, there is ample evidence that non-native and invasive species do have a large, negative impact on populations of native species (Wilcove et al, 1998; Mack et al, 2000; Chornesky and Randall, 2003; Ricciardi, 2003; Cassey et al, 2005). Of particular concern is the question of when do naturalized non-native species become invasive? Most invasive non-native species go through a latent period that may last for decades during which they are non-problematic naturalized species before becoming invasive problems (Mack et al, 2000). At the current time it is very difficult to predict which non-native species will eventually become invasive problems and which will remain minor additions to the local biota. A focus on species richness as the prime measure of ecosystem quality (as in Brown and Sax, 2004, 2005; Sax and Gaines, 2008) misses the point that a native species may continue to exist despite losses of many local populations (Ricciardi, 2003) and the fact the invasive populations may greatly alter ecosystem properties such as nutrient cycling, energy flow, fire regime and hydrology, so that the ecosystem no longer resembles or functions like the previous ecosystem even though local native species persist (Mack et al, 2000). In my experience, a prairie invaded by black locust and autumn olive may still have a majority of native prairie species but it is no longer a prairie; instead it has been converted to shrubby woodland that does not resemble or function like a prairie.

Perhaps most surprising to me is that Brown and Sax (2005) suggest questions of whether a decline in biodiversity or the effects of invasive species are good or bad are not questions to be answered by scientists. In many ways their position is an abdication of public trust. Most scientists have been educated and supported throughout their careers by funds supplied by the government, which ultimately means from taxpayers. Such education and support is provided with the expectation that scientists will function as experts able to make recommendations to the public about what are the best choices to make given a particular situation. The original ecologists in the USA certainly felt a duty to make recommendations about how humans should interact with ecosystems (Worster, 1990). Restoration ecologists must be willing to provide expertise and make judgements about how to restore ecosystems, how to set reasonable goals for restoration projects, and whether proposed management plans are good or bad, although they must be careful to separate their statements about data from their opinion and be willing to accept that the general public may disagree with their opinions and recommendations.

Given our legitimate concerns about the development of a world of domesticated, homogenized ecosystems, the question of whether to restore ecosystems so they contain species previously found there or simply to focus on ecosystem function is a false dichotomy. We could almost certainly maintain basic ecosystem functions of hydrology, nutrient cycling and energy flow with a few well-chosen hardy species but such ecosystems would be lacking in both native species diversity and the dynamics of continued evolutionary change, at least evolutionary change in species rich assemblages of locally native populations. In order for ecological restoration to be truly successful all stakeholders involved in the restoration have to be informed about the restoration and must value both the process and final or ongoing (given that restoration projects never really end) product of restoration (Higgs, 2003; Jordan, 2003; Allison, 2007).

Restorationists highly value restorations that return and maintain local native species, and allow for the continuation of natural processes within those restorations (Higgs, 2003; Jordan, 2003). For most restorationists an ecosystem that lacks many native species would not be particularly valuable even if it maintains desired ecosystem function because one of the most valuable aspects of ecological restoration is the maintenance of native species diversity (Higgs, 2003; Jordan, 2003). In today's world a focus on native species is not usually a form of prejudice against non-native species (Simberloff, 2003), rather it reflects a love and respect for the species that evolved in that place and which have meaning to the local human population (Olwig, 1995; Jordan, 2003).

Because global climate change, human habitat modification, and spread of non-native species will almost certainly continue for the foreseeable future, probably at increasing rates, managing ecological restoration projects so that sites continue to maintain populations of native species will be an increasing challenge. We can assume that in a restoration project we can limit any further direct human habitat modification. However, the effects of global climate change are likely to become greater and may result in changes to the local ecology that are so great that many local species will no longer be able to reproduce and survive on site. Naturalized non-native species usually disperse either on their own or with assistance from other species, so that it is difficult to keep them from arriving at a restoration site. Thus there is likely to be continued and increasing spread of non-native species in the future. As mentioned previously, due to time and budget limitations, restorationists must focus on the most serious problems that prevent working on all real and potential problems for a site. The focus on pressing problem species may allow species initially seen as non-problematic to increase until they too become a large problem. Restorationists must work with proven methods and be sceptical of unproven methods until experimental trials have demonstrated their effectiveness. For example, Donlan has suggested that because American prairies lack large native herbivores, it might be worth introducing large African herbivores to those grasslands to both preserve populations of those herbivores and to allow them to fill the niche once occupied by American native large herbivores. As much as I like the mental image of zebra in the prairies, I would not advocate pursuing Donlan's idea without extensive testing. There are good data which indicate that non-native herbivores have a negative effect on populations of native plants but only limited or even positive effect on populations of non-native plants (Parker et al, 2006). Zebra in Illinois may only make the situation worse for native prairie plants.

There may be ecosystems that are so damaged or changed from their original condition that restoration to the original ecosystem or even something similar to the original will be prohibitively expensive in terms of time, labour and money (Jackson and Hobbs, 2009). In such cases, the most that can be accomplished via restoration may simply be managing or directing the development of new ecosystems that combine native and non-native species so that ecosystem functions are preserved (Jackson and Hobbs, 2009).

Restorationists will have to be flexible in their management of restored sites in the future. Management plans will have to be modified as we see first hand how global climate change and the arrival of new non-native species affect restoration projects. The combination of global climate and invasive species is creating groups of organisms that have never occurred together in the past and which form ecosystems that some ecologists refer to as 'novel ecosystems' (Hobbs et al, 2006). The development of novel ecosystems greatly complicates our understanding of ecological restoration because at least some novel ecosystems appear to be especially robust and potentially stable. Thus we are unlikely to be able to restore ecosystems to an absolutely pristine predamage condition. Given all the environmental changes likely to occur in the next century pristine is probably an impossible goal. It may be time to ask how long a species has to be naturalized in an area before it can be considered native. There is no obvious way to make that determination (Coates, 2007) but many non-native species have evolved local differences once in situ in new habitat (such as adaptation to micro-habitat differences in disturbance in non-native dandelion *Taraxacum officinale* (Solbrig and Simpson, 1976)). Is there a point at which we consider evolution and adaptation to local conditions sufficient to make a population native to a site? Surely there must be because we do not consider species that moved to a new area on their own to be invasive (whether it be opossums moving from South America to North America millions of years ago or more recently cattle egrets spreading worldwide from Africa). Allowing for such changes in classification will be a challenge for restorationists given our usual focus on local native species and the value we place on ecosystems dominated by those species.

At this point my own energies are directed at eliminating or limiting the spread of invasive populations that are especially detrimental to prairies (black locust, autumn olive) because they thoroughly change the character and function of prairies. I cannot imagine a time when either of those species would be considered native or not problematic in prairies, especially considering that some woody natives like grey dogwood and smooth sumac are problematic. Flexibility in light of the changing environment cannot be used as an excuse for letting restored sites that were restored for a particular reason become something completely other than what was originally intended. If climate change researchers are correct that by 2095 Illinois will have a climate that resembles the climate of East Texas today (Kling et al, 2003), then the restored prairies I manage may become more similar to prairies or savannas currently found in Texas, but I think there should be a commitment to maintaining them as prairies rather than allowing them to become forest or scrub.

Conclusions

Do restorationists worry about invasive species too much or too little? Well that depends upon the individual restorationist, but invasive species do merit serious consideration and concern. Because populations of invasive species change ecosystem properties and have a negative effect on native species in sites of ecological restoration, they are a problem that needs to be controlled as much as possible. Global climate change, the continued spread of non-native species, and the potential for increased homogenization of the world's biota indicate that the problem of invasive species will most likely get worse in the future. Restorationists will have to focus on limiting the influence of invasive species that cause the greatest changes to their projects. Even as they maintain such focus, they must be on the lookout for the arrival of new problem species and changes to local ecology as climate change progresses. They will have to be in constant communication with all stakeholders in a restoration to ensure that the restoration continues to fulfil both ecological and cultural goals of the restoration. Our restoration sites may undergo considerable shifts due to climate change but the restoration itself must be maintained in a way that reflects the original plans and possibilities of that site. Restorationists must also use ecological restoration to maintain ecosystems of ecological and cultural value. We must also work to promote restorations that allow for ecological and evolutionary functions and processes that occur outside the realm of human influence. Ideally we will use ecological restoration to prevent the development of a completely domesticated, ecologically boring world that exists only to satisfy basic human needs. If we can do those things, restorationists will play a key role in preserving an ecologically diverse and interesting planet.

References

Allison, S. K. (2002) 'When is a restoration successful? Results from a 45-year-old tallgrass prairie restoration', *Ecological Restoration*, 20, 10–17

Allison, S. K. (2007) 'You can't not choose: Embracing the role of choice in ecological restoration', *Restoration Ecology*, 15, 601–605

Anon. (undated) Illinois Natural History Survey; www.inhs.uiuc.edu/chf/outreach/VMG/wysclover.html

Anon. (undated) Ecosystem Restoration Associates; www.eraecosystems.com/about/story

Blewett, T. J. and Cottam. G. (1984) 'History of the University of Wisconsin Arboretum prairies', *Transactions of the Wisconsin Academy of Sciences, Arts, and Letters*, 72, 130–144

Brown, J. H. and Sax, D. F. (2004) 'An essay on some topics concerning invasive species', *Austral Ecology*, 29, 530–536

Brown, J. H. and Sax, D. F. (2005) 'Biological invasions and scientific objectivity: Reply to Cassey *et al.*', *Austral Ecology*, 30, 481–483

Cassey, P., Blackburn, T. M., Duncan, R. P. and Chown, S. L. (2005) 'Concerning invasive species: Reply to Brown and Sax', *Austral Ecology*, 30, 475–480

Chornesky, E. A. and Randall, J. M. (2003) 'The threat of invasive alien species to biological diversity: Setting a future course', *Annals of the Missouri Botanical Garden*, 90, 67–76

Coates, P. (2007) *American Perceptions of Immigrant and Invasive Species*, University of California Press, Berkeley, CA, USA

Duncan, R. P. and Williams, P. A. (2002) 'Darwin's naturalization hypothesis challenged', *Nature*, 417, 608–609

Hiebert, R. (2001) 'Prioritizing weeds: The alien plant ranking system', *Conservation Magazine*, 2, 1–2

Higgs, E. (2003) *Nature by Design: People, Natural Process, and Ecological Restoration*, MIT Press, Cambridge, MA, USA

Hobbs, R. J. and Harris, J. A. (2001) 'Restoration ecology: Repairing the Earth's ecosystems in the new millennium', *Restoration Ecology*, 9, 239–246

Hobbs, R. J., Arico, S., Aronson, J., Baron, J. S., Bridgewater, P., Cramer, V. A. P., Epstein, R., Ewel, J. J., Klink, C. A., Lugo, A. E., Norton, D., Ojima, D., Richardson, D. M., Sanderson, E. W., Valladares, F., Vila, M., Zamora, R. and Zobel, M. (2006) 'Novel ecosystems: Theoretical and management aspects of the new ecological world order', *Global Ecology and Biogeography*, 15, 1–7

Jackson, S. T. and Hobbs, R. J. (2009) 'Ecological restoration in the light of ecological history', *Science*, 325, 567–569

Jordan, W. R., III. (2003) *The Sunflower Forest: Ecological Restoration and the New Communion with Nature*, University of California Press, Berkeley, CA, USA

Kareiva, P., Watts, S., McDonald, R. and Boucher, T. (2007) 'Domesticated nature: Shaping landscapes and ecosystems for human welfare', *Science*, 316, 1866–1869

Kling, G. W., Hayhoe, K., Johnson, L. B., Magnuson, J. J., Polasky, S., Robinson, S. K., Shuter, B. J., Wander, M. M., Wuebbles, D. J., Zak, D. R., Lindroth, R. L., Moser, S. C. and Wilson, M. L. (2003) *Confronting Climate Change in the Great Lakes Region: Impacts on our Communities and Ecosystems*, Union of Concerned Scientists, Cambridge, MA, USA, and Ecological Society of America, Washington, DC, USA

Mack, R. N., Simberloff, D., Lonsdale, W. M., Evans, H., Clout, M. and Bazzaz, F. A. (2000) 'Biotic invasions: Causes, epidemiology, global consequences, and control', *Ecological Applications*, 10, 689–710

McKinney, M. L. and Lockwood, J. L. (1999) 'Biotic homogenization: A few winners replacing many losers in the next mass extinction', *Trends in Ecology and Evolution*, 14, 450–453

Norton, D. A. (2009) 'Species invasions and the limits to restoration: Learning from the New Zealand experience', *Science*, 325, 569–571

Olwig, K. (1995) 'Reinventing common nature: Yosemite and Mount Rushmore: A meandering tale of a double nature', in W. Cronon (ed) *Uncommon Ground: Rethinking the Human Place in Nature*, W. W. Norton, New York, NY, USA

Parker, J. D., Burkepile, D. E. and Hay, M. E. (2006) 'Opposing effects of native and exotic herbivores on plant invasions', *Science*, 311, 1459–1461

Ricciardi, A. (2003) 'Assessing species invasions as a cause of extinction', *Trends in Ecology and Evolution*, 19, 619

Sax, D. F. and Gaines, S. D. (2008) 'Species invasions and extinctions: The future of native biodiversity on islands', *Proceedings of the National Academy of Sciences*, 105, 11490–11497

Simberloff, D. (2003) 'Confronting introduced species: A form of xenophobia?', *Biological Invasions*, 5, 179–192

Society for Ecological Restoration Science and Policy Working Group (2002) *The SER Primer on Ecological Restoration*, Retrieved 1 July 2007 from www.ser.org/

Solbrig, O. T. and Simpson, B. B. (1976) 'A garden experiment on competition between biotypes of the common dandelion', *Journal of Ecology*, 65, 427–730

Vitousek, P. M. and Walker, L. R. (1989) 'Biological invasion by *Myrica faya* in Hawaii: Plant demography, nitrogen fixation, and ecosystem effects', *Ecological Monographs*, 59, 247–265

Wilcove, D. S., Rothstein, D., Dubow, J., Phillips, A. and Losos, E. (1998) 'Quantifying threats to imperiled species in the United States: Assessing the relative importance of habitat destruction, alien species, pollution, overexploitation, and disease', *Bioscience*, 48, 607–615

Worster, D. (1990) 'The ecology of order and chaos', *Environmental History Review*, 14, 1–18

18

A View from Continental Europe: The Case Study of *Prunus serotina* in France in Comparison with Other Invasives

Aurélie Javelle, Bernard Kalaora and Guillaume Decocq

Introduction: Is the forest a permanent social construction?

In the last two decades, many studies have explored the complex interaction between the characteristics of non-native species that enable them to efficiently invade an ecosystem ('invasiveness') and the properties of the new recipient ecosystems that make them susceptible to invasion ('invasibility') (Alpert et al, 2000). Surprisingly, there has been relatively little research into the social causes and consequences of biological invasions outside a selected set of agricultural pests (Perrings et al, 2002). Most studies ignore the human components of how nature and human societies interact to determine ecosystem invasibility, and how bio-invasions affect people's behaviour. (See Lambert this volume, Chapter 11).

In this study, we focus on the American black cherry (*Prunus serotina* Ehrh) in France as a case study. Our aim is to understand the social conditions enabling its invasion, and to establish what values and perceptions are assigned to this invader by the people. By recording and interpreting local knowledge held by forest users about the ecology and impacts of *Prunus serotina* invasion we should understand the organizational and cognitive relationships that the local culture has with its non-human environment. For this purpose, a constructivist perspective has been adopted. The forest must be conceived, not so much as a natural object or objective reality, but rather as a mental construction, at both local and national levels, where nature and society interact in a twofold dynamic, social and natural. This mental construction reflects stable forms of knowledge, meanings, discourses, practices and activities of the social and institutional actors involved. This way of thinking has led to the deconstruction of the idea of an ontological separation between nature and society; instead, the forest is considered as the outcome of a dynamic process involving natural and social interactions between human societies and ecosystems (Kalaora, 1993).

Hereafter, we report the main results of an ethno-ecological study that has been conducted (Javelle et al, 2006) within the framework of a multidisciplinary research programme dedicated to *Prunus serotina* in northern France from 2003 to 2006 (Decocq, 2006). We then compare this case study to invasions by other plants and by animals in order to highlight the conflicts their presence may generate. These situations are then questioned in order to find patterns in the current thinking about human–nature relationships.

Study species, study area and methods

Prunus serotina is a tree species native to North America, which has been introduced into many European forests for ornamental, timber production and soil amelioration purposes since the 17th century. For at least three decades it has spread throughout temperate forests of Western and Central Europe, particularly on well-drained, nutrient-poor soils, and has become one of the main forest invaders (Starfinger, 1997).

This study was mostly conducted in the forest of Compiègne (14,417ha), located in northern France and currently the area most heavily invaded by *P. serotina* in France. Initially, the species was probably introduced as an ornamental in the gardens of a neighbouring castle in the mid 19th century. In 2003, it was present in the whole forest area and has become the dominant species in many stands. Another part of the study, focusing on foresters, was extended to two other invaded forests: Fontainebleau (southeast of Paris: 25,000ha) and Saint-Amand (northern France: 7675ha).

We used ethno-scientific methods to analyse the empirical knowledge of local users of the landscape facing *Prunus serotina* invasion to assess how new and exogenous knowledge is being incorporated, how populations are culturally inserted in ecosystems through cognitive processes, such as emotional and behavioural responses, and how this local knowledge affects action. Data were gathered in 2004 (Compiègne) and 2005 (Fontainebleau and Saint-Amand) through non-directive interviews. First, 46 users were first selected to represent the whole community associated with the forest, including a range of socio-economic categories (e.g. forest managers, woodcutters, walkers, cyclists, hunters, horse riders). Then 47 other interviews were undertaken with public and private forest managers at all levels in the hierarchy. This placed the topic of biological invasion into the broader context of global change effects on the forest. For purposes of analysis and discussion, the interview results were considered in the light of the data obtained from a bibliographic survey. More details can be found in Decocq and Kalaora (2009).

Social construction of *Prunus serotina*: Emptiness and invisibility

Users of the Compiègne forest, as well as those of other forests, are generally unresponsive to the topic of biological invasions in general and *Prunus serotina* in particular. There was a multiplicity of identities of *Prunus serotina* in the absence of a unified view. Beyond a few individuals who have a sound knowledge of environmental issues, the majority do not know the issues. There was a strong gradient between awareness of *Prunus serotina* as a biological invader and the connection the interviewee had with the scientific world, either as a researcher or as a member of a naturalist association. The assertion of the need to eradicate the invader followed the same gradient.

For most users, *Prunus serotina* is integrated into their everyday life; it is not differentiated from other tree species. However, some walkers have developed a particular hiking trail in autumn to cross heavily invaded areas and admire the golden foliage of the invader (an unusual colour in natives). Others pick fruit to make jam. Some horse riders like to stop occasionally at the edge of invaded stands to eat some fruit. Some people have dug up saplings to plant them in their garden, because of the late flowering and beautiful autumn foliage. Surprisingly, all those users who noticed *Prunus serotina*'s presence in the forest claim that they always have seen it there. Despite the rapid invasion of the forest, where it seems to have spread significantly since the 1970s, the dynamic change is still too slow and gradual to be perceived by forest users. This lack of perception probably results from a 'creeping normalcy': due to the slow, fluctuating change, the fundamental benchmarks of normality gradually evolve in an imperceptible manner (Diamond, 2005). Is there a better example than those users who make jam with the fruit while they do not know the species' name, history or biological characteristics? Yet black cherry jam is gradually becoming absorbed into family rituals!

Forest users who 'note' the presence of *Prunus serotina* are unable to name it. To speak about it, walkers use periphrases such as 'the tree which is everywhere' or 'the tree found everywhere.' For them, *Prunus serotina* is recognized but unknown, identified but without identity. Conversely, other users say that they know *Prunus serotina*, but they do not necessarily recognize it! It is surprising, even disturbing, that many foresters claiming that they have a (good) knowledge of *Prunus serotina*, confuse it with other (native) species such as *Prunus padus* (local forest managers) or *Prunus mahaleb* (National Forest Inventory), or even with unrelated exotic species such as *Betula lutea* (Regional Centre of Private Forests). In the latter case it should be noted, however, that yellow birch is commonly called 'cherry' in Quebec. The lack of vernacular names or the use of wrong ones is evocative of the fuzziness about its semantic definition; this strongly contrasts with the unified, conventional view of oak and beech, which are considered to be the noble heritage of Compiègne forest. Although the species is physically present almost everywhere in the forest, it is, from a sociological point of view, invisible to most people; this has been labelled the 'invisible omnipresence paradox' (Decocq and Kalaora, 2009).

Why such 'social invisibility'? First we must say that we are in a media gap. Although biological invasions were known at least since Darwin, they have been brought to wider attention by the popular media only after the 1992 Rio Conference. In France the problem of invasive plants is hardly publicized. The first articles were published in the 1990s in scientific or professional journals. Only ten years later was the topic seized by major newspapers and popular magazines, even more recently by radio and television. The type of information often follows a gradient corresponding to the scientific knowledge of the authors and/or of the readers: the closer to the general public the magazine is, the more abstract, partial and subjective is the message it delivers. Scientific magazines usually inform of ecological risks, while professional ones focus rather on specific economic or health problems, and tend to temper the risks. In the latter case, articles are often written by non-specialists, who sort out the information they have collected. For the general public, awareness about biological invasions is thus limited.

A second reason is that 'whistleblowers' are facing emptiness. In Compiègne, apart a few knowledgeable persons, some scientists and local naturalists, only field foresters who are directly confronted with *Prunus serotina* invasion act as whistleblowers. But for the latter, the recognition of biological invasions as a problem is recent, related to integration of new biological knowledge. Moreover, for them the problem is primarily economic, because forest regeneration

Figure 18.1 *Old* Prunus serotina *stand*

fails in invaded stands, and thus the production role of the forest is threatened. Conversely, for scientists and naturalists, the threat is definitely ecological: the invasion alters certain equilibria and reduces biodiversity. There is a lack of both a common strategy or a reflection on experience. For instance, the first whistleblower in Compiègne was a field forester who between 1968 and 1971 related the failure of a beech plantation to the unusual spread of an 'atypical Bird cherry' (later identified as *Prunus serotina*) in the stand. At that time he tried to alert other foresters, but met with disbelief from his colleagues. Moreover, the only information he obtained about the species focused on the American context and *Prunus serotina* was presented as a highly valuable tree; of course, no ecological risk was mentioned. This may explain why, in 1986, an experiment was implemented in Compiègne to promote the species through very dynamic forestry; it rapidly failed and was discontinued in 1999. It is noteworthy that in Germany, Belgium and The Netherlands similar experiments had already failed at the beginning of the 20th century! The information did not cross borders.

Biological invasions: From denial to conflict

This poor awareness of *Prunus serotina* invasion in Compiègne forest seems paradoxical. One explanation might come from pressure by the general public on the French National Forest Office, which is often accused of overexploiting the forests. Publicizing the invasion problem would suggest another deleterious effect of already highly controversial silvicultural methods (e.g. monospecific, even-aged plantations, large clear-cutting areas and soil ploughing). The invasion highlights the weaknesses of the human-made forest ecosystem. Similarly, another invasive plant species, the water primrose (*Ludwigia grandiflora* ((Michaux) Greuter and Burdet), is also witness to a historical human impact. In western France, it:

> *tells the story of wetlands during the past century: marsh abandonment by farmers, leading to unmanaged spaces, and local actors watching their closure by the willows, reeds, brambles ... and* Ludwigia grandiflora. *Developments leading to the loss of these marshes, their drying due to intensive cultivation, and now loaded with nitrates and other products of intensive agriculture, have created an environment in which ...* Ludwigia *seems to thrive.* (Menozzi, 2007)

Biological invasions may thus trigger conflicts between social groups when they relate to broader and/or older issues. The conflict may begin when scientists and naturalists recommend the eradication of an invasive plant on behalf of biodiversity conservation, while the advocates of the species criticize this destruction for economic, heritage or aesthetic reasons. For example, tourist or native walkers along the Mediterranean coast are opposed to campaigns to remove Hottentot fig (*Carpobrotus edulis* (L.) N.E. Br.). Their concern is because of the beauty of its flowers, which beautify the coast: 'as they have always seen this species there, it is an ecological crime to destroy its populations' (Médail, 2004). Such conflicts place science against sensibility, and are complex environmental issues that cannot be tackled by science alone (Ravetz, 1998). The level of conflict may move from local to regional and even national, as the number and complexity of interactions between protagonists grows. As shown by Mermet (2002) in another context (the reintroduction of bears to the Pyrénées), specific environmental conflicts may echo local stumbling blocks, the ecological question reflecting the balance of power on a broader scale. Plant invasions, as with the reintroduction of bears to the Pyrénées, are sometimes a pretext for rekindling an existing but lapsed debate. This is one of the weaknesses of environmental issues, which may connect extremely diverse, even sometimes contradictory, representations and objectives and so, instead of creating a new object, remain an instrument of social conflict and add confusion (Alphandéry et al, 1991). However, such conflicts may also be positive in the case of biological invasions when they forge new social ties: 'The disagreement is treated as a vector of relations and social differentiation' (Coser Lewis, 1982 (1953); Simmel, 1995), In this case, conflicts can lead to redefined practices, identities, norms and knowledge, to producing new frameworks, and ultimately, to rethinking the relationship with these species. These changes may also sometimes be used on behalf of particular interests, such as territorial ownership (Mormont, 2006).

Figure 18.2 Prunus serotina *flowers*

It is noteworthy that unlike most animal invasions plant invasions often spark little or no debate. With other case studies in France, invaders such as Japanese knotweed (*Fallopia japonica* (Houtt.) Ronse Decraene), Pampas grass (*Cortaderia selloana* (Schultes and Schultes fil.) Ascherson and Graebner), or South African ragwort (*Senecio inaequidens* DC) are spreading with few constraints. Similarly, the spread of *Prunus serotina* in the forests that we studied is silent and does not cause conflict. As we stated earlier this social invisibility is primarily due to lack of any cognitive apprehension framework, including patterns of knowledge and action, rather than to physical presence or absence. No one is concerned about the presence of *Prunus serotina*; it is here but like a stranger no one recognizes it. So it is without status other than that of intruder who must be ignored or even whose presence is denied. For most people, the species is definitely not perceived as an invader: *Prunus serotina* invasion is just a problem of and for ecologists. A notable exception pertains to a few species, such as, for example, ragweed (*Ambrosia artemisiifolia* L.). Ragweed is well-known but not necessarily recognized by the general public because its pollen causes serious respiratory allergies. As such, it is unanimously considered as a harmful species to be eradicated; it seems that an adverse effect on human health can be a socially unifying and compelling argument.

Doubts over the scientific arguments

Unlike plants, invasive animals often generate more consideration and emotional response. Their eradication raises many more questions and moral issues, including the way of killing animals, even if they are acknowledged as invaders (Mougenot and Mormont, 2009). Being animals themselves, humans probably project their anthropomorphism on animals, making it easier to destroy a plant than to kill an animal. Within the broad field of biological invasions there is a discrepancy so animals are perceived as 'subjects', while plants are 'objects' (Manceron, 2009). As such, the former raise moral, philosophical and ethical issues that the latter do not.

The degree to which arguments about plant or animal invasions are objective and scientifically compelling is also questionable. Recommendations are often given by scientists who have been brought in with the ambiguous role of 'experts' (Roqueplo, 1998). Alternatively, members of naturalist associations that are more or less connected to the scientific world may also act as experts, but their arguments tend to be biased by ideology or a desire for action. In both cases, the guidelines are frequently overstated due to subjective considerations. It should be remembered that the social representations of a given species sometimes change over time, depending on social needs and fashions. From a desirable ornamental in the 18th century, *Prunus serotina* has successively become a valuable timber tree, a soil-improving species, an aggressive forest pest, a controllable weed and a species we have lived with during the 19th and 20th centuries (Starfinger et al, 2003). There is a similar story in South Africa with the Barbary fig (*Opuntia ficus-indica* (L.) Mill.) (van Sittert, 2002). Claeys-Mekdade (2006) demonstrated rather neatly

Figure 18.3 *Pure stand of Prunus*

that, 'emotions tend to carry more weight than scientific discussion' among researchers' discussions about the eradication of invasive species. Furthermore, new facets of a plant species are revealed as their scrutiny highlights the changes taking place. Hence, the way the species is viewed is neither universal nor timeless (Allain, 2001). Although scientific data are given a high intrinsic value by our society, they may still be biased by emotions (see Rotherham this volume, Chapter 15).

We note that interdisciplinary approaches have become more popular in addressing environmental issues, especially biological invasions. This implies bridge-building between human and natural sciences, which is influenced more by the researchers' personalities than by scientific objectivity (Morlon, 1999). Brun et al (2007) show that as much as the maturity of his ideas and his relationship to other researchers, a scientist's way of thinking depends on his personal history, beliefs, social environment, cultural roots and other individual factors. So any decision about invasions is likely to be influenced by personal factors. For this reason, we should beware of how words such as 'native', 'invasion', 'indigenous' and 'exotic' are used, often to justify eradication campaigns, because of their political and cultural implications (Claeys-Mekdade, 2006). A 'xenophobic temptation' may even develop against the free movement of plants worldwide, which has its own advocates, such as Clément (2002), a gardener self-appointed defender of 'refugee plants'. Why should the commitment to eradicate a species be analysed systematically? Is this just anthropocentrism? Are our attitudes to exotic invasive species a transient fad or the insidious effect of moralistic media trying to ease their conscience? Are they really trying to protect nature on behalf of future generations; to correct the excesses of our civilization? Maybe so, but in this context, some species tend to become scapegoats (Bruno, 1990).

Figure 18.4 *Walking at the edge of invaded stands*

Final considerations to overcome the anthropocentric view

Our study of *Prunus serotina* reveals a very heterogeneous distribution of ecological knowledge, perceptions and values among forest users. This suggests that societies adapt poorly to changing conditions and so may not react quickly, if ever, when invasion occurs. This social invisibility of a plant or animal is not an intrinsic attribute of the alien species, but an emergent property of the species with respect to its perception by the local human community connected to the recipient ecosystem. All else being equal, ecosystems may differ in their susceptibility to invasion simply because their associated human communities differ in their perception of new players: invisibility does increase invasibility! Consistent with Perrings et al (2002), we conclude that social or cultural norms of behaviour are relatively insensitive to new risks.

Biological invasions are looked at anthropocentrically (see Lambert this volume, Chapter 11). When it cannot become valuable, a species must neither pre-empt human space nor impair human activities. Otherwise, its invasiveness forces humans to share their territory, and the invader is perceived as 'taking up too much space' (Menozzi, 2007). Cohabitation is even more difficult when the species threatens human health or economy. Within the framework provided by the global biodiversity crisis, our relationship to nature, including cohabitation with invasive species, should be questioned (Maris, 2006). It is argued that we must move beyond a vision of frozen nature, caused by our underlying fear of change (Claeys-Mekdade, 2006). Environmental ethics might open the way, but science still has to accept uncertainty (Ravetz, 1998), as well as a debate encompassing moral values and anecdotal data. Going back to biological invasions, science has shown its limits in alone providing solutions. In most cases, eradication seems neither feasible nor appropriate for the various other users sharing the same natural landscapes as the scientists. At the very least, there is a debate to be had.

References

Allain, Y.-M. (2001) 'Végétal ornemental', *Dossiers de l'environnement*, 21, 39–42

Alpert, P., Bone, E. and Holzapfel, C. (2000) 'Invasiveness, invasibility and the role of environmental stress in the spread of non-native plants', *Perspectives in Plant Ecology, Evolution and Systematics*, 3(1), 52–66

Alphandéry, P., Bitoun, P. and Dupont, Y. (1991) *L'équivoque écologique*, La Découverte, Paris

Brun, E., Betsch, J.-M., Blandin, P., Humbert, G., Lefeuvre, J.-C. and Marinval, M.-C. (2007) 'Postures des scientifiques et interdisciplinarité', *Nature, Sciences, Sociétés*, 15, 177–185

Bruno, J.-B. (1990) 'La sacralisation de la science', in C. Rivière and A. Piate (eds) *Nouvelles idoles, nouveaux cultes*, L'Harmattan, Paris

Claeys-Mekdade, C. (2006) 'A sociological analysis of biological invasions in Mediterranean France', in S. Brunel (ed) *Invasive Plants in Mediterranean Type Regions of the World*, Environmental Encounters, 59, Council of Europe Publishing, Strasbourg, pp209–220

Clément, G. (2002) *L'éloge des vagabondes*. Editions du Nil, Paris

Coser Lewis, A. (1982) *Les fonctions du conflit social*, fifth edition (first published 1953), Presses Universitaires de France, Paris

Decocq, G. (ed) (2006) 'Dynamique invasive du cerisier tardif, *Prunus serotina* Ehrh., en système forestier tempéré. Déterminants, mécanismes, impacts écologiques, économiques et socio-anthropologiques', Programme INVABIO II du Ministère de l'Ecologie et du Développement Durable, rapport final, Amiens, F, www.ecologie.gouv.fr/Dynamique-invasive-du-cerisier.html

Decocq, G. and Kalaora, B. (2009) 'Perceptions des changements en forêt: de l'invisibilité à la normalité rampante. L'exemple de l'invasion par le cerisier tardif', in Galochet, M. and Glon, E. (eds.) *Des milieux aux territoires forestiers*. Artois Presses Université, 293–312

Diamond, J. (2005) *Collapse: How Societies Choose to Fail or Succeed*, Viking Books, New York

Javelle, A., Kalaora, B. and Decocq, G. (2006) 'Les aspects sociaux d'une invasion biologique en forêt domaniale de Compiègne: la construction sociale de *Prunus serotina*', *Natures, Sciences, Sociétés*, 14, 278–285

Kalaora, B. (1993) *Le musée vert*, L'Harmattan, Paris

Manceron, V. (2009) 'Les animaux de la discorde', *Ethnologie Française*, 39(1), 5–10

Maris, V. (2006) 'Le développement durable: enfant prodigue ou avorton matriphage de la protection de la nature?' *Les ateliers de l'éthique*, 1(2), 86–101

Médail, F. (2004) *Facteurs écologiques, évolutifs et sociologiques impliqués dans l'invasion du littoral méditerranéen par* Carpobrotus spp *(*Aizoaceae*). Eléments pratiques pour leur limitation.* Programme INVABIO du Ministère de l'Ecologie et du Développement Durable, rapport final, Marseille, F (www.ecologie.gouv.fr/article.php3?id_article=9457)

Menozzi, M.-J. (2007) 'La jussie, belle plante ou mauvaise envahisseuse?' *La Garance Voyageuse*, 78, 19–22

Mermet, L. (2002) 'Homme ou vie sauvage? Société locale ou bureaucratie centrale? Faux dilemnes et vrais rapports de force', *Annales des mines*, 28, 13–20

Morlon, P. (1999) 'Notes impertinentes sur l'interdisciplinarité', *Natures, Sciences, Sociétés*, 7, 38–41

Mormont, M. (2006) 'Conflit et territorialisation', *Géographie Economie Société*, 3(8), 299–318

Mougenot, C. and Mormont, M. (2009) 'Etats de guerre ou de paix: autour de la prolifération des rats', *Ethnologie Française*, 39(1), 35–43

Perrings, C., Williamson, M., Barbier, E. B., Delfino, D., Dalmazzone, S., Shogren, J., Simmons, P. and Watkinson, A. (2002) 'Biological invasion risks and the public good: An economic perspective', *Conservation Ecology*, 6(1), 1–7

Ravetz, J. (1998). 'Connaissance utile, ignorance utile?', in Theys, J. and Kalaora, B. (eds), *La terre outragée,* Latitudes, Paris, pp89–108

Roqueplo, P. (1998) 'L'expertise scientifique; concensus ou conflit?', in Theys, J. and Kalaora, B. (eds), *La terre outragée,* Latitudes, Paris, pp183–198

Simmel, G. (1995) *Le conflit*, Circé, Paris

Sittert, L. van (2002) '"Our irrepressible fellow-colonist": The biological invasion of prickly pear (*Opuntia ficus-indica*) in the Eastern Cape *c*.1890–1910', *Journal of Historical Geography*, 28(3), 397–419

Starfinger, U. (1997) 'Introduction and naturalization of *Prunus serotina* in Central Europe', in Brock, J. H., Wade, M., Pyšek, P. and Green, D. (eds) *Plant*

Invasions: Studies from North America and Europe, Backhuys Publishers, Leiden, pp161–171

Starfinger, U., Kowarik, I., Rode, M. and Schepker, H. (2003) 'From desirable ornamental plant to pest to accepted addition to the flora? The perception of an alien plant species through the centuries', *Biological Invasions*, 5(4), 323–335

19
Native or Alien? The Case of the Wild Boar in Britain

Martin Goulding

Introduction

Wild boar *Sus scrofa* were driven to extinction in Britain several centuries ago by habitat loss and overhunting. However, this species is unique among Britain's lost mammals as it is the only one to recolonize as free-living self-sustaining populations, albeit accidentally. Since the early 1990s, escaped farmed wild boar established populations in several counties of Britain (see Figure 19.1), and having no natural predators (since the wolf *Canis lupus* and lynx *Lynx lynx* have long gone from Britain) they are predicted to increase significantly (Goulding et al, 1998). The political and social implications of the wild boar's reintroduction are many because the species affects a wide spectrum of issues with particular concerns about agriculture, ecology and public safety. In Britain the association between beast and man runs deep as wild boar feature prominently in British folklore and heraldry and provoke nostalgic thoughts of the greenwoods from times past. As a 'Royal Beast of the Chase' wild boar holds considerable appeal to the hunting fraternity, something shared by those with an appreciation of fine cuisine, as wild boar meat is considered a gastronomic delicacy. Should the wild boar's return to Britain be celebrated as a former keystone woodland species reclaiming ancestral habitat? Or is a more cautious approach necessary due to uncertainty over the returning animals' genetic purity, the species' reputation as an agricultural pest and the threat they may pose to public safety?

A former native species

Six thousand years ago Britain was covered by vast tracts of woodland, and more than one million wild boar were estimated to inhabit woods of oak, ash,

Figure 19.1 *Boar group in the Forest of Dean, England*

lime and hazel (Yalden, 1999). It was their finest hour. Through subsequent ages, woodland clearance to create farmland, increased timber usage and more efficient hunting eventually brought an end to the native wild boar population. It is believed that free-living wild boar became extinct in England at the turn of the 14th century, and during the 16th century in Scotland. References to wild boar in England at dates later than the 14th century are thought to refer to animals introduced from the continent into managed hunting estates as status symbols for landed gentry (Rackham, 1986).

Coincidentally, in the late 1990s, at the same time as the first few escapees were gaining a foothold, the feasibility of reintroducing wild boar to Britain to replace a native animal lost in historical times through man's activities was being mooted by conservationists (Howells and Edward-Jones, 1997; Leaper et al, 1999). Did the wild boar's accidental reintroduction pre-empt a possible official reintroduction? No, it did not. Although British governments have pledged commitment to several biodiversity- and conservation-orientated directives (CEWNH, 1979; ECHD, 1992; UK BAP, 1994), these directives stipulated the desirability of reintroducing animal and plant species in decline and needing strict protection at a European level. Wild boar populations in Europe are far from being in need of protection; quite the opposite in fact, because their numbers are increasing considerably in many European countries. Reasons for the increase are unclear, but are speculated to include: agricultural changes (increased cereal crop planting and larger field sizes); socio-economic changes (depopulation of rural areas); climate change (milder winters, increased oak and beech mast output); hunting factors (supplementary feeding, reintroduction/

translocation of animals); lack of predation (Sáez-Royuela and Tellería, 1986; Pohlmeyer and Sodeikat, 2003). Furthermore, it would be a bold decision for a government to reintroduce a species as controversial as wild boar. However, conservationists have noted that government failure to support the presence of the reintroduced wild boar would expose Britain's international commitment to the preservation of biodiversity as one of 'clear hypocrisy whereby we advocate tolerance by others while failing to do so ourselves' (Gow et al, 2008).

Genetic uncertainty

The original British wild boar farm stock was from surplus zoo animals predominantly of French origin, supplemented with stock imported from European wild boar farms (Booth, 1995). However, wild boar readily cross-breed with domestic pigs, and some British farmers hybridize their wild boar stock with domestic pigs to encourage more frequent farrowing and increase litter size (Goulding, 2001). It is not known whether any of the free-living wild boar escaped from such farms; hence the genetic purity of free-living wild boar is uncertain. Wild boar is a former native species, hybrid animals are not, and an accidentally reintroduced population of wild boar × domestic pig hybrids would have little public (The Mammal Society, 2009), and even less government support (Moore, 2004). Determining the genetic composition of wild boar is not straightforward as no simple test is available to differentiate between pure-bred wild boar and hybrids (Food Standards Agency, 2009). Currently, morphological characteristics are still used to differentiate between the two, although results can be inconclusive (Mayer and Lehr Brisbin, 1991).

The government department ultimately responsible for the wild boar in Britain, the Department for Environment, Food and Rural Affairs (Defra), has researched the purity of the reintroduced wild boar by comparing DNA sequences of reintroduced wild boar against those of several domestic pig breeds (Defra, 2006). Results were 'inconclusive', but the research importantly concluded the muddled genetic purity of Britain's free-living wild boar was comparable with the muddled genetic purity of continental populations (Moore, 2004), thus removing the immediate threat of their swift eradication on the grounds of genetic impurity. Worldwide, the origin and genetic purity of many wild boar populations are unclear as they are a favoured animal for hunting and have been translocated into various localities and countries for this reason. Some localities may have already supported a wild boar population of their own, so populations are often genetically mixed (Tikhonov and Bobovich, 1997). Ancestral cross-breeding with escaped or pannaged domestic pigs has further altered the genetic make-up of some continental wild boar populations (Clutton-Brock, 1999).

The disadvantages of wild boar

Agricultural problems

The returning wild boar entered a British landscape very different to the landscape the original native populations would have known. Nothing of the supposed 'wildwood' remains today, and surviving patches of woodland have been subject to hundreds of years of use, modification and fragmentation. Plant communities have changed in composition, distribution, diversity and species richness. Farming practices have changed considerably too, and so has our use of woodlands. In the Forest of Dean, for example, where there is a large reintroduced population of wild boar, recreational activities abound. People camp, fish, birdwatch, ramble, jog, cycle, exercise dogs, and even play golf within the confines of the forest. For some people wild boar living in forests such as these is not appropriate and they are feared and unwelcome. However, if the wild boar move from the forest into surrounding agricultural land, they are even less welcome.

Wild boar can cause havoc in an agricultural environment by rooting up pasture (see Figure 19.2), consuming cereal crops and breaching stock fences. Indeed, they are considered an important agricultural pest in much of Europe. Crop losses attributed to wild boar are substantial enough for some countries, for example Poland, Italy and France to adopt compensation schemes to reimburse farmers for their economic losses. The most evident form of damage so far with the returned British wild boar is rooting of grassland, although damage to cereal crops has also been recorded (Goulding et al, 1998; Wilson, 2004). To

Figure 19.2 *Damage to grassland by boar rooting, Forest of Dean, England*

date agricultural damage is only minor and localized. However, an expanding wild boar population would equate to an increase in agricultural damage.

A potentially more economically devastating threat to farming interests is that wild boar can play host to transmissible diseases of livestock such as foot-and-mouth, swine fever and bovine TB. There is a very real danger that domestic livestock, particular domestic pigs reared in large outdoor units, will infect the free-living wild boar with an economically significant disease that may become endemic within wild boar. They may act as a reservoir for the disease, reinfecting domestic livestock at subsequent meetings. Furthermore, wild boar can wander several kilometres in a single night, and occasionally undergo long-distance dispersal covering far greater distances. One individual radio-tracked in Britain travelled 18km from its home range (Goulding, 2003). An infected wild boar thus has the potential to carry disease over a considerable area, making disease-containment contingency plans of culling of all animals within a certain area completely impracticable. If the wild boar population continues to increase, domestic pig breeders may eventually be forced to keep pigs indoors to avoid the risk of disease (or interbreeding), or spend considerable sums on providing adequate fencing. Pig farmers in East Anglia, a major outdoor-reared domestic pig-producing area of Britain, are reportedly so concerned about wild boar they have hired a marksman to shoot on sight any wild boar seen in the vicinity of a pig farm (*Farming Today*, BBC, 2009). The morality of culling a former native species because it is incompatible with current agricultural practice appears not to have been considered.

Wildlife conservation issues

And what of the other native species that now find themselves sharing their habitat with the interloping wild boar? We do not know how the wild boar will affect native species already resident, although some clues can be gleaned from the literature. For example, hunting returns have implemented wild boar in causing a fall in woodcock *Scolopax rusticola* numbers in some German hunting estates (Nyenhuis, 1991), and wild boar have reportedly predated the nests of red-legged partridges *Alectoris rufa* in Spain (Leaper et al, 1999). Conversely, corvids will directly associate with wild boar that deliberately encourage the birds' attention. This is presumably to encourage the removal of ticks or parasites from the wild boars' coats (Massei and Genov, 1995). Wild boar may compete with small mammals by digging up caches of buried mast (Focardi et al, 2000) and may opportunistically predate the young. It would be particularly interesting to learn how populations of wild boar and badgers *Meles meles* might coexist in Britain as both species are opportunistic, omnivorous, predominantly nocturnal and root for food among the leaf litter. The badger's diet of earthworms, insects, plant roots, cereals, soft fruit, beech mast, acorns, small mammals and carrion is also similar to the diet of wild boar.

Wildlife-related road traffic accidents (RTAs)

In Europe, it was estimated that 0.5–5 per cent of the spring population of wild boar in each country was involved in RTAs each year (Groot Bruinderink and Hazebroek, 1996), and RTAs have injured or claimed the lives of several wild boar in Britain to date. Wild boar cross roads throughout the night moving from one feeding area to another; and roads that bisect areas of woodland are particularly associated with accidents. Increasing the risks of RTAs further, wild boar often root on roadside verges, or loiter on the edge of the tarmac. In the dark of night, on an unlit country road, spotting a dark-coated wild boar in the road may be difficult. Fortunately in Britain to date no human injuries have been reported, but it is probably only a matter of time. Defra estimated, using data from continental populations and current population estimates, that Britain can expect about six wild boar RTAs annually (Wilson, 2005). Again, an increasing wild boar population will bring a concurrent increase in wild boar-related RTAs.

Advantages

Enough of the problems! Should Britain not rejoice that such a charismatic species has honoured us with its presence once again, and why should a former native species even have to justify its presence? If justification were needed, the statements for the defence are likely to include the wild boar's ecological importance to woodland ecology. Visually analogous to ploughing, the wild boars' strong snouts root through the leaf litter and surface vegetation ripping up the earth as they move forward. Rooting by wild boar has aptly been defined by Sousa (1984) as 'a discrete, punctuated killing, displacement, or damaging of one or more individuals that directly or indirectly creates opportunity for new individuals to become established'. Rooting creates bare patches of soil that act as seed beds allowing dormant seeds to germinate, improves soil fertility by mixing soil horizons, increases decomposition rates, and improves soil aeration (Sims, 2006).

It has been said that a 'woodland without wild boar is a woodland in decline' and rooting does indeed ensure that woodland ecological processes are dynamic and constantly in a state of flux. For example, on one grassy woodland ride I frequently visited in a British woodland, bare soil exposed by rooting was first recolonized by annual plants such as scarlet pimpernel *Anagallis arvensis* and common centaury *Centaurium erythraea*. Later followed perennials such as dog violet *Viola riviniana* and creeping buttercup *Ranunculus repens*. A monoculture of grasses had become interspersed with flowering plants and the biodiversity of that particular woodland ride had increased. Ultimately, perennial grasses will again crowd out the recolonizing flowering plants but by then other areas will have been rooted and the ecological cycle is repeated. And not only flora is affected; all associated invertebrate life that interact with these particular flower species are also involved, as are all other species further up that particular food chain.

The impact of wild boar rooting on Britain's woodland ecology is complex and has still to be fully understood. Furthermore, rooting intensity will vary from year to year due to fluctuating wild boar numbers, unpredictable natural food supplies and climatic conditions (Sims, 2006). Rooting causes a disturbance regime that has been absent from Britain's woodlands for several hundred years, and which is likely to favour some species but not others. In Britain, particular concerns have been raised that wild boars' rooting may destroy bluebells *Hyacinthoides nonscripta* (Goulding et al, 1998). Monocultures of bluebells are a characteristic feature of many English woods and are acknowledged as one of Britain's 'great wild flower spectacles' (Mabey, 1996). Preliminary research results suggest, however, that although wild boar root up and eat bluebell bulbs the impact on plant density is localized and short-lived (Sims, 2006). Also, the wild boars' ecological role in plant dispersal by inadvertently snagging seeds in their shaggy coats as they move through the vegetation should not be overlooked. The seeds are deposited some distance away from the parent plant during the wild boars' characteristic grooming behaviour of rubbing against tree bark to remove parasites, when moulting, or when wallowing in mud. Wallowing also plays an important ecological role in creating ephemeral pools for aquatic insects, dragonflies and amphibians. Several studies have also reported the positive effects of wild boar consuming the larva of invertebrates considered pests of timber production (Genov, 1981; Schmid-Vielgut et al, 1991). Boar are also considered important in the dispersal of important mycorrhizal fungal associates in woodland ecosystems.

Defra has also recognized that the presence of a large and novel wild animal may provide advantageous economic opportunities for localities with wild boar to benefit from wildlife tourism and, at the opposite end of the wildlife–welfare spectrum, through hunting from the sale of sporting rights and carcasses (Wilson, 2005).

Urbanization

If wild boar stepping from forest into agricultural land are not welcome, then those moving from forest into urban areas are often even less so. Wild boar are attracted to human waste and adapt well to a human urban environment (Sáez-Royuela and Tellería, 1986; Goulding, 2008). Urbanization of wild boar occurs worldwide, and an insight into what the future holds for residents of urban areas adjacent to the Forest of Dean, for example, can be seen in Collserola Park, Barcelona, Spain. In comparison to the Forest of Dean, wild boar in Collserola Park were thin on the ground until quite recently, but a recent surge in wild boar numbers has led to a sharp rise in contact between wild boar and people (Cahill and Llimona, 2004). Like the Forest of Dean, Collserola Park attracts visitors who come to walk, jog or cycle in scenic surroundings. Complaints from visitors highlight the fear people have of wild boar regarding their personal safety and that of accompanying children. Some of Collserola's wild boar have become habitualized to people and urban life. This is either by people feeding them, or their becoming accustomed to rooting through lawned gardens, vegetable patches and rubbish bins, often during broad daylight. Their nuisance factor from digging up lawns, jaywalking across roads and scattering rubbish is high, and wild boar are poorly tolerated by most residents. In response, park rangers capture habitualized wild boar using a dart gun and remove the sleeping animal from public view, where it is shot dead. Prevention is determined the key management tool and Collserola Park maintains a public awareness campaign that informs visitors and locals alike to the negative implications of feeding wild boar. Advice is also given on garden fencing and the disposal of rubbish inside specific containers to reduce wild boar attraction. In Britain, the Forestry Commission are emulating these practices and notices advising that wild boar are not to be fed and dogs should be kept on a lead appear in strategic locations in the Forest of Dean, such as picnic areas and campsites, where visitors and locals are most likely to come into contact with wild boar.

Public opinion and decision time

The role of the media in informing the general public on wildlife issues is an important consideration for wildlife management programmes, particularly those involving reintroduction of large and potentially dangerous mammals. A review of 107 British press articles which referred to the returning wild boar showed a predominantly negative media coverage (Goulding and Roper, 2002). The issues raised most often were fears of the risk of attack on humans (60 per cent of articles) and agricultural crop damage (54 per cent). An awareness that wild boar were a former native species was the third most reported specific issue (42 per cent). Although wild boar are indeed a potentially dangerous species, the threat to public safety is minimal, as a British government wildlife agency reports, 'wild boar are normally secretive and nocturnal if they are not interfered with and there are very few documented cases of boar attacking people in Europe or elsewhere' (Wilson, 2005). Therefore, with an informative public relations and education exercise, woodland recreation activities such as those enjoyed by visitors to the Forest of Dean, for example, should not see a drop in visitor numbers because of wild boar (see Figure 19.3).

A government-funded public consultation exercise was initiated in 2006 with the British public invited to offer opinions on the future management of reintroduced wild boar. The government's stated management aim was ambitiously to 'create an acceptable balance between wild boar and the interests of farming, conservation, woodland management and human safety'. Five options were provided for consideration (Defra, 2005), summarized as:

- no direct Government management;
- eradicate all existing feral populations and cull all new escapees;
- eradicate all hybrids but allow managed wild boar populations;
- manage existing wild boar populations on a regional basis by limiting the spread of existing populations and preventing establishment of wild boar in some areas, particularly those with extensive populations of domestic pigs in outdoor units;
- no direct government management of existing established populations but prevention of new populations becoming established.

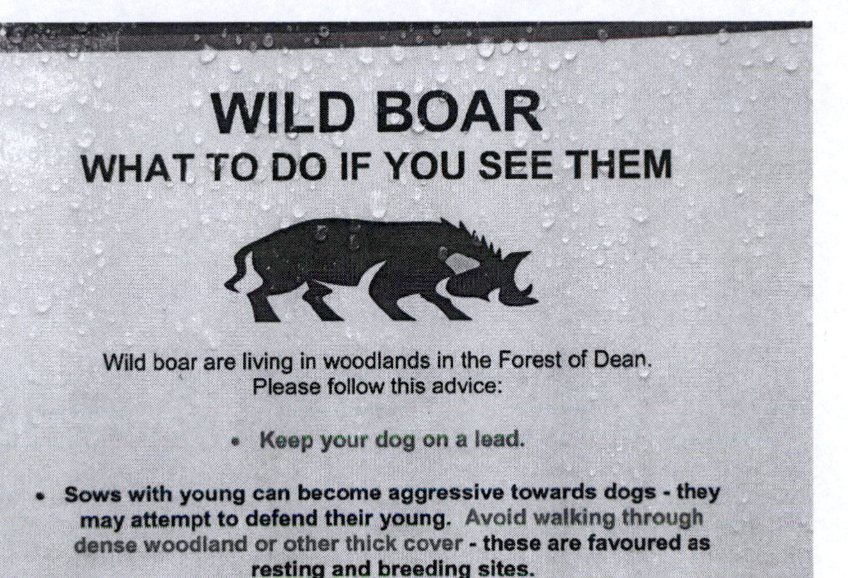

Figure 19.3 *Boar warning sign, Forest of Dean, England*

Responses showed the majority of respondents (56.1 per cent) did not want the feral populations eradicated. However, almost 80 per cent of respondents recognized the need for some management of feral boar populations (Defra, 2005).

In 2008, ten years after the presence of the reintroduced wild boar in Britain was first acknowledged, and having considered the results of scientific research, risk assessments, public consultation and veterinarian advice opinion, Defra published their long-awaited management decision (Defra, 2008). The 'Wild Boar Action Plan' stated that the British government 'considered regional management to be the most appropriate management approach given the current numbers of feral wild boar'. In other words, primary responsibility for the management of reintroduced wild boar lay with local communities and individual landowners. Thus, if you do not want wild boar on your land, you can legally shoot or shoo them away. But if you do want wild boar on your land, no one can tell you otherwise. Crucially for the wild boar's future survival, there was to be no national eradication programme; the wild boar is back in Britain to stay.

Just the beginnings ...

Media coverage of the wild boar's reintroduction showed there was strong awareness among the British public that wild boar was a former native species. There was little recognition of the species as an alien invader, and even less interest in an eradication campaign – once the wild boar's genetic purity was deemed to be on a par with continental populations. However, there still is considerable debate over whether this particular native species is compatible with today's Britain, a debate that will increase in frequency and ferocity as the wild boar population increases.

References

BBC (2009) *Farming Today*, 27 October 2009, www.bbc.co.uk/programmes/b00nd1c8, accessed 22 November 2009

Booth, W. D. (1995) 'Wild boar farming in the United Kingdom', *IBEX Journal of Mountain Ecology*, 3, 245–248

Cahill, S. and Llimona, F. (2004) 'Demographics of a wild boar *sus scrofa* Linnaeus, 1758 population in a metropolitan park in Barcelona', *Galemys*, 16, 37–52

CEWNH (1979) 'Conservation of European Wildlife and Natural Habitats Bern, 19.IX.1979.' European Treaty Series, No. 104. http://conventions.coe.int/Treaty/en/Treaties/Word/104.doc, accessed 21 November 2009

Clutton-Brock, J. (1999) *A Natural History of Domesticated Animals*, 2nd edn, Cambridge University Press, Cambridge

Defra (2005) 'Feral wild boar in England: A consultation by the Department for Environment, Food and Rural Affairs', www.naturalengland.org.uk/Images/wildboar-consultationdoc_tcm6–4510.pdf, accessed 21 November 2009

Defra (2006) 'Feral wild boar in England: A consultation by the Department for Environment, Food and Rural Affairs. Summary of responses', www.naturalengland.org.uk/Images/wildboarconsresponsessummary_tcm6–4509.pdf, accessed 21 November 2009

Defra (2008) 'Feral wild boar in England: An action plan', www.naturalengland.org.uk/Images/feralwildboar_tcm6–4508.pdf, accessed 21 November 2009

ECHD (1992) 'European Community Habitats Directive 1992 (92/43/EEC)', http://eur-lex.europa.eu/LexUriServ/LexUriServ.do?uri=CELEX:31992L0043:EN:NOT, accessed 21 November 2009

Focardi, S., Capizzi, D. and Monetti, D. (2000) 'Competition for acorns among wild boar and small mammals in a Mediterranean woodland', *Journal of Zoology*, 250, 329–334

Food Standards Agency (2009) 'The development and validation of DNA marker methods for the verification of meat from wild boar', www.food.gov.uk/science/research/researchinfo/choiceandstandardsresearch/authenticityresearch/q011ist_meat/q01129/, accessed 10 November 2009

Genov, P. (1981) 'Significance of natural biocenoses and agrocenoses as the source of food for wild boar', *Ekologia Polska*, 29, 117–136

Goulding, M. J. (2001) 'Possible genetic sources of free-living wild boar *Sus scrofa* in southern England', *Mammal Review*, 31(3), 245–248

Goulding, M. J. (2003) 'Investigation of free-living wild boar (*Sus scrofa)* in southern England', Unpublished DPhil thesis, University of Sussex, Sussex

Goulding, M. J. (2008) 'The habituation of Britain's wild boar: lessons from abroad', *ECOS*, 29(3/4), 38–44

Goulding, M. J. and Roper, T. J. (2002) 'Press responses to the presence of free-living wild boar in southern England', *Mammal Review*, 32, 272–282

Goulding M. J., Smith, G. and Baker, S. J. (1998) 'Current status and potential impact of wild boar (*Sus scrofa*) in the English countryside: A risk assessment', Central Science Laboratory Report to the Ministry of Agriculture, Fisheries and Food, York, www.naturalengland.org.uk/Images/wildboarriskassessment1998_tcm6–4642.pdf, accessed 10 November 2009

Gow, D., Minter, R. and Carver, S. (2008) 'A response to Feral Wild Boar in England', www.wildland-network.org.uk/reports_info/DGwildboar_response.pdf, accessed 10 November 2009

Groot Bruinderink, G. W. T. A. and Hazebroek, E. (1996) 'Ungulate traffic collisions in Europe', *Conservation Biology*, 10, 1059–1067

Howells, O. and Edward-Jones, E. (1997) 'A feasibility study of reintroducing wild boar (*Sus scrofa*) to Scotland: Are existing woodlands large enough to support minimum viable populations?', *Biological Conservation*, 51, 77–89

Leaper, R., Massei, G., Gorman, M. L.and Aspinall, R. (1999) 'The feasibility of reintroducing wild boar (*Sus scrofa*) to Scotland', *Mammal Review*, 29, 239–259

Mabey, R. (1996) *Flora Britannica*, Sinclair-Stevenson, London

Massei, G. and Genov, P. V. (1995) 'Observations of black-billed magpie (*Pica pica*) and carrion crow (*Corvus corone cornix*) grooming wild boar (*Sus scrofa*)', *Journal of Zoology*, 236, 338–341

Mayer, J. J. and Lehr Brisbin, I. (1991) *Wild Pigs of the United States: Their History, Morphology and Current Status*, University of Georgia Press, Georgia

Moore, N. (2004) 'The Ecology and Management of Wild Boar in Southern England, Defra Final Project Report, VC0325. http://sciencesearch.defra.gov.uk/Document.aspx?Document=VC0325_2113_FRP.doc, accessed 18 November 2009

Nyenhuis, H. (1991) 'Predation between woodcock (*Scolopax rusticola* L.) game of prey and wild boar (*Sus scrofa* L.)', *Allgemeine Forst und jagdzeitung*, 162, 174–180 (English summary)

Pohlmeyer, K. and Sodeikat, G. (2003) 'Population dynamics and habitat use of wild boar in Lower Saxony, Workshop on CSF', Hannover, Germany, 2003, www.tiho-hannover.de/einricht/wildtier/pdf/sodeikat_workshop, accessed 21 November 2009

Rackham, O. (1986) *The History of the Countryside*, Dent, London

Sáez-Royuela, C. and Tellería, J. L. (1986) 'The increased population of the Wild Boar (*Sus scrofa* L.) in Europe', *Mammal Review*, 16(2), 97–101

Schmid-Vielgut, B., Dopf, M. and Bogenschutz, H. (1991) 'Effect of fenced-in wild boar on May-Bug population Density', *Allgemeine Forst Zeitschrift*, 46, 719–721 (in German)

Sims, N. K. E. (2006) 'The ecological impacts of wild boar rooting in East Sussex', Unpublished D.Phil Thesis, University of Sussex, Sussex

Sousa, W. P. (1984) 'The role of disturbance in natural communities', *Annual Review of Ecology and Systematics*, 15, 353–391

The Mammal Society (2009) 'The re-establishment of wild boar, *Sus scrofa*, in Britain', www.mammal.org.uk/index.php?option=com_contentandview=articleandid=201andItemid=232, accessed 18 November 2009

Tikhonov, V. N. and Bobovich, V. E. (1997) 'Immuno- and cytogenic peculiarities of wild boar sub-species (*Sus scrofa*) from different regions of Eurasia and examination of their genetic purity for conservation of subspecies', *Proceedings of the 1st International Symposium on Physiologie and Ethology of Wild and Zoo Animals*. Berlin, Germany. 18–21 September, *Supplement II*, 229–232

UK BAP (1994) UK Biodiversity Action Plan, www.ukbap.org.uk/Library/PLAN_LO.PDF, accessed 18 November 2009

Wilson, C. J. (2004) 'Rooting damage to farmland in Dorset, southern England, caused by feral wild boar (*Sus scrofa*)', *Mammal Review*, 34(4), 331–335

Wilson, C. J. (2005) 'Feral Wild Boar in England: Status, impact and management. A Report on behalf of Defra European Wildlife Division', www.naturalengland.org.uk/Images/wildboarstatusImpactmanagement_tcm6–4512.pdf, accessed 18 November 2009

Yalden, D. W. (1999) *The History of British Mammals*, T & AD Poyser, London

20
Exotic and Invasive Species: An Economic Perspective

Craig Osteen and Michael Livingston[1]

Introduction

From an economic perspective, an exotic species is invasive when its costs exceed its benefits. A species' costs include damage or harm to production, use, trade and other agricultural and ecological benefits, in addition to costs of excluding and managing the species, where benefits and costs can be monetary and non-monetary. Its benefits include net returns, satisfaction from use and agricultural or environmental benefits, such as pest control (e.g. Hlasny and Livingston, 2008). Invasive species should be excluded or managed when the activity's benefits, derived from preventing or reducing species' net damage or costs, exceed the activity's economic and environmental costs. Evaluations of exotic and invasive species and decisions to exclude or manage species express people's objectives, values and preferences in response to economic incentives, within market and institutional structures. However, biological factors, especially the potential entry, spread, increase and damage of invasive species in agricultural or ecological systems, can affect the market valuation of benefits and costs and create unique economic decision and public policy issues. This chapter examines economic decisions, important economic and institutional factors and public policy issues related to exotic and invasive species, focusing on the USA.

US government definition of invasive species

Executive Order 13112 (Order), signed in 1999, defines an invasive species as alien to an ecosystem and potentially harmful to economic activity, the environment or human health if introduced (National Invasive Species Council, 2006). (We use alien, non-native, exotic, non-indigenous and foreign as synonyms.) A species native to one area can be alien and invasive to other areas of the USA. Exotic species under human control and domestication are not invasive, but the same species uncontrolled can be invasive. Invasive species include exotic pests and foreign animal diseases that harm agriculture and the environment, such as insects, nematodes, weeds and pathogens; zoonotic pathogens that can be transmitted between humans and animals; and seeds, eggs, spores and other biological material capable of propagating pests and diseases. The Order contains an economic definition of invasive species, comparing a species' benefits to its harm or costs. So, exotic species that provide net benefits, such as some crops, ornamentals, livestock, biological control agents or fish and game species, are not invasive.

People's objectives, values or preferences and economic factors, such as prices, income and interest rates, influence the cost and benefit evaluations that determine whether a species is invasive or not. (We relate values to moral, normative or ethical beliefs about right and wrong, acceptable and unacceptable actions, the role of humans and other beings in ecological systems, and the relative desirability of objects, characteristics and conditions.) Some biological scientists and environmental activists consider exotic species to be intrinsically harmful, but Sagoff (2007, 2009) argues that such evaluations reflect value judgements and are not based entirely on science. The reasons for different attitudes and preferences toward exotic and native species are beyond the scope of this chapter, but those differences contribute to public policy controversies. For example, the State of Maryland controls exotic mute swans because they feed aggressively on submerged grasses and might endanger native species in the Chesapeake Bay ecosystem, but some people like the swans, and animal rights groups object to the control programme. In another example, some Potomac River sport fishermen view largemouth and smallmouth bass as highly desirable, but those species, while native in much of the USA, are not native to that river. Also, some people consider northern snakeheads from Asia, aggressive predators that can compete with bass and other game fish, discovered in the Potomac in 2004, as invasive. In 2002, the US Fish and Wildlife Service classified all snakehead species as injurious, prohibiting imports and interstate commerce. However, some Asian Americans value snakeheads as food and guides have promoted snakehead sport fishing.

Some definitions of invasive species emphasize biological factors: the ability to spread in the absence of natural enemies, adapt to a new environment, establish a self-sustaining population or outcompete native species (Office of Technology Assessment, 1993; Florida Department of Environmental Protection, 2003; Plant Conservation Alliance, 2005; Convention on Biological Diversity, 2007). From an economic perspective, the entry, spread, increase and damage of an exotic species and the biological factors that affect them do not determine, by themselves, whether the species is invasive or not. People's evaluations of the species' monetary and non-monetary costs and benefits are necessary. For example, the ability to invade or colonize new habitats is necessary for the survival of species such as salmon, viewed by many as desirable, as well as species viewed as undesirable (Botkin, 2001, 2008). However, the potential of exotic species to enter, spread, increase and cause damage has important implications for evaluation of costs and benefits, classification of species as invasive or beneficial, and the exclusion or management of invasive species.

Economic activities and invasive species

Economic activities, in response to prices and other economic factors, influence the entry, spread and impact of exotic and invasive species in agricultural and ecological systems, as well as the vulnerability of those systems. Imports and foreign travellers can unintentionally introduce invasive species into a country, while domestic commerce and travel can spread them. Some authors argue that globalization of commerce and growth in trade and travel from ecologically diverse countries without commensurate preventative measures enhance invasive species spread (Jenkins, 1996; Kolar and Lodge, 2000; Sutherst, 2000; Mumford, 2002).

Agricultural commodity and livestock imports can introduce non-native pests and diseases that damage agriculture. Hlasny and Livingston (2008) examined relationships between economic variables and identifications of exotic insects in the USA from 1866 to 1990 and found agricultural imports to be the primary variable explaining introductions. Short transit times for fruits, vegetables or nursery products can enhance pest survival (Reichard and White, 2001). However, consumers benefit from imports, which expand food variety and stabilize year-round fresh fruit and vegetable supplies and prices.

Vehicles, vessels, containers, freight and packing material can introduce and spread invasive species. Many ports encourage rapid movement of goods, which could increase entry risks. Changes in transportation technologies, such as containerization or air transport of fresh agricultural products, also may increase those risks (Nugent et al, 2001; Mumford, 2002).

Some individuals or firms intentionally introduce exotic plant or animal species for production, sale, use or other purposes. Such species could become invasive or carry invasive species. For example, imports of live plants can transport pests and pathogens unless treated. Some invasive species or exotic species carrying invasive species may be smuggled into countries where banned (Ferrier, 2009), while some organisms, such as pathogens, could be released as bioterrorism agents.

Alternatively, some invasive species enter a country or spread through natural pathways. Wind currents spread some insects, pathogens and weed seeds. For example, Asian soybean rust apparently entered the USA from South America in 2004 with hurricanes (Johansson et al, 2006). Some insects spread pathogens. For example, the glassy-winged sharpshooter transmits *Xyllela fastidiosa*, a bacterium that causes phoney peach disease in peaches and Pierce's disease in grapes. Domestic and wild animals can spread animal or zoonotic disease pathogens, such as influenza viruses.

The conversion of forest or wild land to agricultural and urban uses can encourage increase and spread of invasive species and susceptibility to damage (Kolar and Lodge, 2000; McNeely, 2000; Westbrooks, 2001). Cultivated lands may be less ecologically and genetically diverse and more vulnerable to pests than are natural ecosystems (Perrings, 2000). About 99 per cent of North American cultivated acreage is planted with non-native crops potentially vulnerable to pests from their native areas (Capinera, 2002).

Agricultural production and resource management practices, such as continuous cropping, decreased mouldboard ploughing, monocultures of genetically uniform crops or overgrazing rangelands, may have encouraged successful pest invasions (OTA, 1993; Weitzman, 2000). Moving agricultural equipment can spread soil pests. Transporting untreated logs and wood products can spread forest pests. Shipping livestock or captive game animals can move diseased animals or encourage disease transmission. Increased concentration and vertical integration of livestock and poultry production might promote rapid spread of and vulnerability to foreign animal diseases, but also facilitate tracing disease outbreaks and more effective responses (Shields and Mathews, 2003).

Invasive species and economic decisions

Entry, spread, increase and potential damage, influenced by biological and economic factors, have implications for the valuation and use of exotic species, the prevention and management of invasive species and the implementation of strategies.

Population dynamics and mobility: the spatial and temporal dimensions

If not excluded or managed, invasive species populations (pest infestations or disease outbreaks) increase and spread until potential habitat is completely infested or susceptible host population is infected. Biological factors (including natural enemies and competitors), climate, economic activity, and exclusion and management practices influence the extent of habitat of susceptible host, population dynamics, mobility and damages. Populations can spread from one location, sometimes assisted by natural events or human actions and cause damage at other locations. Some species move great distances, sometimes across international boundaries. So decisions to exclude or manage invasive species or not can affect damage or costs at other locations, currently and in the future. Rates of increase and spread, time of initial discovery, pathways of spread, and stage of the invasion influence response time and the selection and timing of exclusion, monitoring and management (Livingston and Osteen, 2008).

Risk and uncertainty

Pest mobility, population dynamics, damage, exclusion and management effectiveness and factors affecting them can be stochastic or random, causing variable outcomes and risky returns. Incomplete or sparse information about pest identity, pathways or vectors of spread, the presence and magnitudes of populations at different locations, growth, spread, damage and the availability and effectiveness of control practices create uncertain outcomes and returns. Inspection, monitoring, research and experience can reduce uncertainty. Some decision rules explicitly recognize the effects of risk and uncertainty in selecting exclusion and management strategies.

Irreversible effects

In extreme cases, invasive species can cause rapid and persistent changes that are very costly or impossible to reverse, such as native species extinction, severe ecosystem degradation or large agricultural yield reductions, disrupting associated industries and economies (Perrings, 2000). Excluding, containing or eradicating infestations might prevent irreversible economic or ecological damages.

Externalities and public goods

Exotic and invasive species can create externalities, where people do not account for all the costs and benefits of their decisions to use, prevent or manage such fauna and flora (Nugent et al, 2001). Some economists argue that market incentives for monitoring, exclusion and management of invasive species are inadequate, because those activities have public good characteristics of non-excludable costs or benefits and non-rival consumption, and are not easily traded in markets. (Import bans, quarantine treatments, border inspection and offshore preclearance or management programmes are exclusion options, while containment, control and eradication are management activities.)

Due to the mobility and population dynamics of invasive species and the difficulty of detecting and eliminating them, a producer, firm or individual would not necessarily capture all benefits of preventing or reducing damages or bear the costs of not doing so, thus the costs and benefits of monitoring, exclusion and management would be at least partially non-excludable (Sumner, 2003). As a result, people might misclassify, from a societal perspective, some invasive species as beneficial or allocate insufficient resources to exclusion, monitoring or management (Perrings, 2000). For example, when a pest infestation increases and spreads, producers or land managers might not account for the costs or benefits of their actions incurred by others at different locations or in the future, so they would not monitor or manage pests as intensively as if they had done so. Importers might not bear all economic or environmental costs of inadvertent pest introductions and import more or use fewer preventive practices than if they considered all costs, while consumers might purchase more imports because of lower prices (OTA, 1993; McNeely, 2000). Importers, retailers, growers or consumers might benefit from intentionally introduced exotic species, while others incur costs if species become invasive, which would encourage more introductions than if all costs were considered. Alternatively, some exotic species, such as some biological control agents, could create external benefits that exceed costs, so that market incentives for encouraging introduction and use might be inadequate.

Market incentives for exclusion, monitoring and management can be inadequate for other reasons. Successful exclusion, containment and eradication of pest infestations are non-rival in consumption, that is, available to everyone without reducing anyone's benefit (Sumner, 2003). So, there is no cost for providing the services to additional users in potentially affected areas and it is difficult to collect payment from them. Also, the overall effectiveness of those activities may depend on the least effective individual, producer or firm, called the weakest-link problem, so there would be no market incentive for others to be more effective (Perrings et al, 2002).

Strategies and implementation

When invasive species increase and spread, potentially affected individuals or firms could benefit if they cooperated to attain greater, more consistent or more cost-effective damage prevention or participated in public or group-coordinated exclusion or management strategies. For example, agricultural producers could agree to monitor or manage pests more intensively, fund exclusion programmes or pay growers at other locations to contain or eradicate pests, without government intervention. Many do not do so, because the benefits are too low compared to costs, including cooperation costs from identifying collaborators, negotiating and enforcing agreements, and coordinating multiple actions. Legal, procedural or other differences across local, state or international borders could increase these costs. Also, individuals could evaluate the costs and benefits of species or strategies differently or have different information about the presence and damages of pests, motivating some, but not all, to cooperate or participate. Some might try to negotiate lower cost or higher compensation, try to obtain benefits without participating or oppose a strategy due to its objective, approach or impact. Both cooperation benefits and costs could increase as potential infested areas or the number and populations of host species increase, which can increase the number of affected individuals, firms or government entities.

Some economists propose government exclusion or management strategies, policies that encourage private sector strategies, taxes or subsidies that encourage socially optimal use of practices or fees that raise funds for prevention or response programmes (Dahlsten and Garcia, 1989; Jenkins, 2002; Perrings et al, 2002). In addition to individuals or firms acting independently, approaches for implementing strategies include government programmes, regulation, cooperative government and private sector programmes, and market-based approaches, such as insurance, liability rules, pest damage bonds, private corporations or grower cooperatives. Which government or private sector approaches should implement exclusion or management strategies for given invasive species depends on strategy benefits (net damage and cost reductions), cooperation costs (discussed above), administration costs, economic welfare effects and the responses of individuals or firms to incentives and constraints that influence cost-effectiveness. Government action could be more cost-effective than private sector action for some situations, because regulatory and spending authority could encourage or compel programme participation that improves effectiveness or reduces implementation costs. For other situations, administrative costs or economic welfare losses caused by distortion of market incentives could reduce a government programme's cost-effectiveness relative to market-based or other government approaches or cause net welfare losses (Aquaye et al, 2007).

The spatial and temporal distribution of a strategy's benefits and costs among individuals or firms creates special implementation issues. Producers in infested areas (or owners of infected herds) could incur large costs to prevent pest or disease spread (such as destruction of trees or herds), while others receive net benefits from successful containment or eradication. Conceivably, net benefit recipients could compensate those bearing net costs, but identifying benefit recipients and their cost share might be difficult. Insurance might serve this role, but those subject to low risk might not participate, causing higher premiums. Public agencies in the USA and other countries sometimes use public funds to compensate those bearing large costs and encourage cooperation, which could reduce incentives for other prevention or control practices.

The potential increase, spread, infestation and damage of invasive species influence the selection of implementation approaches, as well as exclusion or management strategies. If a cost-effective strategy is available, public or group implementation would more likely generate greater net benefits than individual action would, as the species' potential rate of spread, extent of infestation and amount of damage beyond the current infestation increase. Benefit–cost criteria would more likely favour public or group over individual implementation for excluding pests or managing new introductions or outbreaks than for managing pests infesting large portions of potential habitat. The reason is that current and future benefits of excluding pests or managing new introductions or outbreaks could be relatively large, costs and environmental effects relatively small, and for new introductions or outbreaks, cost recipients concentrated in infested areas and benefit recipients in areas not yet affected. When managing pests infesting large portions of potential habitat, costs and environmental effects could be relatively large compared to benefits, with cost and benefit recipients interspersed throughout infested areas. In such cases, benefit–cost criteria would favour public or group over individual management when pests reinfest managed areas from unmanaged or poorly managed areas, and mobility between areas and damage are sufficiently great, but eradication might not be feasible or be less cost-effective than long-term control.

Some government strategies create controversies. Strategy opponents might evaluate benefits, costs or health and environmental effects of a pest or strategy differently than supporters would, or object to costs, benefits or other effects they receive. For example, many Florida residential citrus tree owners challenged in court the US Department of Agriculture (USDA) and State of Florida citrus canker eradication programme initiated in 1995 that destroyed commercial and non-commercial citrus trees to protect citrus production. Non-commercial trees were ornamental or provided fruit for personal use and some owners argued they were insufficiently compensated for trees destroyed.

USDA terminated the programme in 2006 after hurricanes spread the citrus canker pathogen to an extent that eradication became not feasible. In another case, some California groups opposed aerial pesticide spraying for the USDA and State of California light brown apple moth eradication programme begun in 2007, arguing that USDA overestimated the pest's potential economic and ecological impacts and that the programme's health and environmental impacts and economic effects on organic growers were not justified (Pesticide Action Network, 2009).

Institutional issues

Economic decisions, including the importation or introduction of exotic species or the prevention and management of invasive species, are made within a market and institutional structure that includes property rights, laws, regulations, treaties organizations (firms, corporations) and government agencies. The structure defines permissible actions and creates decision incentives, which may reflect customs, traditions, legal decisions or political negotiations.

Major treaties affecting exotic and invasive species focus on international trade in goods and commodities, including live animals, plants and propagative materials and on measures to protect human health, agricultural production or the environment from pests or diseases. The goal of the Agreement on Sanitary and Phytosanitary Measures (SPS Agreement) and the related International Plant Protection Convention, World Animal Health Organization and Codex Alimentarius is to establish a transparent and consistent framework through which countries take protective actions against pest or disease threats, while minimizing trade restrictions. Under these treaties, national government decisions to ban imports or require sanitary or phytosanitary measures should be based on risk assessments and scientific evidence. Some authors argue that the SPS Agreement encourages quarantine policies that tolerate higher than minimum pest introduction risks (Smith, 1999; Sutherst, 2000).

US laws, including the Plant Protection Act and Animal Health Protection Act, authorize USDA to use a wide range of measures to exclude or prevent the spread of non-native pests and diseases in foreign and interstate commerce, with some provisions for intrastate regulation, while states have primary regulatory authority within their boundaries. The US Department of Interior regulates the introduction of exotic wild animal species into native ecosystems under the Lacey Act.

Imports of exotic species: What is invasive?

Selecting an approach to determine which exotic species imports or introductions to permit, restrict or prohibit is a contentious policy issue. The USA uses two primary approaches: 'permitted lists' (also called white or clean lists), which allow imports (or introductions) of listed species only and 'prohibited lists' (also called black or dirty lists), which allow imports of species not listed (Schmitz and Simberloff, 1997). The USA uses 'permitted lists' for biological control agents and agricultural commodities (such as grains, fruits, vegetables, livestock, poultry, meat and dairy products), but uses 'prohibited lists' for nursery stock, seed and some live animals including wildlife.

The 'permitted list' approach, which requires evidence that a species is not a pest, emphasizes risk prevention and costs of exotic species more, and potential benefits less, than does the 'prohibited list' approach. The latter approach requires evidence that a species is a pest and could allow introduction before it is discovered to be invasive. Schmitz and Simberloff (1997) argued that the scientific difficulty of predicting whether or not species will be invasive in new locales limits the effectiveness of both approaches in reducing costs and risks.

Some groups or firms desiring to import, grow or sell potentially valuable alien species, such as nursery stock and seed, support the 'prohibited list' or a less restrictive approach, while others desiring to reduce alien pest introductions often advocate a 'permitted list' or more restrictive approach. For example, Fowler et al (2007) criticized the Lacey Act's effectiveness in preventing invasive wildlife introductions, arguing that the Act requires the 'prohibited list' approach and does not authorize containment measures after species enter.

Two particularly interesting controversies about the rules defining invasive species are exotic plants for planting and biological control agents:

Exotic plants for planting

There is a risk that imported exotic plants, including seeds and other propagative materials, could become invasive weeds. (They could also carry insects, pathogens or nematodes, an issue not discussed here.) Australia changed its exotic plant import rules from a 'prohibited list' to 'permitted list' approach in the late 1990s and prohibited imports unless weed risk assessments showed low risk or species appeared on a permitted seeds list (Australian Government, 2008). USDA's Animal and Plant Health Inspection Service (APHIS), as of 2009, used a 'prohibited list' approach to regulate imports or interstate movement of exotic plants under the noxious weed provisions of the Plant Protection and Federal Seed Acts. That year, the agency proposed a new category of plants for planting not authorized for importation pending pest risk analysis (NAPPRA). A plant taxon would be listed if scientific evidence demonstrated it to be a potential quarantine pest or host of a quarantine pest and could be removed from the list if warranted by pest risk analysis.

Biological control agents

These biocontrol agents are pest enemies such as predators, parasites, pathogens and weed feeders. Government agencies, universities and private firms often import and release them to manage exotic pests. The agents may have fewer adverse environmental effects than synthetic pesticides have, but, as exotic species, they could become invasive if they attack desirable, non-target species.

The USA uses a 'permitted list' approach for biocontrol agents, conducting pest, environmental and/or biological risk assessments before permitting import or release. APHIS regulates most agents for plant pests under the Plant

Protection Act, but the US Environmental Protection Agency regulates some pathogens, such as *Bacillus thuringiensis*, as pesticides. If APHIS finds no significant effect, the agent is not subject to further import regulation. However, if an agent has potential adverse effects on non-target species, the Endangered Species Act or National Environmental Policy Act may require further assessments. Some pest control researchers argue that approval procedures and handling requirements slow the implementation of effective biocontrol programmes (Messing, 2005). Individual states can regulate them with more restrictive rules than APHIS does. Indigenous biocontrol agents generally are not regulated.

Conclusions

The classification of exotic species as invasive or not and decisions to exclude or manage invasive species depend on cost–benefit comparisons, which reflect people's objectives, values and preferences and other economic, institutional and biological factors. The entry, spread and increase of exotic species do not directly determine whether species are invasive or not, from an economic perspective, but do create unique economic and policy problems. Economic activity and biological factors influence entry, spread, increase and damage, which, in turn, influence the costs and benefits of species, as well as the costs, effectiveness and benefits of exclusion and management strategies. Due to the increase and spread of some species, individuals or firms might not account for the full costs and benefits of their decisions, which might distort the market valuation of species' costs and benefits, the private sector classification of species as invasive or beneficial, and the allocation of resources to prevention and management. As a result, governments or private groups might implement prevention and management programmes, develop institutional arrangements to reduce the distortion of costs and benefits or create incentives to encourage prevention and management for some invasive species. Disagreements about the magnitudes or objections to the distribution of costs and benefits of exotic species, their exclusion and management can create public policy conflicts.

Note

1 The authors are agricultural economists with the Resources and Rural Economics Division, Economic Research Service, US Department of Agriculture, Washington, DC. The views presented are those of the authors and do not represent the official views of any agency or organization.

References

Aquaye, A. K. A., Alston, J. M., Lee, H. and Sumner, D. A. (2007) 'The economics and spatial dynamics of eradication policies', in Lansink, A. O. (ed) *New Approaches to the Economics of Plant Health*, Wageningen UR Frontis Series, 20, Springer, Dordrecht

Australian Government, Department of Agriculture, Fisheries, and Forestry, Biosecurity Australia (2008) 'Development of weed risk assessment', www.daff.gov.au/ba/reviews/weeds/development, accessed 17 August 2009

Botkin, D. (2001) 'Naturalness of biological invasions', *Western North American Naturalist*, 61(3), 261–266

Botkin, D. (2008) 'Naturalness of Biological Invasions', Presentation at 2008 PREISM Workshop, 23 October, Washington, DC

Capinera, J. (2002) 'North American vegetable pests: The pattern of invasion', *American Entomologist*, 48(1), 20–29

Convention on Biological Diversity, Programmes and Issues (2007) 'What are Invasive Alien Species', www.cbd.int/invasive/WhatareIAS.shtml, accessed 17 August 2009

Dahlsten, D. L. and Garcia, R. (1989) *Eradication of Exotic Pests: Analysis with Case Histories*, Yale Press, New Haven, CT

Ferrier, P. (2009) *The Economics of Agricultural and Wildlife Smuggling*, Economic Research Report No. 81, Economic Research Service, US Department of Agriculture, www.ers.usda.gov/Publications/err81/, accessed 30 September 2009

Florida Department of Environmental Protection (2003) *Statewide Invasive Species Management Plan for Florida*, www.dep.state.fl.us/lands/invaspcc/2ndlevpgs/pdfs/2003%20Strategic%20Florida%20Plan%20-%20Fedpdf, accessed 17 August 2009

Fowler, A. J., Lodge, D. M. and Hsia, J. (2007) 'Failure of the Lacey Act to protect US ecosystems against animal invasions', *Frontiers in Ecology and the Environment*, 5(7), 353–359

Hlasny, V. and Livingston, M. (2008) 'Economic determinants of invasion and discovery of nonindigenous insects', *Journal of Agricultural and Applied Economics*, 4(1), 37–72

Jenkins, P. T. (1996) 'Free trade and exotic species introductions', *Conservation Biology*, 10(1), 300–302

Jenkins, P. T. (2002) 'Paying for protection from invasive species', *Issues in Science and Technology*, (Fall), 67–72, National Academy of Sciences, Washington, DC

Johansson, R. C., Livingston, M. J., Westra, J. and Guidry, K. (2006) 'Simulating the US impacts of alternative Asian soybean rust treatment regimes', *Agricultural and Resource Economics Review*, 35(1), 116–127

Kolar, C. and Lodge, D. (2000) 'Freshwater nonindigenous species: Interactions with other global changes', in Mooney, H. and Hobbs, R. (eds) *Invasive Species in a Changing World*, Island Press, Washington, DC, pp3–30

Livingston, M. and Osteen, C. (2008) *Integrating Invasive Species Prevention and Control Policies*, US Department of Agriculture, Economic Research Service, Economic Brief number 11

McNeely, J. (2000) 'The future of alien invasive species: Changing social views', in Messing, R. H. (2005) 'Hawaii as a role model for comprehensive US biocontrol legislation: The best and the worst of it', *Second International Symposium on Biocontrol of Arthropods*, pp686–691, www.bugwood.org/arthropod2005/v012/14b.pdf, accessed 17 August 2009

Mooney, H. and Hobbs, R. (eds) *Invasive Species in a Changing World*, Island Press, Washington, DC

Mumford, J. D. (2002) 'Economic issues related to quarantine in international trade', *European Review of Agricultural Economics*, 39(3), 329–348

National Invasive Species Council (2006) *Invasive Species Clarification and Guidance White Paper*, www.invasivespeciesinfo.gov/docs/council/isacdef.pdf, accessed 17 August 2009

Nugent, R., Benwell, G., Geering, W., McLennan, B., Mumford, J., Otte, J., Quinlan, M. and Zelazny, B. (2001) 'Economic impacts of transboundary plant pests and animal diseases', in *The State of Food and Agriculture*, FAO, Rome

Office of Technology Assessment, US Congress (1993) *Harmful Non-Indigenous Species in the United States, OTA-F-565,* Washington DC, US Government Printing Office, September

Perrings, C. (2000) 'The economics of biological invasions', Paper presented at the workshop on *Best Management Practices for Preventing and Controlling Invasive Alien Species*, 22–24 February, South Africa/US Bi-national Commission

Perrings, C., Williamson, M., Barbier, E., Delfino, D., Dalmazzone, S., Shogren, J., Simmons, P. and Watinson, A. (2002) 'Biological risks and the public good', *Conservation Ecology,* 6(1), 1, www.consecol.org/v016/iss1/art1, accessed 28 August 2009

Pesticide Action Network North America (2009) *Light Brown Apple Moth in California*, www.panna.org/resources/lbam, accessed 17 August 2009

Plant Conservation Alliance (2005) 'Weeds Gone Wild', www.nps.gov/plants/alien/bkgd.htm, accessed 17 August 2009

Reichard, S. and White, P. (2001) 'Horticulture as a pathway of invasive plant introductions into the United States', *BioScience,* 51(2), 103–113

Sagoff, M. (2007) 'Are non-native species harmful?' *Conservation Magazine,* 8(2) (April–June), 21–22

Sagoff, M. (2009) 'Environmental harm: Political not biological', *Journal of Agricultural and Environmental Ethics,* 22(1), 81–88

Schmitz, D. C. and Simberloff, D. (1997) Biological invasions: A growing threat', *Issues in Science and Technology*, Summer 1997, www.issues.org/13.4/schmit.htm, accessed 17 August 2009

Shields, D. and Mathews, Jr., K. H. (2003) *Interstate Livestock Movements, Economic Research Service*, USDA, Economic Research Service, Washington, DC, LDP-M-108–01, www.ers.usda.gov/publications/ldp/jun03/ldpm10801/ldpm10801.pdf, accessed 28 August 2009

Smith, J. F. (1999) 'Case studies: International sanitary and phytosanitary laws; national and international nursery regulations', in Coppock, R. H. and Kreith, M. (eds) *Exotic Pests and Diseases: Biology, Economics, Public Policy*, University of California Agricultural Issues Center, University of California, Davis, CA, pp218–221

Sumner, D. A. (2003) 'Economics of policy for exotic pests and diseases: Principles and issues', in Sumner, D. A. (ed) *Exotic Pests and Diseases: Biology and Economics for Biosecurity*, Iowa State University Press, Ames, IA, pp9–18

Sutherst, R. (2000) 'Climate change and invasive species: A conceptual framework', in Mooney, H. and Hobbs, R. (eds) *Invasive Species in a Changing World*, Island Press, Washington, DC, pp211–241

Weitzman, M. L. (2000) 'Economic profitability versus ecological entropy', *The Quarterly Journal of Economics*, 115(1), 237–263

Westbrooks, R. (2001) 'Potential impacts of global climate changes on the establishment and spread of invasive species', *Transactions of the 66th North American Wildlife and Natural Resources Conference*, Wildlife Management Institute, Washington, DC

21
Satisfaction in a Horse: The Perception and Assimilation of an Exotic Animal into Maori Custom Law

Hazel Petrie

Introduction

Perceptions of exotic species affect not only those in the wild but domestic stock too, and the cultural assimilation may run deep. Prior to the arrival of Europeans, New Zealand's only mammals were the dog and the native rat brought by the Maori from their original Polynesian homelands. Early voyagers left pigs and ships' rats behind, but within some three decades of the arrival of the first horse, those animals not only acquired spiritual significance but also featured in 'customary' legal practice, especially as satisfaction for acts of adultery. Interestingly adultery was a very serious crime, especially if the woman involved was the wife of a high-ranking chief, but the idea that horses represented customary satisfaction might seem rather curious given that the animals were not indigenous to New Zealand.

Like other innovations, animate or inanimate, the circumstances surrounding the arrival of the first horse on Maori shores have entered into legend. One story recalls the arrival of a monster from the sea (see, for example, Pomare, 1989), but written records tell us that the first arrivals were a stallion and two mares, which arrived in the Bay of Islands with the first missionaries on 22 December 1814. One of the mares was a gift from the governor of New South Wales to Ruatara, the Ngapuhi chief who invited the missionaries to live under his *mana* at Rangihoua in the Bay of Islands. Missionaries gave his people access to Western knowledge and manufactured goods, so having them under his patronage and protection greatly enhanced his *mana* and prestige.

Mana, or spiritual authority and power, is the fundamental basis of chiefly leadership and the driving force behind tribal well-being. Because they represented the most direct lines of descent from the ancestor gods, the first-born children of noble families inherited great *mana* and typically succeeded to leadership, but *mana* can also be acquired through achievement. However, gaining and maintaining *mana* and a loyal following requires the distribution of wealth so it is also dependent on the control of resources.

The connotations of *mana* attached to horses were evident when Ruatara's successor Hongi Hika returned from England in 1821, having met King George IV. Hongi complained that he had not been treated with the dignity befitting the noble leader he was. He had not been given the grey horse he desired, nor a square of scarlet cloth as had been presented to other chiefs (Francis Hall to Marsden, 20 October 1821; Williams, 1864). Red fabric was a well-known symbol of rank throughout Polynesia and Hongi's remarks suggest that he also saw horses as a similarly appropriate gift for a noble personage.

The horse in New Zealand

The dissemination of horses began slowly and it was not until 20 years after the arrival of the first horse at the Bay of Islands that Moetara Motu Tongaporutu became the first to own one in the Hokianga region; a mere 100 kilometres to the west. In February 1834 he was presented with a sword, cloak and letter of appreciation from the lieutenant governor of Van Diemen's Land (Tasmania) for retrieving the property of a Northland-based British trader following an inter-tribal battle over trading rights. Also aboard the ship carrying these gifts was the agreed payment for some land: a mare and foal. Wearing the cloak and carrying the sword, Moetara rode the mare the following month at a great feast he gave to accompany the scraping of the bones of members of his Ngati Korokoro people who had died in the battle. This was a ritual that traditionally took place some time after death to prepare the deceased for laying in their final resting places and Moetara's grand appearance would surely have enhanced his *mana* in the eyes of the 4000 or more people who attended (Ballara et al, 2007).

Because horses were so rare and expensive during the earliest years of European settlement, they were owned only by the most powerful families (Orbell, 1999). In January 1840 – just before British annexation – a French missionary wrote that a settlement possessing a boat and a horse was a *kainga rangatira* (chiefly settlement); whereas one that had only a canoe and a walking stick was a *kainga ware* (lowly settlement) (Turner, 1986). So horses were replacing the walking stick as Western-style vessels were replacing canoes, but their rarity and their value in facilitating overland travel ensured that their acquisition enhanced their owners' power and status and signified *mana*.

The Hokianga chief Eruera Maihi Patuone, a gifted entrepreneur, became a successful horse breeder, even supplying the government for military purposes. But he is also remembered for gifting horses to tribal leaders from other parts of the North Island. One of these was an old Tuhourangi chief named Kohika

Figure 21.1 *The smart attire worn by these people in 1856 suggests that horse ownership was an indicator of wealth ('A group I once saw in Maori Land', New Plymouth, 1856. Alexander Turnbull Library, Strutt, William 1825–1915: Reference number: E-453-f-005)*

who, in about 1844, asked the missionary Seymour Spencer to help him by writing to Patuone to request a steed. The iron-grey mare that arrived about a year later was the first that Tuhourangi owned. Patuone's biographer Charles Davis noted that the Ngati Tuwharetoa leader Hohepa Tamamutu named one of his sons Kohika, which he thought was due to a temporary loan of the animal and in order that Tamamutu's people might claim a share of its offspring (Davis, 1876). Personal names may invoke imperatives indicating tribal expectations but the *mana* attached to horses was clear when the people of an important Ngai Tuhoe chief named Te Maitaranui bought their first steed, probably in the 1840s, and named it Tuhoe. As Patu Hohepa explained in an essay on the introduction of muskets to Maori society, the naming of items considered to have exceptional value would give them the potential attributes of *mana* (Hohepa, 1999).

The diplomatic significance of gifting horses was evident when the son of Mananui te Heuheu, paramount chief of Ngati Tuwharetoa, was sent to live with Patuone's brother, Tamati Waka Nene. Sending young men destined for leadership to other parts of the country to become acquainted with the principal chiefs of other tribes was a customary practice and so it was that this young man spent over two years with Nene in the Hokianga and subsequently took the name of his esteemed host: Te Waaka. In 1842, before he was escorted home to Taupo by a large party of Nene's Ngapuhi people, Nene presented Te Waaka with a horse (Grace, 1992). This chiefly gift had made a great impact as the travelling artist George French Angus who visited Taupo in 1844 noted:

> *The extraordinary excitement produced by the arrival of so large and singular an animal … gave rise to numberless charcoal drawings of men on horseback, that cover nearly every flat board within the settlement.* (Angas, 1847)

It surely strengthened ties between Ngapuhi and Tuwharetoa as well. That horses were associated with *mana* was again evident when Angas and his companion Thomas Forsaith needed to cross the Whaingaroa Harbour that same year. Because mudflats, exposed by the tide, would make their crossing difficult, their host offered them horses that they might 'ride over like Rangiteras' (*rangatira* means chiefs or noblemen). When Forsaith observed that they had no horses, the chief pointed to some of his men sitting nearby, saying: 'on … their shoulders you shall ride across the flats' (Angas, 1847). This proud leader was not to be shamed by a lack of quadrupeds!

But as well as symbolizing *mana* (which has a spiritual source), it is evident that horses had a *wairua* or spirit of their own. This was evident over 30 years after Tuwharetoa received their first horse and were well stocked with the animals when a party of 70 Ngati Raukawa visited them from further north. All came on horseback except poor Paraone Taupiri and his wife who arrived on foot because their horses had recently died. Evidently a canny man, Paraone composed a lament along the way which he sang on arrival at Taupo:

> *E muri ahiahi, takoto ki te moenga,*
> *Ka rarua aku mahara, ngaro noa te kamia*
> *I te hikitanga wae no Ngati Raukawa.*
> *Tera pea korua kei nga wi ka hau*
> *I roto nga Roto Takawha,*
> *Kai atu ra nga rori ka tuwhera i roto Atiamuri.*
> *Tera pea korua kai nga titahatanga i roto Whakaheke,*
> *Ka kitea mai korua e Ngati Te Whetu.*
> *Ma wai e rangaranga to korua mate i te ao?*
> *Ma Te Hemopo, ma Hohepa Tamamutu,*
> *Ma Te Papanui, ma Te Heuheu,*
> *Ma Paurini Karamu, ma Kingi –*
> *Tongariro e!*
> *K hoki mai ki ahau!*
>
> *Grieving in the evening, I lie on my bed,*
> *With troubled thoughts, my horses gone*
> *From Ngati Raukawa's expedition*

Perhaps you are in the sounding tussock
By the Takawha Lakes
Passing along the roads that lie open at Atiamuri.
Perhaps you are on the winding paths at Whakaheke,
Where Ngati Whetu will see you.
Who will make good your deaths in this world?
Te Hemopo will do so, and Hohepa Tamamutu,
Te Papanui, Te Heuheu,
Parini Karamu and Kingi – O Tongariro!
And you will return to me! (Translation by Orbell, 1999)

So he sang to the dead horses, tracing the journey of their spirits, before addressing six of the leading chiefs with language usually employed to seek support in avenging a military defeat. The song indicated that, by replacing his horses, the chiefs would avenge their deaths and symbolically bring them back. The six duly responded by presenting Paraone and, effectively, his people, with seven horses – one for each of them and one for their sacred mountain, Tongariro, who he had courteously identified with the chiefs (Orbell, 1999).

The significance of horses and their possession of a spirit are also indicated by the funerary arrangements accorded Robert Tahua's horse in the early 1800s. Missionary Richard Taylor sketched a large conical construction on stilts which is reminiscent of the platforms Maori erected to hold the bodies of deceased people before their removal to a final resting place. Taylor left Northland for Whanganui in May 1843, so the sketch is assumed to have been made at an earlier date.

Figure 21.2 *The burial place of Robert Tahua's horse at Rarakarea's farm (Richard Taylor, Reference number: E-296-q-157–2, Alexander Turnbull Library, Wellington)*

Later in the 19th century

From the 1860s, when wars divided Maori from European and Maori from Maori, ways in which biblical symbolisms pertaining to horses were assimilated into syncretic religions gave them other spiritual dimensions. The Taranaki prophet Titokowaru had a grey horse named Niu Tirene (New Zealand), but white horses were more often associated with Maori prophets (Belich, 2007). Te Kooti, the famous 19th-century military leader and founder of the Ringatu religion, had a white horse with spiritual powers — the white horse of Revelation that would ensure he eluded capture by government forces during his time as a fugitive (Binney, 2007). The Tuhoe prophet Rua Kenana, who claimed to be Te Kooti's successor, also had a white horse as did Hipa Te Maiharoa in the South Island. Judith Binney has written extensively on the Maori prophet tradition, explaining how Te Kooti and others interwove traditional sources of authority with the Bible's prophetic heritage to evoke new concepts of future fulfilment and redemption (see, for example, Binney, 1984, 2007). However, Christianity is less evident in the spiritual powers attributed to horses by Tuwharetoa's Ngati Tama Whiti *hapu* (tribal group). Four *kaitiaki* or spiritual guardians protect the four corners of their lands, horses being assigned to the northern boundary. In the 1990s, when the government's Department of Conservation announced its intention to begin a programme of culling wild horses in their area, in order to reduce damage to threatened flora, Ngati Tama Whiti lodged a claim to the Waitangi Tribunal, which inquires into breaches of the Treaty of Waitangi, that such action would violate their right to exercise *rangatiratanga* or chiefly control over their ancestral lands and the horses on them.

Yet despite a body of literature relating to the adaptations to Maori spirituality that have occurred in the wake of European contact, little attention has been paid to the way in which horses featured in custom law, especially in connection with *muru* and the crime of adultery. *Muru* may be defined as ritual plunder or a process of restorative justice, a means of exacting compensation for injuries sustained, and a mechanism for preserving *mana*. The process typically involved the removal of property belonging to the offender's kin group because, under custom law, the perpetrator's wider tribal group was implicated in the crime and liable to a joint penalty. Since it was irrational to seek satisfaction from a party with little *mana*, the payment of a high price exulted the standing of the offender's people and the 'victims' of a large *taua muru* or plundering party gained considerable social prestige (Adams et al, 2009). In earlier times, when long-distance travel tended to be undertaken via the sea or waterways, canoes were often seized. The most prestigious carved war canoes were potent symbols of tribal *mana* but although horses, a new mode of transport, could not carry large numbers of people, they may bear meaningful comparison. That canoes and horses shared similar status is also indicated by Maori grammar, which has two categories of possessives, 'a' and 'o' – the 'o' category being used when the possessor lacks control over the relationship or is subordinate or inferior to what is owned. Both canoes and horses require the 'o' category of possessive pronouns.

With the passing of time, canoes were being replaced by western vessels and horses, although the prominence of the latter as the objects of *muru* was not immediate. In 1836, during a period of hostilities between tribal groups and following the murder of a Waikato girl by the Bay of Plenty's Ngati Whakaue, a British catechist stationed in her home area was subjected to the retributive removal of all his belongings except for his horse and two shirts. When a Select Committee of the British House of Lords asked why they had not taken the horse, he replied: 'I think they had never seen a horse before' (evidence of John Flatt, 6 April 1838). Nevertheless, horses were soon being demanded or removed in response to a variety of offences. Eight horses were taken when pigeons were stolen from another tribal group's reserve (evidence of Tamati Arena Napia, 15 March 1934), and they were also demanded when a chiefly woman was accidentally cut by a police cutlass in 1844. Hugh Carleton, biographer of missionary Henry Williams, who mediated in that case, remarked that horses were then the 'current coin of the realm' (Carleton, 1874–77). 'Current coin' they may have been, but the fact that they were specifically requested implies an association with *mana,* which is supported by the way that horses feature heavily – as something of a standard 'payment' – in *muru* related to adultery.

One relatively well-recorded incident took place in the Bay of Plenty in 1849, when a man from a Christian Maori settlement called Kenana (Canaan) near Te Puke, which was led by one of missionary Thomas Chapman's teachers, committed adultery with the wife of a man named Te Hura from another community (Stafford, 1991). Tohi Te Ururangi, a prominent chief of the region's Te Arawa tribe, sought to obtain *utu* or compensation for the crime, but the offender refused to give up his horse. So Tohi demanded the horse on three separate days – which was, in his words, 'according to the right of the law'. Repeating an act three times is a common theme in Maori ritual, indicating confirmation and/or completion and was evidently a requirement in a number of aspects of custom law. In this case, because the horse was not forthcoming after three requests, Tohi allowed the Kenana people's trading ship to be removed to the injured husband's settlement. Subsequently, after what Chapman described as 'the usual routine', the horse was handed over, but the ship owners refused to take their vessel back without monetary compensation for damage and loss of use while it was withheld. On the basis that the adulterer had contributed only £5 towards the total purchase price of £195, they argued the justice of innocent people suffering for the misconduct of one. The compensation they demanded: four acres of land and £1 5s for each day the ship was detained, was surely a response to Christian teachings as was the individualization of responsibility. Describing themselves as being 'for the most part slaves' or war captives, they stated that because of their religion and loyalty to the Queen, they sought redress through British law. Their refusal to accede to custom law was symptomatic of changes occurring across Maori society as previously subordinate people opted for the advantages of egalitarianism promised by Christianity and British government. In something of a quandary, Tohi reported his actions to the Governor, seeking reassurance that he had acted correctly but also requesting guidance as to British law regarding women, land and murder.

Ann Parsonson has suggested that unprecedented death rates among children and adolescent girls may have contributed to an unusually high number of cases of adultery in the 1840s. She also suggests that, being in high demand, women were less likely to willingly comply with betrothals arranged in infancy (Parsonson, 1981). However, regardless of their frequency, horses continued to be a regular payment for the offence beyond that time and obstructing the process could lead to violence. For example, in one 1852 case, the Ngapuhi chief Repa, who charged others with 'having taken improper liberties with one of his wives', demanded a horse as satisfaction. A newspaper report said this was 'according to native usage'. Unfortunately, however, the other parties denied the accusation and refused to hand over the horse, precipitating a fight in which five people were killed, including Repa himself (*Daily Southern Cross*, 1852). Five years later, when Pita Hongihongi of Te Ati Awa committed adultery with Ihaia Te Kirikumara's wife Hariata, Pita was obliged to forfeit his horse as well as a gun (Parsonson, 1981). Both horses and cows were taken in an 1878 incident in Whangarei which might indicate a nascent interest in dairy farming (*Te Wananga*, 25 May 1878). Nor was a white man married to a Maori woman immune to custom law, as at least one is known to have had horses removed for a similar misdeed (Best, 1903).

As the Kenana people demonstrated, new values including individual responsibility, individual wealth and the commodification of time were creeping into Maori society as increasing numbers converted to Christianity, so other penalties sometimes took the place of *muru* during the 1840s and 1850s, although agreed justice sometimes incorporated elements of both old and new. For example, when an East Coast missionary advised a man, who had lived with another man's wife for five years and had three children with her, to return the woman and pay a fine to the wronged husband, the *runanga* (local council) decreed that paying the fine entitled him to keep the woman (Oliver and Thomson, 1971). This was one of several instances where the payment of compensation was considered to settle a divorce.

Runanga, or meetings of heads of communities, were a regular way of settling disputes – at least until the early 1860s when Governor Sir George Grey attempted to meld them into more formalized institutions. But the influences of Christianity and British legal systems were already evident, as they were when a Wairarapa man was found guilty of adultery in 1859. He was sentenced to a cash fine of £50, which he did not have, so a woman apparently unconnected with the offence offered to pay it for him. However, the *runanga* objected saying that it was not proper for someone other than the offender to pay. So, they took his horse, a racehorse named Dick Turpin (Evidence of Hamuera Tamahau, 12th November 1891).

There were some exceptionally extravagant cases, too. In 1862, Resident Magistrate C. Hunter Brown reported that an Opotiki chief whom he described as a very great man, 'especially in his own estimation', volunteered 15 horses together with two pieces of land, a canoe, a gun and a greenstone (nephrite jade) hatchet as compensation for a similar sin. Somewhat taken aback by the generosity of the offer, the offended husband referred it to the local *runanga* to consider what might be fair. Their judgement was for 12 horses but the chief refused to take back any part of the self-imposed damages. Brown suggested that his motive may have been to gain *mana* from a munificent payment or perhaps, more sanguinely, because he knew of an offence by someone else and that his large payment would necessitate a still greater one to his party for the other's crime (Brown, 1862).

A time of change

However, from the mid 1850s, a number of significant changes to New Zealand's political scene had increased tensions between Maori and settler, which encouraged some individuals and groups to return to custom law. The shift to a settler parliament instead of government from London was particularly damaging to Maori interests as provisions requiring freehold land ownership in individual title effectively disenfranchised most Maori who still held their land communally. Political disempowerment, denial of chiefly authority, and an apparently insatiable demand for their land led to a rise in Maori nationalism. As a response to these pressures, a number of tribes from the Waikato region had installed the first Maori king in 1858. In this milieu, the pressure to conform with government's preference for cash fines to the Crown rather than compensating aggrieved parties increased.

Waikato Magistrate Frances Dart Fenton endorsed the policy saying that, 'sometimes the woman is so worthless that no money should be paid to her husband' – which statement belies the great gulf between Maori and settler perspectives (Resident Magistrate F. D. Fenton to the Colonial Treasurer, 25th February 1858). Sir William Martin, formerly New Zealand's Chief Justice, concurred. Among rules he offered to chiefs attending an 1860 government conference in Auckland was one giving magistrates the power to punish adultery by way of a fine to the Crown. It was said to be 'an evil thing that the wife's infidelity should be a means of making money for the husband' (*Maori Messenger*, 1860). But the Maori perception was quite different. The protest of one hereditary leader was translated in 1848 as asking: 'Am I a slave to have money given me for the use of my wife?' (*New Zealand Spectator & Cook's Strait Guardian*, 1848). It is likely that the Maori word he used and which was translated as 'slave' was the same or similar to that employed by Te Keene in response to Martin's suggestion which implies a very significant loss of *mana*. One of several chiefs at the 1860 meeting, Te Keene warned that wronged husbands would not be satisfied by payments to the Queen. 'If the husband were a chief', he said, he would 'fall back upon the customs of Maori law, and the slave man who has committed adultry [sic] with his wife [would] perish by his hand.' '[T]he sin connected with women' was 'a great offence' and the proposed law would 'greatly increase the grief of the husband's heart'. His word, *tutua*, was translated as slave but more correctly means a low-born person of little *mana*.

The Crown certainly faced an uphill battle gaining acceptance of law change. Indeed, when war was raging in the Waikato in 1864, a Northland

Figure 21.3 *Unknown artist, sketches of a Maori* muru *at Parawera; appearance of a member of the Force. Between 1860 and 1890?, Alexander Turnbull Library: Reference No. A-081–005*

resident magistrate's attempts to prevent a wronged spouse from receiving at least five horses awarded by his Maori assessors was the catalyst for establishing a '*Runanga Kei Waho*' or 'outside committee'. The magistrate's practice had been to fine the guilty man and woman but the clearly stated intention of the *runanga kei waho* was to subvert the authority of the magistrate's court and return to custom law (White to Native Minister, 11th April 1864).

At Parawera near Te Awamutu in the Waikato, sometime between 1860 and 1890, a Maori who witnessed a *muru* related to adultery drew six sketches of the proceedings. They are exceptional, not only because naturalistic drawing did not feature in Maori art prior to European contact, but also for their episodic format (Neich, 1994). The first shows the arrival of a party of relatives of the offended people and a horse is evident to one side of the scene. Subsequent scenes illustrate stages in the proceedings, including the arrival of a policeman who announced that at least one of several horses intended as payment had been stolen. It is not clear whether the policeman was a government officer or one of the King Movement's policemen.

These drawings highlight the performance aspects of Maori custom law which requires that justice is done in a public forum. The large numbers of people present on this occasion are indicated by the sketches which identify each group by their tribal names. One records the arrival of Ngati Raukawa,

Figure 21.4 *Unknown artist, sketches of a Maori* muru *at Parawera; the co-respondents, confronted by the injured husband and wife, while the giddy dance proceeds in front of the marae. Between 1860 and 1890? Alexander Turnbull Library: Reference number: A-081–004*

while the one below indicates the relative positions of the home people and Ngati Haua, who are seated in a U-shaped formation while members of Ngati Koroki dance in the centre.

The *muru* at Parawera was between groups associated with the Maori King Movement. The local people were Ngati Apakura, King Tawhiao's tribe, whereas those seeking compensation were Ngati Raukawa, Ngati Haua and Ngati Koroki. Probably unconnected with this event is a letter written to a government-sponsored Maori newspaper in 1875 by Hoani Meihana Te Rangiotu, a Christian teacher and chief of high rank, who was concerned about the continuation of *muru*. His perception of this as taking the law into their own hands and as an unwillingness to adopt a doctrine of forgiveness reflects the shift towards British systems, especially by Christian converts, and the tendency for dissatisfied or disillusioned groups – such as those associated with the King Movement – to revert to custom law. Meihana reported that he had called a 'synod' to discuss the suppression of unenlightened Maori customs and explained that his people were baptized Christians who had promised to forsake the old ways and submit to British law so that the innocent would 'not suffer for the guilty'. Men were still being beaten for adultery and made to give horses or even children to *taua muru* or plundering parties, he said. That horses were highly prized by Maori is evident in their being juxtaposed with children as a means of gaining satisfaction for perceived wrongdoing (*Te Waka Maori o Niu Tirani*, 30th November 1875).

Conclusion

As customary justice systems were aimed at restoring *mana* to victims and their kin groups, it should not be surprising that horses featured so prominently as *utu*, especially during the 1840s and 1850s when their rarity and value not only symbolized *mana* but imbued them with its qualities and power. However, the continuation of *muru* as a justice system from the late 1850s is more likely to represent a highly visible rejection of British law with its preference for cash fines to government. Disillusioned Maori, who were consciously asserting their *mana* and chiefly authority and thumbing their noses at presumptuous colonial administrators, were insisting on their own ideas of justice. In that context, horses might be seen as a 'traditional' or 'customary' means of gaining satisfaction with all the practical advantages of modern convenience. This account presents a rich example of how exotic species are perceived within a culture and also of how they may be absorbed into the customs, laws and lives of native peoples.

References

Adams, T., Benton, R., Frame, A., Meredith, P., Benton, N. and Karena, T. (2009) 'Te Matapunenga: A Compendium of References to Concepts of Māori Customary Law', p53, www.lianz.waikato.ac.nz/publications-internal.htm, accessed 8 October 2009

Angas, G. F. (1847) *Savage Life and Scenes in Australia and New Zealand*, vol II, London

Ballara, A., Klaricich, J. and Tate, H. A. (2007) 'Moetara Motu Tongaporutu', *Dictionary of New Zealand Biography*, updated 22nd June 2007, www.dnzb.govt.nz/

Belich, J. (2007) 'Titokowaru, Riwha', *Dictionary of New Zealand Biography*, updated 22 June 2007, www.dnzb.govt.nz/

Best, E. (1903) 'Maori marriage customs', *Transactions and Proceedings of the New Zealand Institute*, vol 36, p54

Binney, J. (1984) 'Myth and explanation in the Ringatū Tradition: Some aspects of the leadership of Te Kooti Arikirangi Te Turuki and Rua Kenana Hepetipa', *Journal of the Polynesian Society*, 93, 347 and passim

Binney, J. (2007) 'Te Kooti Arikirangi Te Turuki', *Dictionary of New Zealand Biography*, updated 22 June 2007, www.dnzb.govt.nz/

Brown, C. H. (1862) 'Report from C Hunter Brown, Esq, of an Official Visit to the Urewera Tribes', June 1862, *Appendices to the Journals of the House of Representatives* (AJHR), E-9 Sec. IV, p31

Carleton, H. (1874–77) *The Life of Henry Williams*, Volume II, *Upton, 1874–1877*, New Zealand Herald Office, Auckland, pp82–83

Daily Southern Cross, 6 February 1852, p2

Davis, C. O. B. (1876) *The Life and Times of Patuone, the Celebrated Ngapuhi Chief*, Auckland, pp99–100

Grace, J. Te H. (1992) *Tuwharetoa: The History of the Maori People of the Taupo District*, Auckland, pp238–239

Hohepa, P. (1999) 'My Musket, My Missionary, and My *Mana*', in Alex Calder and Bridget Orr (eds) *Voyages and Beaches: Pacific Encounters, 1769–1840*, Honolulu, p197

Maori Messenger: Te Karere Maori, 31 July 1860, pp5,9

Neich, R. (1994) *Painted Histories: Early Maori Figurative Painting*, Auckland, pp161–172

New Zealand Spectator and Cook's Strait Guardian, 14 June 1848, p2

Oliver, W. H. and Thomson, J. M. (1971) *Challenge and Response: A Study of the Development of the Gisborne East Coast Region*, Gisborne, pp64–65

Orbell, M. (1999) 'Paraone's horses: A letter from Hōhepa Tamamutu, 1875', *Kotare: New Zealand Notes and Queries*, May 1999, 2(1), pp3–4

Parsonson, A. (1981) 'The Pursuit of *Mana*', in W. H. Oliver and B. R. Williams (eds) *The Oxford History of New Zealand*, Auckland, p144

Pomare, M. (1989) 'The Taniwha – the landing of the first horse in New Zealand', in James Cowan (ed) *Legends of the Maori*, vol 2, Auckland, pp119–120

Stafford, D. M. (1991) *Te Arawa: A History of the Arawa People*, Auckland, p305

Unpublished sources

Petitjean to Epalle, 31st January 1840, cited in Turner, 'The Politics of Neutrality: The Catholic Mission and the Maori 1838–1870', Unpublished MA thesis, University of Auckland, 1986, p151

Unpublished letters Francis Hall to Marsden, 20th October 1821, Missionary Letters *etc*, F. Hall 1821, H. Williams 1864, MS-0053, Hocken Library, Dunedin, New Zealand

Evidence of John Flatt, 6th April 1838, Minutes of Evidence before the Select Committee on New Zealand, *Great Britain Parliamentary Papers. Colonies: New Zealand*, 1837–1838, Vol 1, p45

Evidence of Tamati Arena Napia, Motukauri block, 15th March 1934, Native Land Court, Northern Minute Book, 65, p153

Evidence of Hamuera Tamahau, 12 November 1891, Native Land Court, Wairarapa Minute Book 18

Te Waka Maori o Niu Tirani, 30 November 1875, J. Grindell, Turanga, pp283–284 and personal communication, Professor Mason Durie, 3 November 2009

White to Native Minister, 11 April 1864, BAFO, 10852, 2a, p470 and White to Native Secretary, 2 May 1864, BAFO 10852, 2a, pp303–306, Archives New Zealand, Auckland

Resident Magistrate F. D. Fenton to the Colonial Treasurer, 25 February 1858, AJHR, 1860, E N0.1c, p30

22

Fire and Loathing in the Fynbos: Notions of Indigenous and Alien Vegetation in South Africa's Western Cape, *c.*1902–1945

Simon Pooley

Introduction

In the late 19th and early 20th centuries, botanists in South Africa's Western Cape felt hard-pressed to popularize and protect the unique indigenous Fynbos flora of the region. They saw themselves ranged against the extensive transformations of the landscape being undertaken by farmers and foresters, the expansion of urban areas and infrastructure, and the depredations of flower pickers. The introduction of a suite of invasive alien plants into the region in the 19th century, notably a range of Australian species well suited to the poor nutrients and rainfall and fire regimes of the region, presented a physical but also a symbolic focus for their advocacy. In the early 20th century this was played out in the context of political attempts to build unity among the English and Afrikaner populations after the South African War ended in 1902. The new science of ecology was consciously used as an integrative influence. However, the ecological theory imported with the experts arriving from Britain in the period of reconstruction, as influential a biological invasion as the earlier wave of alien plant imports, had unfortunate consequences for scientifically informed research and management of the local flora.

Naturalizing a 'state without a nation'

Beinart used the phrase 'a state without a nation' to describe the period of South African history from 1888 to 1948 (Beinart, 1994, p7). After the South African

War of 1899–1902, Dubow argues, the ideology of South Africanism 'emerged to inhabit the space left by a retreating imperialism and a temporarily broken republicanism' (Dubow, 2006, p5). In this context, Dubow argues that science was used to express universality and progress within the imperial 'chain of civilization', and was regarded as a means of uniting moderate Afrikaners and British colonials in a positive and progressive patriotism (Dubow, 2006). Preservationist and conservationist movements led to the founding of the Kruger National Park in 1926, and a patriotic pride in the South African landscape was evoked, with Table Mountain as the ultimate symbol of national unity.

In 1913, the National Botanical Gardens were founded at Kirstenbosch in Cape Town, on the southern slopes of Table Mountain, with Harold Pearson as the first director. The purpose of the garden was to combine useful research with the cultivation and display of the newly unified country's indigenous flora. The Botanical Society of South Africa was founded in the same year. Kirstenbosch's second director, (Robert) Harold Compton, would use the rhetoric of patriotism to mobilize support for the protection of the country's 'natural wealth', its indigenous flora. One of the reasons he gave for encouraging South Africans to grow indigenous flowers in their gardens 'is because they *are* South African' (his emphasis) (Compton, 1927, p3). Van Sittert has described how what he calls 'floral nativism provided both a sense of identity for emerging White settler nationalism and a justification for evicting the underclass from the commons' (van Sittert, 2003, p113).

It is worth recalling that the National Botanical Survey, which was initiated in 1918 under Illtyd Buller Pole Evans, director of the Department of Botany at the Department of Agriculture, and whose patrons included the first Union Prime Minister, Louis Botha, and then his successor Jan Smuts, was made possible by the establishment of the Union of South Africa in 1910. Botany became professionalized in this period, and was explicitly seen by Smuts as a way of promoting unity in the country. As Anker puts it:

> *Unity through diversity was the new approach, and the means to achieve it was ecology. The ecologists could – armed with the science of botanical relations and succession – unite environments that formerly were divided by topography and climate.* (Anker, 2002, p65)

Botany, patriotism and the politics of national unity were closely bound up, and in his guide to undertaking ecological work, the influential Natal-based botanist and ecologist John William Bews (1884–1938) made these links explicit, recommending that ecologists use the language of sociology to describe relationships in the plant world (Anker, 2002, p62).

Northern hemisphere ecology invades

Bews, one of the most influential South African ecologists of the first half of the 20th century, was at this time importantly influenced by, and disseminating, the Nebraskan ecologist Frederic Clements' idea of succession. Clements' ecology, influential since the publication of his *Research Methods in Ecology* (1905) and summed up in his book *Plant Succession* (1916), might best be described as 'dynamic'. Bews discussed Clements' research methods and his influence on plant ecology in this period (Bews, 1935, p3). Clements argued that vegetation progresses linearly through a series of increasingly complex stages towards a stable climax community, which is in equilibrium with the prevailing environmental conditions. In the initial stages of succession, soil is more important, but ultimately it is climatic factors, rainfall, temperature and winds, which determine the nature of the climax community. Climate thus drives a linear unidirectional succession towards a single vegetation type (monoclimax), though a variety of disturbances (notably fire) may inhibit or temporarily reverse this progression (see Worster, 1994, pp208–236).

While Anker has noted the 'South Africanisation' of botany in the early decades of the century, imputing nationalistic motives, it is also simply the case that the field was dominated by British men trained in temperate climes who were influenced by soil-oriented notions of phytogeography. As Bews, Professor of Botany at Natal University College (1910–1938), pointed out in an address to the South African Association for the Advancement of Science in 1920, botany was mostly taught by imported teachers using syllabi drawn up on northern hemisphere lines and using foreign textbooks: 'even now we have no satisfactory South African textbook of Botany' (Bews, 1921, p65).

In 1916, six years after his arrival in Natal from Edinburgh to take up the province's first professorship in botany, Bews had sought to remedy the lack of an ecologically informed overview of the region's vegetation in his paper 'An account of the chief types of vegetation in South Africa, with notes on the plant succession' (Bews, 1916). Bews frames his overview of the main vegetation formations of the region within the concept of vegetative succession advanced by the American ecologists Henry Cowles and Frederic Clements. It followed from their work that if climatic and geological conditions remain stable then climax vegetation formations should be resistant to invasion (Bews, 1916, p131).

Bews bases his description of the 'Sclerophyllous Formations' (now collectively known as the Fynbos Biome) of the winter rainfall region of the Western Cape on his own fieldwork and the publications of the famous Cape botanist Rudolf Marloth. These winter rainfall formations show a succession, he argues, through lichen associations, lithophilous mosses, chomophytic vegetation (on the surface of rocks), to heath. What he terms 'Macchia' (Fynbos vegetation) is 'another sclerophyllous formation, much more extensive', which is 'the climax type for its own climatic habitat' (Bews, 1916, p137). Unfortunately, this influential linear, synchronous explanation of the dynamics of Fynbos vegetation proved to be disastrously misleading.

There was little room, in the interpretation of Clementsian ecology adopted in South Africa, for the role of disturbances such as fire in the functioning of the 'natural system'. Rather, there was a strong tendency to interpret fire as an anthropogenic disruption of the 'natural balance'. This is particularly unfortunate as Fynbos is a highly fire-adapted vegetation type. It is not possible to understand Fynbos ecology without investigating the complex range of plant strategies for coping with and taking advantage of stochastic variations in fire frequencies and behaviour. The idea that fire degrades the natural vegetation of the region led to some regrettable misinterpretations which had negative consequences for environmental management for decades to follow.

Further, the idea of the climax community of plants as an integrated organism, as argued by Clements, further obscured the ecology of Fynbos plants. According to recent interpretations, it is in fact the different (unsynchronized) strategies evolved by various plant species to a variety of kinds and frequencies of fire events that accounts, in part, for the sustained diversity of plants in the vegetation type. Variation in fire regimes prevents the long-term dominance of any one species, thus maintaining the coexistence of numerous species. The structure and species dominance apparent in a Fynbos community at any one time is likely to be highly influenced by the intensity of the last fire, and the time elapsed since it occurred, confounding attempts to recognize a stable climax community (Rebelo et al, 2006, pp82–85).

A remarkable (though not unique) feature of the period was the preference of botanists for studying ecosystems in a 'natural' and 'undisturbed' state, notably excluding the influence of human beings on the landscape. This resulted in a tendency to denigrate 'secondary' vegetation types perceived to have been interfered with by humans, and vigorous campaigns to preserve 'virgin' indigenous vegetation against invasion by alien interlopers and damaging practices such as veld burning.

On the question of plant invasions, Bews remarks that in the formations of vegetation he identified in South Africa, no one species dominates. He observes that the Australian wattle species *Acacia mollissima*, extensively planted in Natal, had spread naturally and where it had established itself, 'hardly any native species is able to exist ... An introduced species therefore is apparently able to assume complete dominance, while our native species of trees are not.' However, Bews argues that on the whole, 'the vegetation of South Africa ... is resistant to invaders' (Bews, 1916, p157). Under 'natural' circumstances, then, the advanced, complex, climax stage of indigenous vegetation must be resistant to invasion as it is the best-adapted possible vegetation formation for the circumstances. It is only if this natural state of affairs is disturbed by human activity that invasion can occur. This was a view that was to persist until the 1980s, and is central to Oxford ecologist Charles Elton's important text on biological invasions, *The Ecology of Invasions by Animals and Plants* (1958). Such confusions of 'natural' (or 'native') with 'best adapted' have subsequently been eloquently dismantled by evolutionary theorist Stephen Jay Gould, who noted that in Darwin's conception:

> *Natural selection is only a 'better than' principle, not an optimizing device ... many native plants, evolved by natural selection as adaptive to their regions, fare poorly against introduced species that never experienced the local habitat.*

Further:

> *Since organisms (and their areas of habitation) are products of a history laced with chaos, contingency, and genuine randomness, current patterns (although workable, or they would not exist) will rarely express anything close to an optimum.* (Gould, 1998, pp6, 7)

Indigenous and alien invasions

Late 19th-century and early 20th-century agricultural literature on invasive plants in South Africa reveals a focus on the invasion of productive agricultural land (particularly grazing lands) by unproductive or poisonous plants. These plants prospered on disturbed lands, and as a result of selective grazing (livestock removed their more palatable competition). Although the bulk of the research covered alien plants, notably prickly pear (*Opuntia* spp), Port Jackson willow (*Acacia saligna*), blackwood (*Acacia melanoxylon*), *Lantana camara*, tumbleweed (*Salsola kali*), various species of *Hakea*, and water hyacinth (*Eichhornia crassipes*), the question of whether these invasive species were alien or indigenous appears not to have been a focus. (Moran and Moran (1982) provides an overview of what was researched.) It was, rather, their negative effects which were of consequence.

Indigenous species such as sweet thorn (*Acacia karroo*), silver clusterleaf (*Terminalia sericea*), sickle bush (*Dichrostachys cinerea*), Rhenosterbos (*Elytropappus rhinocerotis*) and *Helichrysum argyrophyllum*, or even vegetation formations such as rhenosterveld, were identified as problematic invaders of productive farmland, just as harmful as alien invasive species (see for instance Scott, 1935, p105; on Rhenosterbos and *H. argyrophyllum*, see Schönland, 1923, pp102–104). In his highly influential *Veld Types of South Africa* (1953), agricultural researcher John Acocks (1911–1979) would identify even the much prized indigenous Fynbos as an '*invader* that threatened the local grassveld' (Acocks, 1953, pp14, 17). He did so because he was writing a description of the vegetation of the country from the perspective of livestock farmers wanting to assess its grazing potential. Acocks' vegetation units were conceived as units of vegetation homogeneous enough to facilitate equivalent farming potential. He divided Fynbos (which he called Macchia) into lowland and montane Macchia, and he divided the latter into 'mountain' and 'false' Macchia. 'False Macchia' included all the Fynbos on the Cape Mountains from the central Swartberg and Langeberg all the way to the eastern limits of these ranges near Grahamstown. It was 'false' because he argued that it was derived from natural forest or grassland through human intervention: overgrazing, clearing or burning. Fynbos ecologist Richard Cowling has shown that this influential classification is inaccurate, and in the long term has proved harmful in its influence on the beliefs and land management practices of the agricultural community in the region (Cowling and Holmes, 1992, pp41–42).

Many non-native species were recommended by the agricultural authorities for specific qualities, which in some cases made them more suitable to the farmers' purposes than indigenous plants, notably hardy succulent species such as agave and prickly pear (*Opuntia* species) which could be used for fencing, shelter, soil stabilization and fodder in arid and semi-arid areas of the country (for Agave, see du Toit et al, 1923, p67, and for *Opuntia* species, see Beinart, 2003, pp266–303). Even Port Jackson willow was recommended as a 'sluit stopper' to help prevent erosion (see Benke, 1908, p652). (Many of these species would subsequently be declared alien invasive plants (ARC, 2010).)

Agriculturalists did however draw attention to the alien nature of invasive vegetation in the case of exotic timber plantations. This was partly a turf war with the Department of Forestry; many of these trees were being planted in river catchment areas, deemed most suitable for the imported species favoured at the time. Pastoral farmers wanted access to these mountain pastures as back-up in times of drought, and further argued that the trees were drying up water supplies. Eucalypts in particular were regarded as 'thirsty' trees which used more water than indigenous species. Early conflicts over this issue came to a head in the British Empire Forestry Conference held in South Africa in 1935 (see below) (Fourth British Empire Forestry Conference, 1936, pp3–6).

Many of the species regarded as harmful invasive aliens were imported by foresters from the 1880s onwards. South Africa has very little indigenous timber, and the indigenous hardwoods are slow-growing. For a country with a large and growing mining industry, developing its railway and telegraph networks, this was a serious lack. The drought commission report of 1923, while noting some potential drawbacks of planting exotic tree species en masse, pointed out that some 50 per cent of South Africa's irreplaceable imports ('raw materials needed for continuous expansion and progress'), valued at £3,045,008, consisted of 'timber, wooden manufactures and paper'. The timber famine experienced during World War I had emphasized the fact that the country had no option but to afforest (du Toit et al, 1923, p67). Foresters worked assiduously to see which tree species were best adapted to cultivation in the local conditions, in the process experimenting with numerous species of pine from Europe and the Americas, and eucalypts, wattles and others from Australia (Brown, 2003, p350). The stabilization of dunes and drift sands has a longer history, with numerous species (notably species of *Acacia*, *Hakea* and *Casuarinas*) being imported (the majority from Australia) to the Cape from the 1840s onwards (Avis, 1989, p57).

The afforestation, particularly of mountain catchment areas, with exotic plantations, proved controversial. In addition to the concerns of farmers, botanists complained of the invasion of indigenous plant communities by introduced species, notably species of pine, eucalypts, *Acacia* and *Hakea*. South Africa's Chief of Forest Management, J. D. M. Keet, speaking at the British Empire Forestry Conference held in South African in 1935, acknowledged that there had 'been a large agitation recently in South Africa against afforestation' (Fourth British Empire Forestry Conference, 1936, pp48, 109, 115). He noted a variety of arguments advanced against afforestation, including the view that it destroys the natural flora, and that it spoils the natural beauty of the country. He relates that, 'It is our plantations, especially, that stand suspect. They are accused of being ecologically foreigners to our climate ...' (Fourth British Empire Forestry Conference, 1936, p117). Keet grumbled that, 'there are South Africans ... who cannot see any beauty in an exotic tree' (Fourth British Empire Forestry Conference, 1936, p49). He made the point that afforestation with exotics paid for the protection of much larger areas of indigenous flora (Fourth British Empire Forestry Conference, 1936, pp48–49). At the same conference, South Africa's P. C. Kotze noted that General (Jan) Smuts had called for 'a compromise between the forester and a nature lover'. He recalled seeing a picture in:

> *... one of the illustrated papers [in which] the General was shown on the slopes of Table Mountain on a hot day, sitting down on a bed of pine needles under the shade of* Pinus pinaster. *In the picture one could see what appeared to be a patch of wag-'n'bietjie [the thorny* Acacia caffra *(Thunb)] and protea bush, but the General did not take advantage of the seat of the wag-'n'bietjie, nor of the shade of the Fynbos, but sat under the pine trees. What better compromise could have been reached?* (Fourth British Empire Forestry Conference, 1936, p178)

This is certainly a powerful image of such a compromise and eloquent plea for tolerance, showing Smuts the well-known botanist who liked to pose as the philosopher on the mountain, and is also a sly poke at the selective nature of complaints against exotics.

The Jonkershoek Forest Influences Research Station was established in the upper catchment of the Eerste River near Stellenbosch in 1935, on the recommendation of the Fourth Empire Forestry Conference. The Conference concluded that the proposed studies of 'the effects of forests on climate, water conservation and erosion' would be 'of inestimable advantage both to it and the world at large' (Wicht, 1948, p4). This research was also deemed necessary in order to address the fears of 'the public', quite understandable in light of the narrative of desiccation of the country long propounded by experts, that 'plantations of exotic pines, eucalypts and wattles might dry up water supplies, exhaust the soil and even promote erosion' (Wicht, 1948). These concerns are still prevalent, and form the rationale for the nationwide Working for Water Programme in South Africa today. (See Noemdoe, 2001, pp121–125.)

It was through these burning experiments, conducted as part of the forestry hydrological research programme, that foresters became particularly interested in the indigenous Fynbos. Further, in the 1930s large areas of mountain veld were bought and put under the control of the Department of Forestry, to conserve the vegetation and water resources (Wicht and Kruger, 1973, p8). This legitimated research on indigenous vegetation within forestry, alongside their focus on exotic timber trees.

Defending the realm: the Cape Floral Kingdom

What was the nature of this indigenous Fynbos vegetation that made it worthy of such sustained botanical interest? The Cape Floral Kingdom (CFK) is the smallest of the world's six plant kingdoms, comprising the southwest corner of South Africa, a winter rainfall region experiencing long, hot, dry and windy summers, and relatively short, wet winters.

The CFK is astonishingly biodiverse, supporting nearly 9000 species of plants, 69 per cent of them endemic – thus 44 per cent of southern Africa's plant species, on 4 per cent of its area (Rebelo et al, 2006, p91). Most of these species occur in the Fynbos ('fine bush') biome. Fynbos has evolved (and continues to evolve) in the context of nutrient-poor soils, complex topography, pollinator specialization, fire, and in response to the Mediterranean-type climate dominant in the region since the establishment of modern circulation patterns between 33 and 3 million years ago (Rebelo et al, 2006, pp67, 93). The vegetation is highly adapted to fire, the natural occurrence of which was significantly boosted by fire-stick farming (to encourage natural fields of carbohydrate-rich geophytes) since the late Pleistocene epoch (<2 million years ago) (Deacon et al, 1992, pp15, 20; Deacon, 1992, p261).

As noted above, a great deal of confusion resulted from Cape botanists' attempts to make the Fynbos fit Clements' theory. Fire is certainly the major source of difficulty, being regarded as a disturbance of the natural plant succession, a threat to indigenous Fynbos, and a friend of invasive introduced plants. I will illustrate how this worked out through looking at just two prominent Cape botanists in the period.

In 1923, Robert S. Adamson moved to South Africa to take up the Harry Bolus Professorship of Botany at the University of Cape Town. Born in Manchester, Adamson was educated in Edinburgh and Cambridge, taught botany at the University of Manchester, and collaborated on ecological studies with the originator of the concept of the 'ecosystem', Alfred Tansley, in England. Adamson's *The Vegetation of South Africa* (1938) was the only one of a series of monographs on the vegetation of the British Empire, planned by the British Empire Vegetation Committee, to be completed. He also co-edited *The Flora of the Cape Peninsula* (1950).

In his 'preliminary account' of the vegetation of Table Mountain (1927), Adamson argues that the extremely diverse plant communities and their apparent lack of correlation with their habitats is a result of a high level of human disturbance. Burning, planting, grazing and the felling of timber have changed the nature of the vegetation almost completely, he argues, and of these factors, 'fire has been very much the most extensive and far-reaching in its effects' (Adamson, 1927, pp284–285). Adamson regarded fire as an anthropogenic disturbance, rather than a natural agent of vegetation change integral to Fynbos ecosystem functioning.

In his discussion of the spread of alien plants, Adamson notes that fire aids the rapid spread of *Pinus pinaster* in particular, arguing that it reduces the invader's competition (indigenous plants). Species of *Hakea* are also assisted in their spread by fire: 'The extremely resistant fruits open after burning and the seeds germinate at once' (this is known as serotiny) (Adamson, 1927, p303). So, although Fynbos requires fire to regenerate, a number of introduced species regenerate more quickly than Fynbos after fire, gaining a competitive advantage. The reasons are complex and for a discussion, see Richardson et al, 1992, pp271–308.

In concluding, Adamson finds evidence of widespread 'destructive change' to what he imagines to have been the original indigenous climax vegetation of Table Mountain (Adamson, 1927, p304). He is of the opinion that: 'by far the most effective agent of change has been the frequent burning of the vegetation. The effects of this need not be repeated; they result in the retrogression of the normal succession and an impoverishment of the flora' (Adamson, 1927).

In 1919, Cambridge-educated Robert Harold Compton (1886–1979) took over as director of Kirstenbosch, with the concomitant position of Harold Pearson Professor of Botany at the University of Cape Town (a university from 1918), and began editing the *Journal of the Botanical Society of South Africa*. Compton became an influential figure in South African botany, in 1937 founding what would in 1952 be named the Compton Herbarium (today South Africa's second-largest).

Compton's antagonism to 'promiscuous fire-making' was fuelled by the danger that runaway fires lit by those camping and hiking on Table Mountain posed to the flora of his botanical garden (and that of the mountain in general) (Compton, 1924, p3). His most vehement statement on veld burning was delivered in his presidential address to the South African Biological Society on 27 November 1924. Compton took as his subject 'Veld Burning and Veld Deterioration' (Compton, 1926).

Compton writes that in pre-European times, the indigenous peoples of the region burned the veld 'with that sublime indifference to ultimate consequences that characterises uncivilised races' (Compton, 1926, p5). Writing of the arrival of Europeans and the cross-fertilization of farming knowledge, he opines that 'it is remarkable how, when two races come into contact, each adopts the worst features of the other: so that while the blacks acquired the habits of wearing trousers and drinking spirits, the white learnt the gentle art of ruining the veld in the most certain and rapid way conceivable' (Compton, 1926). (One wonders what he had against trousers.) The point he's working around to is that white farmers are mistaken in believing that they have inherited the practice of burning the veld from their European forefathers, and his strategy is to assert that 'veld-burning is Kafir pastoral practice, and is unworthy of Europeans' (Compton, 1926). (He was right that they learned this practice from Khoikhoi pastoralists at the Cape (Kolb, 1738, pp62–63), but pastoralists all over the world have used this approach to veld management for millennia).

Compton presents a morality tale about the burning of Fynbos by Cape stock farmers to force up new shoots for their stock to graze on, an approach which he argues enjoys a little short-term success but inevitably leads to deterioration of the veld, a paucity of plant species, and in the final stage, 'bare sandy rubble, devoid of humus, soil and vegetation' (Compton, 1926, p6). He refers to Bews' idea that 'grazing or burning puts back the succession' of vegetation, and that 'putting back the succession' signifies the production of an earlier or more primitive type of vegetation than previously existed' (Compton, 1926, p12). The parallels he draws between 'primitive vegetation', 'uncivilized races' and 'backward and retrograde farmers' pursuing 'Neo-Hottentot pastoral practice' (i.e. burning the veld) are clear (Compton, 1926, pp5, 11).

As he concludes his address to the Biological Society Compton turns to a matter he feels will be close to their hearts: the destruction of indigenous species. Curiously for one taking an evolutionary view of human civilization, and who clearly sees European civilization as the most advanced, he proceeds to lament 'our destructive civilisation', which has 'entailed the loss or decimation of so many of our most beautiful and interesting plants and animals' (Compton, 1926, p18). He complains that:

> *The mountain slopes are burnt bare for the imagined benefit of a few goats or cows; that is we lose our unique and irreplaceable species of plants in order to gain a little butter disparagingly called 'farm,' and a few bottles of milk of inferior quality. In these and many other ways biologists are being robbed of their subject matter, and the most intensely and intrinsically national of all the country's possessions, its flora and fauna, are being replaced by a deadly and pernicious monotony of alien weeds.* (Compton, 1926)

The 'natural' and the 'wild'

Identification of the indigenous flora as 'natural' or 'wild' was complicated by issues arising around flower picking and landscape gardening. Compton, like his predecessor Pearson, was concerned about the impacts of flower picking. Flower shows were popular, poor people sold cut flowers in the city, and Cape flowers were being exported. (For a discussion of the origins of the wild flower trade, shows and the protection societies see van Sittert, 2003, pp119–120, 123–124.) The impacts of (mostly Cape Coloured) flower pickers on the protected species growing on the slopes of Table Mountain were a concern, also because some of them burned the veld to induce flowering (they were more sanguine about the effects of burning on Fynbos than the botanists). The *Minute of His Worship the Mayor* (1940) recorded flower pickers burning 'for ... the removal of over-ripe veld' (p33). However, it was the larger commercial operations run by some farmers which were of more concern to Compton. Revised regulations to Cape Ordinances on wild flowers recommended by the Wild Flower Protection Society in 1922 were frustrated by legal disputation over definitions of 'wild' and 'cultivated'. In Caledon, a magistrate ruled that 'flowers which are protected from fire, pruned and weeded, and generally protected by farmers on whose property they grow, may be regarded as cultivated'. As Compton complained, 'How are the police to know the cultivated from the uncultivated wild flowers? How can they know from what area they have been gathered?' (Compton, 1924, pp4–5).

Landscape gardening with indigenous plants made it difficult to be sure of the 'natural' distributions of some indigenous plants around Table Mountain. Showy species were planted along roadways and in picnic areas. In 1911, it was suggested that in the Corporation Plantation on Signal Hill 'steps should be taken to clean out the greater part of the sugar bush which was very unsightly ... leaving only those shrubs which are in isolated positions and of good shape' (*Minutes of His Worship the Mayor*, 1911, pp47–48). In 1929, the city's Director of Parks and Gardens noted that 'seeds of various indigenous trees and shrubs have been sown including 12 different species of protea planted near foreman's house as an experiment and if found suitable seed will be sown on different parts of Lion's Head to improve our native flora' (van der Houten, 1929, p3).

Such dilemmas notwithstanding, concern over the destruction of Fynbos resulted in a 1945 Royal Society commission into the preservation of the indigenous vegetation of the Western Cape chaired by the director of research at Jonkershoek, C. L. Wicht (Wicht, 1945). Professors Compton and Adamson were both members. The final report argued that, 'the flora of the Cape is its chief beauty ... the flora is unique, it is one of the richest, most varied and beautiful in the world' (Wicht, 1945, p7). The authors were concerned that: 'this considerable asset is being lost, through the ravages of fire, browsing and erosion; the invasion of undesirable exotic species; the illegal gathering of flowers; and the indiscriminate conversion of the veld to other uses' (Wicht, 1945, p8).

In the view of the assembled experts, writing in the closing year of World War II, the strongest argument for the conservation of the flora was a recognition of its 'amenity values', which 'touch the deepest sources of mental and spiritual refreshment, both conscious and unconscious' (Wicht, 1945, p9). After the general destruction – and politically divisive effects of the war in South Africa – they saw the preservation of the Cape vegetation as a means to healing, and called for 'co-operation between State departments, universities, local public bodies, and the people' to preserve this 'part of our natural heritage' (Wicht, 1945, p53).

Conclusion

Anker and van Sittert have drawn attention to the ideological dimension of some of the pioneering plant ecological work undertaken in this period. I have given further examples and elaborated ideas about vegetation in the writings of the botanists considered here. However, I wish to conclude by arguing that ideas about indigenous vegetation were most significantly shaped in this period by botanists' struggle to adapt ecological ideas developed in the northern hemisphere to the complex, fire-adapted Fynbos vegetation of the Cape Floral Kingdom.

Though gifted and committed scientists, these botanists held value-laden conceptions of indigenousness and what is 'natural'. These were bound up with the social concerns and ecological orthodoxy of their time, leading them to believe that the diversity of the Fynbos was threatened by one of the chief drivers of this diversity, namely fire. They defended a conception of a complex, advanced, equilibrial climax state of vegetation against what they saw as a disturbance-driven retrogression to a simpler, more primitive state.

Perhaps this should give us pause for reflection on the values and theoretical trends that inform our current thinking about alien and indigenous vegetation in the Cape, and how it should be managed. There are certainly powerful resonances with some contemporary agendas, which I will touch on briefly here. While state and municipal foresters had come around to doing prescribed block burns and invasive plant control from the 1960s, the collapse of state forestry in the late 1980s led to a vacuum in the control of invasive species over much of the country until the founding of Working for Water programme in 1995. (For a profile of Working for Water, see Woodworth, 2006). This successful programme combines poverty relief with environmental management, specifically the clearing of invasive alien plants from waterways and catchment areas. This programme is founded on antagonism to harmful introduced species, and advocates routinely use emotive language referring to 'cleaning' areas of 'infestations' of alien plants which 'threaten to engulf and exterminate the unique indigenous fauna and flora'. It has even been claimed that 'criminals use the vegetation as cover to pounce on innocent victims' (Noemdoe, 2001, pp121, 122). These kinds of claims have been made since the 1870s (see Beinart, 1994, p273; *Report of the City Engineer*, 1978, p19).

Events such as the large fires on the Cape Peninsula in January 2000 have been used to advocate widespread eradication of undesirable exotics from the region. The founding of the Working on Fire programme in 2003 was in part premised on the perceived role of alien invasive species in increasing the frequency and intensity of fires (a matter of some debate). However, some recent research suggests that it is weather conditions that trigger the largest fires, rather than fuel load of vegetation. It is these large fires that consume most of the vegetation that gets burned in any year (Forsyth and Van Wilgen, 2008, pp3–9; Southey, 2009). If validated, the consequences for the rationale of current fire management practice – which focuses on prescribed burning and control of alien invasive plants – may be profound.

In a stimulating article, anthropologists Jean and John Comaroff have drawn parallels between the emotional public response to the Cape 2000 fires with its focus on invasive alien plants, and xenophobia in the 'new' South Africa. They interpret both reactions as symptoms of the 'second postcolonial epoch' (post-1989) in an era of 'neoliberal global capitalisms', emphasizing the challenges these developments pose to the power of states and notions of national identity (Comaroff and Comaroff, 2001, pp632–633). However, I have argued that debates over invasive introduced and indigenous plants, hybridized by debates over autochthony and national identity, have a much longer history in the region. They cannot be fully comprehended without taking cognizance of longer term trends in ecological thinking and land management, and the ways these have been 'naturalized' in the landscapes of the Cape Floral Kingdom.

Note

1 A previous version of this chapter was published in the *Journal of Southern African Studies*, 36(3) (September 2010), 599–618.

References

Acocks, J. P. H. (1953) *Veld Types of South Africa*, Department of Agriculture, Pretoria, South Africa

Adamson, R. S. (1927) 'The plant communities of Table Mountain: Preliminary account', *Journal of Ecology*, 15, 278–309

Adamson, R. S. (1938) *The Vegetation of South Africa*, British Empire Vegetation Committee, London

Adamson, R. S. and Salter, T. M. (1950) *Flora of the Cape Peninsula*, Juta and Co., Cape Town

Anker, P. (2002) *Imperial Ecology: Environmental Order in the British Empire: 1895–1945*, Harvard University Press, Cambridge, MA

ARC, 'Legislation on Weeds and Invasive Plants in South Africa', Agricultural Research Council, www.arc.agric.za/home.asp?pid=1031, accessed on 19 January 2011

Avis, A. M. (1989) 'A review of coastal dune stabilization in the Cape Province of South Africa', *Landscape and Urban Planning*, 18, 55–68

Beinart, W. (1994) *Twentieth-Century South Africa*, Oxford University Press, Oxford

Beinart, W. (2003) *The Rise of Conservation in South Africa: Settlers, Livestock, and the Environment 1770–1950*, Oxford University Press, Oxford

Benke, J. D. (1908) 'Port Jackson willow as a sluit stopper', *Agricultural Journal: Department of Agriculture of the Cape Colony*, 32

Bews, J. W. (1916) 'An account of the chief types of vegetation in South Africa, with notes on the plant succession', *The Journal of Ecology*, 4, 129–159

Bews, J. W. (1921) 'Some aspects of botany in South Africa and plant ecology in Natal', *South African Journal of Science*, XVIII(1 & 2), 63–80

Bews, J. W. (1935) *Human Ecology*, Oxford University Press, London

Brown, K. (2003) 'Trees, forests and communities: Some historiographical approaches to environmental history on Africa', *Area*, 35(4), 343–356

Clements, F. E. (1905) Research Methods in Ecology, University of Nebraska Publishing Company, Lincoln, USA

Clements, F. E. (1916) *Plant Succession: An Analysis of the Development of Vegetation*, Carnegie Institute of Washington, Washington, DC

Comaroff, J. C. and Comaroff, J. L. (2001) 'Naturing the nation: Aliens, apocalypse and the postcolonial state', *Journal of Southern African Studies*, 27, 627–651

Compton, R. H. (1924) 'News and notes', *Journal of the Botanical Society of South Africa*, 10

Compton, R. H. (1926) 'Veld burning and veld deterioration', *South African Journal of Natural History*, 6(1) 5–19

Compton, R. H. (1927) 'News and notes', *Journal of the Botanical Society of South Africa*, 13

Cowling, R. M. (ed) (1992) *The Ecology of Fynbos: Nutrients, Fire and Diversity*, Oxford University Press, Cape Town, South Africa

Cowling, R. M. and Holmes, P. M. (1992) 'Flora and vegetation', in Cowling, R. M. (ed) *The Ecology of Fynbos: Nutrients, Fire and Diversity*, Oxford University Press, Cape Town, South Africa, pp23–61

Deacon, H. J. (1992) 'Human settlement', in Cowling, R. M. (ed) *The Ecology of Fynbos: Nutrients, Fire and Diversity*, Oxford University Press, Cape Town, South Africa, pp260–270

Deacon, H. J., Jury, M. R. and Ellis, F. (1992) 'Selective regime and time', in Cowling, R. M. (ed) *The Ecology of Fynbos: Nutrients, Fire and Diversity*, Oxford University Press, Cape Town, South Africa, pp6–22

Dubow, S. (2006) *A Commonwealth of Knowledge: Science, Sensibility, and White South Africa 1820–2000*, Oxford University Press, Oxford

du Toit, H. S. D., Gadd, S. M., Kolbe, G. A., Stead, A. and Van Reenen, R. J. (eds) (1923) *Final Report of the Drought Investigation Commission* (U.G.49–'23) Government Printer, Cape Town, South Africa

Elton, C. (1958) *The Ecology of Invasions by Animals and Plants*, Methuen, London

Forsyth, G. G. and Van Wilgen, B. W. (2008) 'The recent fire history of the Table Mountain National Park and implications for fire management', *Koedoe*, 50, 3–9

Gould, S. J. (1998) 'An evolutionary perspective on strengths, fallacies, and confusions in the concept of native plants', *Arnoldia*, Spring, 3–10

Gunn, M. and Codd, L. E. (1981) *Botanical Exploration of Southern Africa*, A. A. Balkema, Cape Town, South Africa

Kolb, P. (1738) *The Present State of the Cape of Good-Hope*, trans. M. Medley, 2nd edn, Vol. 1, W. Innys & R. Manby, London

McNeely, J. A. (ed) (2001) *The Great Reshuffling: Human Dimensions of Invasive Alien Species*', IUCN, Gland, Switzerland

Minutes of His Worship the Mayor for the Mayoral Year ending 13th September, 1911 Chancellor Oppenheimer Library, University of Cape Town, Government Publications, G682 VC2 s.2247

Minutes of His Worship the Mayor for the Year ending 6th September, 1940, Chancellor Oppenheimer Library, University of Cape Town, Government Publications, G682 VC2 s.2247

Moran, V. C. and Moran, P. M. (1982) 'Alien invasive vascular plants in South African natural and semi-natural environments: Bibliography from 1830', *South African National Scientific Programmes Report No. 65*, CSIR, Pretoria, South Africa

Mucina, L. and Rutherford, C. (eds) (2006) *The Vegetation of South Africa, Lesotho and Swaziland*, SANBI, Pretoria, South Africa

Neely, A. H. (2004) '"Blame it on the Weeds": Politics, Fire, and Ecology in the New South Africa', Unpublished MSc thesis, University of Oxford, Oxford

Noemdoe, S. (2001) 'Putting people first in an invasive alien clearing programme: Working for Water Programme – are we succeeding?', in McNeely, J. A. *The Great Reshuffling: Human Dimensions of Invasive Alien Species*', IUCN, Gland, pp121–125

Proceedings of the Fourth British Empire Forestry Conference (1936) London, pp3–6

Rebelo, A. G., Boucher, C., Helme, N., Mucina, L. and Rutherford, M. C. (2006) 'Fynbos Biome', in Mucina, L. and Rutherford, C. (eds) *The Vegetation of South Africa, Lesotho and Swaziland*, pp53–219

Report of the City Engineer (1978) Chancellor Oppenheimer Library, University of Cape Town, Government Publications, G682 VC2 s.2247, p19

Richardson, D. M., Macdonald, I. A. W., Holmes, P. M. and Cowling, R. M. (1992) 'Plant and animal invasions', in Cowling, R. M. (ed) *The Ecology of Fynbos: Nutrients, Fire and Diversity*, Oxford University Press, Cape Town, South Africa, pp271–308

Schönland, S. (1923) 'The deterioration of the veld in the S. E. area of the Botanical Survey', Appendix 6, in du Toit, H. S. D., Gadd, S. M., Kolbe, G. A., Stead, A. and

Van Reenen, R. J. (eds) *Final Report of the Drought Investigation Commission*, pp102–104

Scott, J. D. (1935) 'Some problems in the restoration of veld', *Farming in South Africa*, 10(March), 103–105

Southey, D. (2009) 'Wildfires in the Cape Floristic Region: Exploring Vegetation and Weather As Drivers of Fire Frequency', Unpublished MSc thesis, University of Cape Town, Cape Town, South Africa

van der Houten, A. W. (1929) 'Report of the Director of Parks and Gardens for the Year ended June 30th, 1929', in *Minute of His Worship the Mayor for the Mayoral Year ending 8th September, 1929*, Chancellor Oppenheimer Library, University of Cape Town, Government Publications, G682 VC2 s.2247

van Sittert, L. (2003) 'Making the Cape Floral Kingdom: The discovery and defence of indigenous flora at the Cape *c.*1890–1939', *Landscape Research*, 28(1), 113–129

van Wilgen, B. (2005) 'Managing fires: The science behind the smoke', *Quest*, 1, 26–33

Wicht, C. L. (1945) *Preservation of the Vegetation of the South Western Cape*, Royal Society of South Africa, Cape Town, South Africa

Wicht, C. L. (1948) 'Hydrological research in South African forestry', *Journal of the South African Forestry Association*, 16, 4–21

Wicht, C. L. and Kruger, F. J. (1973) 'Die Ontwikelling Van Bergveldbestuur in Suid-Afrika', *South African Forestry Journal*, 86, 1–17

Woodworth, P. (2006) 'Working for water in South Africa: Saving the world on a single budget?', *World Policy Journal*, (Summer), 31–43

Worster, D. (1994) *Nature's Economy: a History of Ecological Ideas*', 2nd edn, Cambridge University Press, Cambridge

23
Biological Invasion and Narratives of Environmental History in New Zealand, 1800–2000

James Beattie

Introduction

In the past, successive New Zealand settlers and governments spent millions of pounds *bringing in* introduced species. Now New Zealand spends millions of dollars on *removing* introduced species. The reason for this reversal lies mainly in the massive ecological impacts of introductions. The reason why New Zealanders feel so strongly about this is also because of the close associations between nationalism and New Zealand's flora and landscape that developed over the course of the 20th century. As the Department of Conservation (DOC), the government body responsible for managing conservation land, outlined to its incoming minister: 'New Zealand's indigenous biodiversity – our native species, their genetic diversity, and the habitats and ecosystems that support them – is of huge value to our economy, our quality of life, and our sense of identity as a nation' (DOC, 2008).

New Zealanders identify themselves as Kiwis. Wearing the coveted black shirt and silver fern of the All Blacks, the national rugby team, symbolizes most childhood dreams. Tourism – the country's largest export earner (NZTE, 2009) – relies on New Zealand's endemic flora and landscape and fictive images of a clean and green country, deeply ironic since milk and meat production is sustained on thousands of hectares of introduced pasture. In these narratives of environmental purity and nationalism a line is drawn between the exotic and the indigenous. The former is equated with the bad days of colonization, the latter with the emergence of New Zealand as a nation shaking off the shackles of empire. A narrative of biological invasion which pits introduced against indigenous nature is central not only to New Zealand's late 20th- and early 21st-century environmental policy, but also to its national identity.

This chapter examines the emergence of this narrative in the 19th century before analysing recent historical interpretations of settler attitudes towards exotic and indigenous nature and towards wider processes of environmental change. Much scholarship on environmental history falls into a simple binary trap, projecting present views of nature, divided into indigenous versus exotic, onto the past. First, many histories of conservation arrange in teleological fashion the rise of native species conservation as the inevitable endpoint of earlier conservation trends. Second, many writers assume that prior to protection of native environments in the 1880s and 1890s settlers must have disliked indigenous nature because they were removing it. According to this view, appreciation for introduced and indigenous nature cannot exist simultaneously with acclimatization and large-scale environmental change. To set the historical context, this chapter provides an overview of environmental change in New Zealand before examining the rise of indigenous conservation and its historical interpretations; it ends with examples that complicate historical interpretations of settler attitudes towards the indigenous and exotic.

Historical and environmental settings

Perhaps nowhere else in the world was the environmental change unleashed by colonization so profoundly documented or as rapid as that which took place in New Zealand from the early 19th century. European and Maori written and oral accounts, along with state statistics, photographs, maps and paintings declaim the processes of swamp drainage, deforestation, pasture sowing, and plant and animal introductions ushered in by colonization. The balance in power relations between Maori and European tipped towards the ever-increasing numbers of Europeans arriving from the 1860s. Earlier, especially before the 1840s, missionaries and traders existed only on the sufferance of tribal chiefs, while many Maori tribes also became successful – and wealthy – traders, exporting their produce of newly introduced food crops as far away as Australia.

Formal colonization commenced in 1840 when the British Crown and many Maori chiefs signed the Treaty of Waitangi. Within 40 years of its signing, New Zealand's European population had grown massively. From a population of Europeans probably fewer than 2000 before 1840, by 1881 well over half a million lived in New Zealand. This figure rose to 1 million early in the 20th century. Rapid environmental transformation, war and Maori land-loss characterized colonization (Pawson and Brooking, 2002).

New Zealand's environmental transformation witnessed the introduction of plants and animals from around the world. A variety of motives triggered acclimatization: nostalgia, economic necessity, aesthetic considerations, religious beliefs and even health concerns. In the early years of a European settlement, animals for labour and plants for food often made the difference between life and death. So too did the generosity of Maori, who supplied food to struggling early European settlements. Given the importance of food crops, a thriving trade in edible plants – and also ornamentals – developed. From the 1840s to the 1860s, Australia provided an important source of many varieties, having an already established network of botanical gardens and nurseries. Settlers also brought out seeds from home or had them sent. Technological innovations such as the Wardian case aided acclimatization. Effectively a mini-glasshouse, this contraption enabled live plants to be shipped across the globe. Later, the introduction of more regular and faster steamer services further opened up New Zealand to the world. Lastly, settlers themselves feverishly established acclimatization societies, releasing (sometimes successfully, sometimes not) a variety of plants and animals into New Zealand, in the process unleashing a series of large-scale and often unintended environmental transformations (Beattie, 2008).

19th and 20th century interpretations of ecological change

Soon after contact, settlers and Maori were commenting upon the rapidity with which introduced species were establishing themselves in New Zealand. William Swainson spoke for many gardeners, noting that 'all sorts of European produce flourish, and all sorts of live stock thrive to an amazing degree' in New Zealand (Swainson, 1840). Gargantuan pumpkins weighing 43 pounds and various other oversized vegetables testified in tangible terms to the productivity of New Zealand's soil and climate (Beattie and Stenhouse, 2007). To some, they also demonstrated the superiority of European species to native ones.

In the late 1860s, William Travers, a politician and naturalist, presented a series of popular public lectures exploring the impact of 'civilised races' on New Zealand's environment. A devotee to Darwin, Travers observed that introduced plants and animals were often displacing indigenous ones, citing as one example the replacement of the native rat by the more aggressive European one. Favouring introduced organisms because they helped improve land lying waste, Travers also sounded a note of caution. An introduced aphid had already damaged crops across the country, he warned (Travers, 1870). Others at the time welcomed the transformation less circumspectly. For imperial tourist Charles Dilke, 'the English fauna and flora are peculiarly well fitted to succeed at our antipodes [and] ... have not even to encounter the difficulties of acclimatization in their struggle against the weaker growths indigenous to the soil' (Dilke, 1868). From the 1870s, acclimatization ushered in several ecological problems. In parts of New Zealand rabbits reached near-plague levels, forcing (against many scientists' advice) farmers to introduce mustelids to control them. The resulting great depredations on New Zealand's native bird populations have elicited criticism ever since (Pawson and Brooking, 2002).

By the 20th century, the first scientific monographs on introduced species in New Zealand appeared. In 1919, G. M. Thomson, a largely self-trained science teacher and enthusiastic acclimatizer, produced a work examining introduced species in New Zealand (Thomson, 1922). A generation later in 1949, North American historical geographer Andrew Hill Clark published the influential *The Invasion of New Zealand by People, Plants and Animals* (Clark, 1949). Its title captures Clark's approach, which tells a history of invasive plants and animals outcompeting indigenous ones. Not only is it a classic of invasion biology literature; it is also a classic example of what imperial environmental historian John MacKenzie later termed the 'apocalyptic school of imperial environmental history', one concerned with narrations of environmental disaster (Mackenzie, 1997). It is within the framework of studies by Thomson and Clark that Alfred Crosby's work, a generation later, appeared.

Alfred Crosby's groundbreaking work *Ecological Imperialism: The Biological Expansion of Europe 900–1900* (1986) shook the historical profession, opening historians' eyes to the power of environment in colonization. Crosby delineated in broad brush strokes and delicate 'pontillism' the impact of settler portmanteau biota, plants, pathogens and animals on southern Africa, Australasia and the Americas. European species, he charged, swept aside more vulnerable indigenous organisms, in the process fashioning 'neo-Europes'; areas whose biotic populations came to closely resemble that of Europe.

Crosby's highly influential work shaped the attitudes and writing of a generation. Only now are environmental historians starting to challenge Crosby's assumptions (Beinart and Middleton, 2004; Beattie, 2008; Coates, 2011). Crosby's work appealed in colonized countries like New Zealand because of the deep concern felt at the loss of indigenous flora coupled with a powerful identification of indigenous flora as a source of nationalism.

The rise of indigenous protection

A key development in European attitudes towards the New Zealand environment and its conservation occurred from the late 1880s. Urban-living European New Zealanders, many of whom were colonial born, began to see in the remnant indigenous flora and landscape symbols of a nation, profitable sites of tourism and retreats from busy towns. Reacting to pressure from both within and without, parliament began to preserve aspects of this nature, leading to the creation of New Zealand's first national park (Tongariro, in 1894) and the establishment of reserves through legislation such as the Land Acts (of 1877 and 1892), and the later Scenery Preservation Act (1903). The conservation estate grew over the ensuing decades thanks to pressure from outdoor enthusiasts and scientists. From the 1920s, decisions to concentrate solely on exotic plantations, rather than native timbers, reinforced images of a weaker indigenous nature that were important to later conservation movements. From the 1960s, high-profile campaigns including attempts to prevent the destruction of iconic landscapes and to make New Zealand nuclear-free, heralded the beginning of the modern environmental movement and a recapitulation of New Zealand's clean, green image. Britain's decision in 1973 to join the European Common Market shocked many New Zealanders, releasing a new wave of nationalism, which also drew on the country's unique natural environment. By the later 20th century, with an increasingly diverse population, New Zealand's native nature became a useful focal point of this nationalism. At the same time very real concerns emerged about loss of native biodiversity. New Zealand's DOC initiated a series of programmes aimed at preserving aspects of New Zealand's unique biota. Most notably and perhaps most famously, the tiny remaining population of black robin, on the brink of extinction, was relocated to a predator-free offshore island (Pawson and Brooking, 2002).

While the government needed to convey its message to New Zealanders, the very real depredations made by introduced species on indigenous New Zealand biota ignited a somewhat extreme dualism in government conservation circles, excoriating the exotic and lauding the indigenous. Two examples of such views will suffice. The DOC's *Protecting and Restoring Our Natural Heritage: A Practical Guide* (2001) rigidly defines pure (and authentic) nature against exotic (and by inference inauthentic) nature. It attempts to return to a fictive 'wilderness' state; to wind back the clock to before Europeans came and then freeze-frame that ecosystem as representative of 'pure' nature. 'Nature', it declares, 'puts plants in the right place', so a 'restored system', it goes on to say, 'can never be as authentic as a natural system.' Imbued by such principles the guide models 'restoration planting' 'on a previous baseline', which was usually the pre-European vegetation of that site. Authenticity is reflected in the plants to be used: restorers are urged to source plants locally, thereby avoiding genetic pollution since native plants 'are best adapted to growing in their local conditions' (DOC, 2001). This work presents a series of problematic assumptions. First, with nature placing plants in the right place, who determines what that right place is? The guide assumes that the workings of ecological processes automatically correspond to aesthetic values, but this is clearly not always the case. One can argue that introduced species naturally spread through New Zealand, but that problems arise when aesthetic assumptions presume they are out of place. Second, the guide freeze-frames extremely dynamic ecological processes to an arbitrary date before 1840. The view precludes the acceptance of environmental change before 1840, whether by non-human forces, Maori or Europeans. In colonial times, the fiction of a primeval New Zealand sanctified the opening up of Maori land for settlement through sale as well as confiscation. Now that same logic sanctifies the protection of New Zealand native plants against introduced ones. The ends are different but the morally imbued language, and its condescension to the complexity of the past and those who inhabited it, remains.

Commenting on New Zealand's Ministry for the Environment Draft Conservation Strategy's description of 'invasive pests' threatening to 'demolish our native species', historian Tom Isern (2002) notes that, 'A bellicose, national-security rhetoric that pits virtuous natives in moral contest against an evil empire of alien invaders may be useful as a stimulus [to conservation], but it also raises impossible expectations', offering no realistic resolution and complete disregard for the reality of the introduced species already naturalized into New Zealand.

Historical interpretations of environmental change

Assumption of a division between indigenous and exotic is central to New Zealand's conservation policy. While important in examining its history, many scholars oversimplify earlier settler attitudes towards nature. Some authors place New Zealand environmental history within a distinct teleology: for them nationalism marks a break with the profligate environmental damage of the colonial period and leads inexorably to the eventual triumph of indigenous species conservation. Earlier celebratory works adopt this approach, but so too do some recent ones. While David Young's (2004) commissioned history of conservation in New Zealand sensitively describes 19th-century settler environmental attitudes, its narrative – and 19th-century conservation – inevitably moves towards the emergence of biodiversity and protection of native species.

For other historians, the late 19th century marks a temporal break between settler likes and dislikes of native nature, an attitude shift supported by the underlying assumption that destruction of native forest cannot occur simultaneously with its appreciation. As noted, massive environmental change took place in the 19th century at the same time as settlers introduced millions of new organisms into New Zealand. To historical geographer Alan Grey, 'the alien nature of the dark and seemingly forbidding bush gave further impetus to its clearing in favour of something more familiar – grassy hills' (Grey, 1994). Similarly, Paul Shepard posits that colonists' view of native flora as barren and unproductive justified its removal (Shepard, 1969). Ploughed land, well-established homesteads and fenced paddocks undoubtedly represented settler ideals, but it would be wrong to think that settlers did not appreciate native species. In the remaining pages, I wish to challenge those interpretations and demonstrate that settlers employed different criteria to appreciate nature and the resulting hybrid landscapes which developed. Native species were incorporated into European gardens and aesthetics just as Maori themselves incorporated introduced species into theirs. At the Mission Station of Waimate North (northern New Zealand) for instance, in 1843 Reverend William Cotton established a native garden, planting nikau palms (*Rhopalostylis sapida*), tree ferns, ti-plant (*Cordyline australis*) and creating an environment apparently much to the liking of his pet kiwi (Raine, 1995). In Dunedin (southern New Zealand) in the 1850s, Jane Bannerman (1855) 'took great pride in watching the development' of her father's manse through bush clearance, but also appreciated native flora, while her brother Arthur travelled by boat to collect native shrubs to plant in their garden. In the 1840s and 1850s, with the fern craze gripping New Zealand, many women collected native ferns, their efforts and specimens preserved in the albums still held by many museums (Beattie and Stenhouse, 2007).

As New Zealand species proved popular with settlers in New Zealand, so they did overseas. Travelling in New South Wales, Australia, in 1834 George Bennett noted that *Cordyline terminalis* 'grows and flowers well' and 'is frequently seen planted in front of the dwelling houses in and about

Sydney'. On Mr H. McArthur's property Vineyard, Bennett found karaka (*Corynocarpus laevigatus*) in 'thriving condition, having reached the elevation of from six to nearly fourteen feet, and borne fruit'. In the 19th century, New Zealand species also proved popular in parts of Europe. In 1852, William Hooker, Director of Kew Gardens, wrote to New Zealand settler David Monro thanking him for sending 'me one or two rare little Beeches ... [and] little Evergreens. These and other Middle Island plants', he predicted, 'would no doubt succeed well in the open air in England if we could get them alive' (Hooker, 1852). Further afield, in 1880 Scottish gardener William Gorrie (1811–1881) examined those New Zealand species introduced into Scotland and which had survived the harsh winter of 1878/9, such as kowhai (*Sophora grandiflora*) and speargrass (*Aciphylla colensoi*), thus providing a useful overview of New Zealand varieties growing in Scotland in this period (Beattie, 2008).

Settlers also held more complex attitudes towards nature than the simple formula 'native = good, exotic = bad' assumed by the two historical geographers discussed above. Landscape artist, garden designer and writer Alfred Sharpe (1876, 1880) appreciated plants first of all for their aesthetic qualities. Following John Ruskin, Sharpe particularly admired older trees, whether introduced or native, but also plants displaying contrasts of colour, light and darkness. Such considerations led him to champion the planting of many Eurasian and American species, to exhort the protection of unique European trees and to decry the destruction of anything he considered beautiful (Beattie, 2006).

This included criticism of the replacement of 'the fine old oaks so wantonly destroyed in Government House grounds' with karaka. For Sharpe, the oak was always picturesque 'whether in winter, with its gnarled and twisted branches; in spring, with its lovely green frondage; in summer, with its massive leafage and shade, and in autumn, with its rich colouring of russet and yellow'. Oak was, moreover, unique because nothing like it existed in New Zealand. To replace oaks 'with the never varying, stiff, awkward looking, dark green karaka', he concluded, 'is an absurdity' (Sharpe, 1876). At the same time, Sharpe criticized the importation of 'vermin and vegetation' for corrupting the originality of the New Zealand bush. He also stressed the need for artists to record native nature before it changed and for government to protect some important native forest from destruction. While recognizing the impact of introduced species in New Zealand, attitudes towards preservation were ultimately determined by aesthetic criteria, not whether a plant was indigenous or exotic.A tangible measure of Sharpe's complex aesthetic still exists in 2009 in several city parks in and around Newcastle, New South Wales, where he moved in the late 1880s. Asked to design or redesign several city parks, Sharpe introduced hundreds of New Zealand species such as pohutakawa, karaka and puriri, setting them alongside European species and incorporating them into his picturesque pleasure garden.

Even with the increasing popularity of Asian plants, primarily ornamental, but also some fruit evident from the 1870s onwards, New Zealand species were planted alongside newly introduced ones, with some native plants even incorporated into Japanese-style gardens. The popularity of rock gardens in New Zealand, evident from the Edwardian period, also integrated New Zealand alpines with other species, although often with an emphasis on planting representative ecological zones (Beattie et al, 2008). Even with mounting concerns over native bird loss, there was no necessary shunning of exotics. In a 1933 letter to Lord Bledisloe, a key player in native conservation, health reformer Sir Frederic Truby King related how he intended 'to convert my Estate into a perpetual Sanctuary and Paradise for New Zealand Native Birds, and such [other] friendly visitors from other lands as elect to join and live amicably alongside those already with us from England and abroad'. Noting that originally he intended 'to depend for shelter almost solely on N.Z. Native shrubs and trees', his mind changed 'due to the comparatively recent discovery that our Tuis [*Prosthemadera novaeseelandiae*] and Mockas actually prefer (as do also the Sparrows already here by the million from England) the beautiful, quick-growing and everlasting flowering gums of Australia, and the Cape-Gooseberries from South Africa, to our Native N.Z. flowers and berries'. For King, as for many others, there was no sharp vilification of the endemic at the expense of the exotic, rather an accommodation of the two so far as possible (King, 1933).

Conclusions

New Zealand's present hybridized landscapes, comprising introduced and native plants (Pawson, 2008), belies the dynamism of its transformation and the divergence in interpretations of that process. Rapid 19th-century environmental change contributed to European identification with, and conservation of, some native nature and introduced species. But in interpreting that history and present management, some historians and government attitudes take a bipolar perspective, pitting exotic against indigene, or identifying a temporal shift between one view and another. The reality was a far more complex set of settler attitudes, drawn from different concepts than simply that of native against exotic nature.

References

Bannerman, J. (1855) *Reminiscences of Her Life to 1855*, typescript, in J. C. Wilson Family Papers, MS 0536–2, Hocken Library, Dunedin, New Zealand

Beattie, J. (2006) 'Alfred Sharpe, Australasia, and Ruskin', *Journal of New Zealand Art History,* 1(27), 38–56

Beattie, J. (2008) 'Acclimatisation and the "Europeanisation" of New Zealand, 1830s–1920s?', *ENNZ: Environment, Nature and New Zealand*, 3(1), 1–25

Beattie, J. and Stenhouse, J. (2007) 'Empire, environment and religion: God and Nature in nineteenth-century New Zealand', *Environment and History*, 13(4), 413–446

Beattie, J., Heinzen, J. and Adam, J. P. (2008) 'Japanese gardens in New Zealand, 1850–1950: Transculturation and transmission', *Studies in the History of Gardens and Designed Landscapes*, 28(2), 219–236

Beinart, W. and Middleton, K. (2004) 'Plant transfers in historical perspective: A review article', *Environment and History*, 10, 3–29

Bennett, G. (1834) *Wanderings in New South Wales, Batavia, Pedir Coast, Singapore, And China; Being the Journal of a Naturalist in those countries during 1832, 1833, and 1834*, vol I, Richard Bentley, London

Clark, A. H. (1949) *The Invasion of New Zealand by People, Plants and Animals*, Rutgers University Press, New Brunswick, NJ

Coates, P. (2011) 'Over here: Undesirable American animals in Britain', this volume, Chapter 3

Crosby, A. (1986) *Ecological Imperialism: The Biological Expansion of Europe 900–1900*, Cambridge University Press, New York

Department of Conservation (DOC) (2001) *Protecting and Restoring Our Natural Heritage: A Practical Guide*, DOC, Christchurch, New Zealand

DOC (2008) 'Briefing to the new Minister of Conservation 2008 – Major policy and implementation issues' Biodiversity', www.doc.govt.nz/publications/about-doc/briefing-to-the-new-minister-of-conservation-2008/major-policy-and-implementation-issues/major-policy-and-implementation-issues-biodiversity/, accessed 9 September 2009

Dilke, C. (1985 [1868]) *Travellers' Tales of Early Australia & New Zealand: Greater Britain, Charles Dilke Visits Her New Lands, 1866 & 1867*, Blainey, G. (ed) Methuen Haynes, North Ryde, NSW, Australia

Fox, W. (1971) *The Six Colonies of New Zealand*, Hocken Library Facsimile, Dunedin, New Zealand

Gorrie, W. (1879–80) 'Notes on New Zealand plants that withstood the severe winter of 1878–79 at Rait Lodge, Trinity, near Edinburgh', *Transactions of the Edinburgh Botanical Society*, 14, 52–64

Grey, A. (1994) *Aotearoa and New Zealand: A Historical Geography*, Canterbury University Press, Christchurch, New Zealand

Hooker, W. J. to David Monro, 4 October 1852, Royal Gardens, Kew, 'Letters from Sir Wm J. Hooker to Dr Sir David Monro, 1852–4', original MS lent by Miss Linda C. Monro, Palmerston North, copied 1951, typescript, MS-1030, Alexander Turnbull Library, Wellington, New Zealand, folio 3

Isern, T. (2002) 'Companions, stowaways, Imperialists, invaders: Pests and weeds in New Zealand', in Pawson, E. and Brooking, T. (eds) *Environmental Histories of New Zealand*, Oxford University Press, South Melbourne, Australia

King, Frederic Truby, to Lord Bledisloe, 8 June 1933, Melrose, Wellington, New Zealand, in *Royal New Zealand Plunket Society, Headquarters: Records*, 'Gardening, Melrose, etc.', pt 1, Hocken Library, MS-1783/083

MacKenzie, J. M. (1997) 'Empire and the ecological apocalypse: The historiography of the imperial environment', in Griffiths, T. and Robin, L. L. (eds) *Ecology and Empire: Environmental History of Settler Societies*, Washington and Edinburgh

NZTE (New Zealand Trade and Enterprise) (2009) 'Tourism', www.nzte.govt.nz/access-international-networks/explore-opportunities-in-growth-industries/growth-industries/pages/tourism.aspx, accessed 21 September 2009

Pawson, E. (2008) 'Plants, mobilities and landscapes: Environmental histories of botanical exchange', *Geography Compass*, 2(5), 1464–1477

Pawson, E. and Brooking, T. (eds) (2002) *Environmental Histories of New Zealand*, Oxford University Press, South Melbourne, Australia

Raine, K. (1995) '1815–1840s: The first European gardens', in Bradbury, M. (ed) Viking, Auckland, New Zealand

Sharpe, A. (1876) *New Zealand Herald*, 24 August, p6

Sharpe, A. (1880) 'Hints for landscape students in watercolour', in Blackley, R. (1992) *The Art of Alfred Sharpe*, Bateman and Auckland Art Gallery, Auckland, New Zealand, p141

Shepard, P. (1969) *English Reaction to the NZ Landscape Before 1850*, Pacific Viewpoint Monograph No 4, Victoria University of Wellington, Department of Geography, Wellington, New Zealand

Swainson, W. (1840) *Observations on the Climate of New Zealand*, Smith, Elder and Co, London

Thomson, G. M. (1922) *The Naturalisation of Animals and Plants in New Zealand*, Cambridge University Press, Cambridge

Travers, W. T. L. (1870) 'On the Changes effected in the Natural Features of a New Country by the Introduction of Civilized Races', Part III, *Transactions and Proceedings of the New Zealand Institute*, 3, 326–336

Young, D. (2004) *Our Islands, Our Selves: A History of Conservation in New Zealand*, Otago University Press, Dunedin, New Zealand

Part IV
The Way Ahead: Conclusions and Challenges

24

Good Science, Good History and Pragmatism: Managing the Way Ahead

Ian D. Rotherham and Robert A. Lambert

Here we pull together issues and ideas and draw some critical conclusions. Importantly, we highlight the implications of these discussions for future conservation management, for economic development, and in terms of pressing academic research needs. The issues of alien and exotic species, of invaders, of deliberate introductions and of reintroductions, are complex and frequently misunderstood. While the economic, ecological and even human consequences of invasive and often exotic species cannot be disputed, the underlying causes and responses often remain unrecognized. An objective of this book has been to raise awareness and to broaden the discussion to embrace a range of academic disciplines and insights.

Introduction

At the dawn of the 21st century, the most celebrated communicator in natural history broadcasting, Sir David Attenborough, mused in a one-off television documentary about *The State of the Planet* (BBC, broadcast 2000). He identified five ways in which human activities were destroying or eroding biodiversity values, in no particular order of magnitude or challenge:

- the over-harvesting of animal and plant resources (both marine and terrestrial);
- the transportation and introduction of species around the world; especially the problem of alien species on fragile oceanic islands, but also at home (i.e. in the UK);
- the destruction of habitats, both our own and for flora and fauna;

- the problem of 'islandization': as we set about creating isolated islands of relatively undisturbed habitat in a sea of totally modified land, rich and complex ecosystems are being replaced by monoculture, uniformity and geometric lines;
- climate change and air pollution: more flooding, wetter winters, warmer summers and more weather extremes.

For many with environmental interests this was a stark and bleak documentary that urged reflection, thought and action. It challenged viewers to see the Earth as a whole, as a grand, functioning and gorgeous ecosystem under threat, while at the same time urging vigilance and commitment on a regional and local scale. Attenborough is a national icon in the UK, cherished as representing the very best of well-informed and accessible public service broadcasting in an age of fluffy sound bites. For him to talk publicly about the challenges and threats posed by alien species on a global scale, but also domestically, was in effect, a clarion call to take up the issue. Indeed, for many in the huge British wildlife constituency (a recent social survey counted 30 million people with some sort of 'engagement', however active or passive, with the natural world (Rollins, 2006, p31)), this may have been their first genuine exposure to the issue. Alien and especially invasive plants and animals raise many issues and generate serious problems. Yet the processes of naturalization and invasion present society and politics with upheaval and challenges beyond the mere environmental and economic. 'The world is globalising, and nature is no exception', mused a Dutch team of a biologist, aquatic biologist and medical epidemiologist in 2007, coining the term 'biological globalization' to describe plants and animals crossing natural barriers and settling in new areas, whatever the driving force (Weijden et al, 2007, p5). In a multicultural modern technological society, perceptions of and attitudes to alien fauna and flora are deeply embedded in sociocultural mental landscapes and are thus based on subjectivity rather than on science and objectivity. Informing the responses to exotic invaders requires clear and balanced science and well-documented case studies, but it also demands a multidisciplinary approach blending natural sciences with the social sciences, the arts and humanities. Only through these insights can a fuller, richer and more complex understanding be arrived at of how nature and people interact over time – each causing and responding to invasions – and long-term effective solutions to ecological crises be found. As Philip Grime (2005) has pointed out, separating invasion ecology from the rest of ecology weakens our comprehension and leads to muddled and ineffective responses. Furthermore, when action requires expenditure of considerable sums of money and the coordination of national and international programmes of control, the public will rightly ask for justification and transparency. With the global economic downturn (2008 onwards) financial constraint is likely to become very serious for at least the next decade, so to justify expansive and costly actions the arguments and cases will need to be much more openly and effectively constructed. Ignoring fundamental issues of values, perceptions and attitudes will ultimately be open to challenge, and this contestation may undermine much-needed conservation management.

There have been some remarkable successes in the reintroduction of regionally or nationally lost species in both Europe (including Great Britain) and North America (Jones, 2002), most especially charismatic birds of prey (Brown and Waterston, 1962; Love, 1983; Carter, 2007). However, the benefits of re-establishing ecological keystone species, and especially large carnivores and herbivores, will always provoke a strong response from some farming, fishing and game management interests. It is likely that finance for more innovative and dynamic approaches may be limited for the foreseeable future, and this may compound suspicions that most reintroductions favour high-profile 'iconic' and attractive species that can generate wealth, through tourism potential for example (RSPB, 2006), rather than ones necessary for healthy ecological functioning. But tourism and environment research since the 1980s has shown that the economic valuation of wildlife (as distasteful as that might be to romantics or scientific purists) as a tourist attraction, be it for the nature tourism or ecotourism industries, now carries real clout at the policymaking table. Case studies of success (for conservation projects, education and rural economies) both domestic and international abound in the academic literature (Page and Dowling, 2002; Diamantis, 2004; Newsome et al, 2005; Fennell, 2008).

Attitudes and perceptions

What is acceptable and what is alien vary with time. In the 1930s, the little owl or 'Frenchie' was perceived by some land managers as a serious threat to native British species, most especially to game bird chicks. An inquiry was launched to investigate its diet (Hibbert-Ware, 1938). Yet today it is a highly valued and admired member of the nation's avifauna, and occasional avian star of TV natural history documentaries. Eagle owls are also now naturalized in small numbers in northern England, and it seems likely (though open to debate across the scientific and birdwatching community) that they were native in prehistoric times; are they welcome or not? Even more problematic are internationally rare animals such as Chinese water deer or golden pheasants – Britain hosts a significant proportion of the total world population as exotic species. Of course, neither brown hare nor rabbit are native (Sheail, 1972); yet the latter is a keystone species of many British ecosystems, and the former is a national Biodiversity Action Plan species. It could hardly be argued that the acceptance of these is based on their minimal ecological impact. Other animals that come under the spotlight are beavers and wild boar. It can be argued that without these two important terrestrial and aquatic mammals, our wildlife habitats lack a major element of their natural functional composition; indeed this absence may be damaging to other key indicator species, and ultimately to wider ecosystem health. A lowland wood without wild boar lacks critical microdisturbance and the dispersal of important fungi. So then how do we respond to attempts either to reintroduce these species, or to tolerate escapees? In neither case are they wholeheartedly welcomed. There is public outcry and a clamouring for control over the toleration for escapees. But if wild boar is native then it is protected; only if we view it as exotic should we consider controls on re-establishing British populations. In Germany, France and parts of Eastern Europe, people and wild boars seem to get along fine (Dennis, 1998, p6), so why not in Britain? There are issues of real impacts on ecology and even the archaeology of wooded landscapes of a large mammal re-established after several centuries. However, these should be dealt with as management of a problem rather than control or eradication of a perceived 'alien'.

Realism and conservation management, not blind eradication

A key issue is that we need to address 'problem' species and not necessarily 'alien' species. This also means the tacit acceptance that we value and even celebrate some exotic plants and animals and rightly so, and also that much management is a subjective, even emotional or sentimental, decision. City dwellers, who value, cherish and take genuine delight in observing what they perceive to be wild nature (be that colourful parakeets or diminutive deer), must be taken seriously, and included in the wider debates, because together (as a sizeable wildlife constituency) they have influence, have a right to be involved and a role in any future policymaking. It is simply unthinkable in a modern progressive democracy for environmental management decisions to be taken by research scientists, politicians and country folk alone. We decide for a variety of reasons that we don't like a certain plant or animal in a particular place. This may be because of real or perceived effects on other pants and animals or for a variety of other reasons. However, while this decision may be informed by objectively gathered and interrogated science, this choice is ultimately subjective, cultural, and political even. This does not necessarily mean that the decision is wrong, but simply that we should be honest about how we arrived at it, which raises difficulties for government agencies and NGOs charged with protecting and enhancing biodiversity and controlling the tide of exotic and invasive species. This basis for decision-making does mean that in a democracy there need to be full, open discussions and stakeholder debates about how such decisions are arrived at. These important challenges affect wider society beyond the realm of environmental managers and research scientists. In an increasingly cosmopolitan international community, some arguments need to be more robust, the logic more transparent, and the accountability for actions more direct.

We may forget the simple fact that many people love exotic species, and indeed in the pet trade are active in their transportation around the globe; a legacy we can trace back to early European intercontinental voyaging (Crosby, 1986), to the rampant 'acclimatizers' (animal and plant transporters) of the 19th and early 20th centuries, who had no real thought for biological effects on native ecosystems, save the aesthetic, useful or comfortingly familiar (Rolls, 1969; King, 1984; Imber, 1998). Direct inheritors of that love for the exotic 'other', many modern British birdwatchers derive real pleasure from seeing ruddy ducks, golden pheasants and ring-necked parakeets in the wild (which they can add to their competitive British checklist), and secretly hope that one day birds such as monk parakeet, black swan and eagle owl will be deemed acceptable by some birding authority or other, and thus countable on national avifauna checklists (Dudley, 2010). One keen birder and naturalist, Mark Hows, has taken his passion for the aliens a step further; he actively seeks them out all over Europe in targeted trips labelled 'Plastic Fantastic I-V', his aim to see and photograph all European 'exotic and introduced species'. You can follow his remarkable hunting exploits, his treatment of alien and invasive species as an observational challenge, on his website: www.hows.org.uk/inter/birds/exotics.htm.

With Himalayan balsam and *Rhododendron ponticum* in Great Britain for example, Rotherham (2001a, b) traced such informal grassroots activities for over 160 years as enthusiasts deliberately spread these plants around their own country and even to other countries hundreds of kilometres away. There is a further issue (which would surely benefit from cutting-edge social science anthropological and cultural research), namely that for people from the same regions as these exotic plants and animals, there are deep cultural and practical ties to them, links between nature and people firmly embedded in society, in generations of family history. In a modern democracy, we will need to understand, confront, explain and, where necessary, justify our actions more convincingly than we have done to date in the face of such traditions. There is good evidence of what clearly amounts to barely disguised racism and xenophobia in some recent 'ecological' writings and in eradication policies in some Western European countries. This is a worrying trend that distances science even further from people, and promotes distrust and unease; but it has been flagged up as a disturbing development by some bolder sociocultural academics (Coates, 2003, pp135–136; Smout, 2003, pp17–19). An understanding of perceptions of exotic and alien, of native and natural, is important to achieving a balanced approach to problem resolution (Coates, 2006). Research on the environmental histories of Australia and New Zealand, those most invaded and homogenized of ecosystems, demonstrates how the cultural depth and impacts of such translocated Anglo-perceptions mould attitudes not only to 'wild' fauna, but to domesticated animals too (Bolton, 1981; Lines, 1991; Griffiths and Robin, 1997; Dunlap, 1999; Garden, 2005). Deeply held cultural attitudes and values affect all that we do, behaviour and responses, and they come with a lot of baggage.

The genie out of the bottle

At present experts around the world are trying to assess the likely costs of alien invaders. Figures banded about suggest a cost of perhaps 5 per cent of the global economy. In Britain, we are doing calculations to assess the cost of eradicating perceived aliens and invasives from the British landscape even though there is little chance of the necessary funding being forthcoming, and there are pragmatic reasons why removal will fail. Also in Britain, the 1970s and 1980s campaign, to first document, then study, then eradicate through concerted and focused trapping the somewhat elusive (at times invisible) South American coypu from East Anglia, remains our only significant long-term management success (Gosling and Baker, 1989). However, in the 1930s, we had to resort to prohibitive legislation in order to remove quickly the North American muskrat (Public Acts, 1932). It now seems in 2010, that aspirations for complete removal of invasive creatures or plants can never be fulfilled since in most instances the species are far too well entrenched and absorbed into native ecosystems, are too widely dispersed, or have many human supporters ready to rally to the species' defence. The calculated costs are way beyond anything that society via its government will pay in these times of new austerity in the early 21st century. This presents the most serious dilemmas because some ecological impacts and conservation problems are indeed very severe and invaders will undoubtedly have further dramatic effects on highly valued fauna and flora. We suggest that the real issue is how to address problem species through long-term sustainable land management, rather than just tackling alien invaders. A more coherent approach will have greater chance of winning support and of delivering long-term controls.

We suggest that present approaches:

- generally lack scientific rigour in their justification;
- fail to inform and engage and call upon all stakeholders, most especially the wider public who form a large and powerful 'wildlife constituency';
- rarely provide a holistic (for example catchment-wide) context or strategy;
- almost always lack financial or human resources to be long-term effective;
- fail to discriminate between genuine and perceived problems;
- have no realistic long-term targets and if they do, no effective monitoring towards achievement;
- are separated from the context of landscape, ecological and social history which underpins many 21st-century landscapes. In other words, they often take place in an 'ahistorical' vacuum;
- do not provide effective management or containment of critical problems at local and regional levels where they occur.

These are serious charges, based on many years of research engagement, critical reflection and first-hand practical personal observation. In accepting the most worrying impacts of certain invasive and often alien species, these concerns beg the question of what can we do differently or better in the 21st century.

A nature conservation argument

First, it is necessary to accept the potential ambiguous status of these plants and animals that we wish to manage and the nature of the threats which they pose. This means reassessing our motives for control and our desired outcomes. In many cases around the world, it is clear which species are exotic and the impacts are obvious. The sad demise of many island ecosystems following the various waves of human colonization over thousands of years is testimony to the potential for utter disruption of otherwise functioning natural systems. Nowhere is this more evident than on isolated oceanic islands such as Hawaii or New Zealand. But there is also the argument that we cannot separate humanity from nature and in these ecocultural systems it is no longer possible to extract the effects of human influence. Over millennia, people have modified global ecosystems, and environmental history (Hughes, 2006; Mosley, 2010) records both species losses as well as the absorption of new arrivals into 'native' ecology. In Great Britain we have gained the rabbit as a hugely influential alien, but at the same time have lost all our major carnivores and keystone mammals such as wild boar. In this context, claims of pristine native or natural ecology seem especially thin.

It seems sensible that in nature reserves and other protected areas, such as many national parks, containing vulnerable native communities and species, most exotic plants and animals should be removed or at least monitored. Even here though, there is room for a sensitive questioning of what is necessary. If a British nature reserve includes areas of Victorian or early planting of sweet chestnut, Austrian pine, European larch or even sycamore for example, then it would be highly detrimental and a complete nonsense to remove these, unless for a major scheme to revert to a more pristine habitat. However, an example of where this might be justifiable would be a mobile sand dune or a relict peat bog. But in most cases these trees would be a part of a rich palimpsest of the cultural and ecological landscape, so contributing to the overall conservation and worth of the site. They might well also have human admirers and friends in the local community whose views cannot simply be ignored. Understanding and valuing sites for their cultural landscapes and their archaeological interest is important in this re-evaluation. Such insight is also helpful to avoid ill-founded attempts to return sites to some presupposed pristine native state, which in reality probably never existed. Some of this runs contrary to much current conservation thinking, though it sits quite comfortably with where conservation began about a century or more ago.

Second, it is important to develop strategies to deal with problems where they arise and to facilitate and target the necessary resources for delivering measurable outputs. In Britain, for example control of problem species is at present undertaken mostly through the planning process in isolation of any wider implementation, and in the long-term is ineffective beyond the parochial level. Some further control is initiated by conservation NGOs and agencies and a very few local authorities. It is heartening to see cross-agency conferences being held on these issues, for example, the British Ornithologists' Union (BOU) autumn 2008 scientific meeting in Peterborough, 'The Impacts of Non-Native Species', aimed to provide an overview of present knowledge and action domestically and, importantly, included British overseas territories (Burton et al, 2010, p654); this event was seen as building on the BOU/UK Joint Nature Conservation Committee conference in 1995 on the introduction and naturalization of birds (Holmes and Simons, 1996). While we applaud these initiatives, the meetings were, as billed, 'scientific meetings'. As environmental historians and social scientists, we must therefore openly bemoan the lack of involvement and interaction in these events for those who could provide deeply valuable insights the sociocultural perspective. Our fervent hope is that this will change over the coming years. In the context of national and regional strategies for problem species there need to be mechanisms in place for the long-term management rather than eradication of target problem species, both animals and plants. This will require long-term, ongoing commitment and funding; there is no one-off panacea. Controlling problem and invasive species is not just for today, it is for ever! For fragile ecologies of emerging economies around the world, vulnerable to bio-invasion, the application of skills and resources to undertake long-term control measures is a huge challenge.

Realistic responses to the problems of invasive species, both native and exotic, present major difficulties but there are possible ways forward. First, to some extent in many countries, national mechanisms are there, though the democracy and accountability of the processes might be questioned. At regional and local levels, responsibility needs to be placed within sensible coherent spatial and administrative areas – such as 'catchment-wide', 'river basin', or National Park designation units. There are usually networks of potential stakeholders who might be empowered to take long-term effective action. This needs to cast the net much wider than current attempts often do, and it is also likely that funded partnerships of interested stakeholders will be essential, raising the issue of financial resources (in lean times) again.

Good reliable information is also central to any successful control programmes. In many cases, research and monitoring of impacts and outputs of strategy implementation can be done by responsible agencies in partnership with appropriate regional universities and, we would suggest, with engaged 'amateur naturalists'. Again of course, there is the question of long-term finance to implement and to coordinate such programmes. This is the 'elephant in the room' for conservation management and the answer is stark; it needs to be paid for with money from somewhere. The costs of not doing anything will far outweigh the expense of implementation, and that is the political argument to claw down the necessary money. However, this aspiration needs to be realistic, especially so in the current austere financial climate. The potential mechanisms exist in local authorities and agencies to implement controlled regional programmes of management at a catchment scale and the costs would probably decline over time as the most intransigent problems are tackled. The key is to make the delivery of such controls and strategies a statutory responsibility and function of local authorities in partnership with agencies. Most costs could be delivered in the long-term through a strategic realignment of existing resources combined with fees for delivering controls on land in other partner ownerships. Agencies would lead on the strategic overview and coordination, and local authorities on the practicalities of management. The aim would be control and management, not eradication. Such an approach would be long-term effective and sustainable and would deliver locally accountable targets acceptable to local people. This would move management away from a scattergun approach of planning-led initiatives and one-off conservation projects, to catchment-wide programmes delivering a high-quality environment reflecting the changing world in which we live. Observation, monitoring and review would help ensure flexibility and responsiveness as new challenges arise, as they certainly will over the coming decades of climate change.

Finally, it is also important to recognize the potential importance and wide-ranging benefits to be gained from deliberate reintroduction programmes as part of positive conservation. These benefits go well beyond the ecological or biological, into political, economic and sociocultural spheres, embracing such intangible concepts as community and regional identity, driven by a strong sense of place. Such intervention ecology is by its very nature subjective and mostly aimed at high-profile species with maximum 'wow' factor and economic returns from nature tourism. As long as this is recognized and acknowledged from the outset, then the dialogue with the public and other stakeholders will be more informed and carry more weight in decision-making.

Conclusion

This volume has raised and addressed a wide range of issues in relation to the importance of attitudes and perceptions. The broad range of multidisciplinary chapters provides insights into the issues and problems that face both researchers and practitioners seeking to address the effects, perceived or real, of exotic, alien and reintroduced species. Traditional nature conservationists may find some of the questions raised here discomforting, and full awareness of the vagaries of perceptions and attitudes over time may be especially challenging. However, the mix of subjective attitude and objective science in conservation is a facet which needs to be addressed and considered, particularly so for human relations with invasive and other exotic species. This is a controversial field where science, history and cultural studies mix (as do politics, economy and society), so judgements of problems and actions are not wholly clear or well-defined. There are blurred boundaries and competing insights and values. Acknowledging these complications will help to better focus debates on the need for specific conservation action to achieve what are generally agreed to be desirable outcomes. Yet in many interventions conservation practice hides behind a veneer of pseudoscience and certainly often challenges democratic processes. This is neither healthy nor likely to be sustainable in the long term. Another lesson of history is that our human attitudes and perspectives towards animals (indeed, towards nature) have change over decades and centuries, most dramatically from 1750 onwards (Thomas, 1984; Coates, 1998; Smout, 2000; Coates, 2006). There is no reason to assume that future approaches to these species and their impacts will not change again. Alien and invasive species raise emotions and generate strong feelings, both for and against. In this context it is important that researchers and practitioners recognize the lessons of history and the limitations of science as they blend into a wider political, economic and sociocultural story. The underpinning science must seek to be objective and dispassionate but conservationists' decisions are often subjective and often personal. This is no bad thing, as long as we are able to recognize it and to debate the merits either way. In a western democracy that is how it must be.

The approaches we suggest will not come free but they will be both more effective and less expensive than other options currently touted around. With regional strategies subjected to local scrutiny, they would resonate more effectively with local people than central visions imposed from above without local democracy and accountability. This might also help avoid the expensive and unnecessary removal of aliens, which detracts from real issues of long-term conservation management. The savings of otherwise wasted effort and money could help support the targets of realistic, achievable and pragmatic regional strategies delivered locally. It might be suggested that this softly, softly approach could never work, but in Great Britain both Swansea City Council and Cornwall County Council provide ample evidence of a mechanism that does just this and which is financially sound. To resonate with Charles Elton's war-induced philosophy of ecology (Elton, 1958) and as Winston Churchill pleaded: 'Give us the tools and we will finish the job.' There is also a strong argument to involve the powerful and diverse wildlife constituency in the recording and monitoring of invasive species, 'citizen science' if you will. Schemes run by the British Trust for Ornithology for example are very successful and there is the potential to make use of voluntary human effort as part of building 'the big society' so favoured by the Conservative–Liberal Democrat coalition government formed in 2010. One such hugely positive development, which draws on the rich amateur tradition for observing and recording wildlife in Britain (going back to at least the late 19th century), is the Recording Invasive Species Counts (RISC) Project launched in March 2010. This is a joint initiative between the Biological Records Centre within the Centre for Ecology and Hydrology, the National Biodiversity Network, Anglia Ruskin University, and the GB Non-Native Species Secretariat, funded by Defra. It has a simple aim: social inclusion in the recording and mapping of invasive species, using popular engagement and participation and media publicity in the project to generate wider and wider societal interest in the issues. At the launch of the project a manageable total of six species was chosen for public recording: Muntjac deer, zebra mussel, Chinese mitten crab, creeping water primrose, Tree of Heaven and American skunk cabbage. More species will be added later, in blocks of four. The launch of the RISC project, at the Linnean Society of London, was covered by the *Guardian*, *Daily Mail*, Radio 5 Live, Anglia TV, and BBC Radio Scotland.

It is important to recognize the subjectivity of decision-making processes and the cultural and historical origins of many of today's problem species. We simply have not done this in the past. Therefore, in the UK today the great majority of plants and animals that now cause consternation among conservationists and have huge effects on 'native' ecology were deliberate introductions. Some helped themselves and escaped from the confines of domestication, but many were released intentionally into the wild (for aesthetic, economic, or sporting reasons), and are today still aided in their spread by enthusiasts. The historic context that we ask for in this volume will help us develop a more balanced perception in terms of future controls, both their desirability and realistic outcomes. We unashamedly ask for more historians of nature conservation to offer new, fresh and updated interpretations of the past (Sheail, 1976; Evans, 1992; Marren, 2002; Smout, 2009). Without this there is much banter and foot-stamping, some excellent scientific talking-shops (albeit rather one-dimensional), some limited and local successes, but overall very little real impact on the spread of perceived alien and problems species. With climate change and other environmental stresses the problems of invasive species and of exotic invaders in particular are bound to increase. Achieving a pragmatic and balanced approach to human interactions with these plants and animals will remain a great challenge. A good starting point is surely Jim Dickson's rallying call in June 1996 before a multidisciplinary audience at Battleby Scottish Natural Heritage for, 'good science, good history and pragmatism' (Dickson in Lambert, 1998, p1) to inform what we do in confronting invasive species management issues in the 21st century.

References

Bolton, G. (1981) *Spoils and Spoilers: Australians Make their Environment, 1788–1980*, George Allen & Unwin, Sydney, Australia

Brown, P. and Waterston, G. (1962) *The Return of the Osprey,* Methuen & Co, London

Burton, N. H. K., Baker, H., Carter, I., Moore, N. and Clements, A. (2010) 'The Impacts of non-native species: A review of the British Ornithologists' Union's Autumn 2008 Scientific Meeting', *Ibis*, 152(3), 654–659

Carter, I. (2007) *The Red Kite,* Arlequin Press, Chelmsford, UK

Coates, P. (1998) *Nature: Western Attitudes Since Ancient Times,* Polity Press, Cambridge

Coates, P. (2003) 'Editorial Postscript: the naming of strangers in the landscape', *Landscape Research*, 28(1), 131–137

Coates, P. (2006) *American Perceptions of Immigrant and Invasive Species: Strangers on the Land,* University of California Press, Berkeley, CA

Crosby, A. (1986) *Ecological Imperialism: The Biological Expansion of Europe, 900–1900,* CUP, Cambridge

Dennis, R. (1998) 'The reintroduction of birds and mammals to Scotland', in Lambert, R. A. *Species History in Scotland: Introductions and Extinctions Since the Ice Age,* Scottish Cultural Press, Edinburgh, Chapter 1, pp5–7

Diamantis, D. (2004) *Ecotourism: Management and Assessment,* Thomson, London

Dudley, S. P. (2010) 'Non-native bird species and the British List', in *The Impacts of Non-native Species, British Ornithological Union Proceedings, British Ornithological Union, Peterborough*

Dunlap, T. R. (1999) *Nature and the English Diaspora: Environment and History in the USA, Canada, Australia and New Zealand,* CUP, Cambridge

Elton, C. S. (1958) *The Ecology of Invasions by Animals and Plants,* Methuen & Co, London

Evans, D. (1992) *A History of Nature Conservation in Britain,* Routledge, London

Fennell, D. (2008) *Ecotourism,* Routledge, London

Garden, D. (2005) *Australia, New Zealand and the Pacific: An Environmental History,* ABC Clio, Oxford

Gosling, L. M. and Baker, S. J. (1989) 'The eradication of muskrats and coypus from Britain', *Biological Journal of the Linnean Society*, 39(1), 39–51

Griffiths, T. and Robin, L. (1997) *Ecology and Empire; Environmental History of Settler Societies,* Keele University Press, Edinburgh

Grime, P. (2005) 'Alien plant invaders: Threat or side issue?', *ECOS*, 26(3/4), 33–40

Hibbert-Ware. A. (1938) *Report of the Little Owl Food Inquiry 1936–37,* H. F. & G. Witherby, London

Holmes, J. S. and Simons, J. R. (eds) (1996) *The Introduction and Naturalisation of Birds*, The Stationery Office, London

Hughes, J. D. (2006) *What is Environmental History?* Polity Press, Cambridge

Imber, M. (1998) 'Extinctions in the fauna of New Zealand and the impact of introduced vertebrates', in Lambert, R. A. (ed) *Species History in Scotland: Introductions and Extinctions Since the Ice Age,* Scottish Cultural Press, Edinburgh, Chapter 9, pp129–141

Jones, K. R. (2002) *Wolf Mountains: A History of Wolves Along the Great Divide,* University of Calgary Press, Calgary, Canada

King, C. (1984) *Immigrant Killers: Introduced Predators and Conservation of Birds in New Zealand,* OUP, Auckland

Lambert, R. A. (1998) *Species History in Scotland: Introductions and Extinctions Since the Ice Age*, Scottish Cultural Press, Edinburgh

Lines, W. J. (1991) *Taming the Great South Land: A History of the Conquest of Nature in Australia,* University of Georgia Press, Athens, GA

Love, J. A. (1983) *The Return of the Sea Eagle*, CUP, Cambridge

Marren, P. (2002) *Nature Conservation: A Review of the Conservation of Wildlife in Britain, 1950–2001*, Harper Collins, London

Mosley, S. (2010) *The Environment in World History,* Routledge, Abingdon, UK

Newsome, D., Dowling, R. K. and Moore, S. (2005) *Wildlife Tourism*, Channel View, Clevedon, UK

Page, S. J. and Dowling, R. K. (2002) *Ecotourism*, Prentice Hall, Harlow, UK

Public Acts (1932) *Destructive Imported Animals Act,* Chapter 12, 22 and 23 George V, UK Parliament

Rollins, J. (2006) 'It's official: Half the UK loves wildlife', *Natural World*, Spring, 31

Rolls, E. C. (1969) *They All Ran Wild: The Enthralling Story of Pests on the Land in Australia*, Angus and Robertston, Melbourne

Rotherham, I. D. (2001a) 'Himalayan Balsam: The human touch', in Bradley, P. (ed) *Exotic Invasive Species: Should we Be Concerned?* Proceedings of the 11th Conference of the Institute of Ecology and Environmental Management, Birmingham, April 2000, IEEM, Winchester, UK, pp41–50

Rotherham, I. D. (2001b) 'Rhododendron gone wild', *Biologist*, 48(1), 7–11

RSPB (2006) *Watched Like Never Before: The Local Economic Benefits of Spectacular Bird Species,* RSPB Economics Department, Sandy, Beds, UK

Sheail, J. (1972) *Rabbits and their History,* Country Book Club, Newton Abbot, Devon, UK

Sheail, J. (1976) *Nature in Trust: The History of Nature Conservation in Britain,* Blackie, London

Smout, T. C. (2000) *Nature Contested: Environmental History in Scotland and Northern England Since 1600,* EUP, Edinburgh

Smout, T. C. (2003) 'The alien species in twentieth-century Britain: Constructing a new vermin, *Landscape Research*, 28(1), 11–20

Smout, T. C. (2009) 'The alien species in twentieth-century Britain: Inventing a new vermin', in Smout, T. C. (ed) *Exploring Environmental History: Selected Essays,* Edinburgh University Press, Edinburgh

Thomas, K. (1984) *Man and the Natural World: Changing Attitudes in England, 1500–1800*, Penguin Books, London

Weijden, W. van der, Leewis, R. and Bol, P. (2007) *Biological Globalisation*, KNNV Publishing, Utrecht, The Netherlands

Index